Advances in Industrial Control

AF786289

Springer

London
Berlin
Heidelberg
New York
Hong Kong
Milan
Paris
Tokyo

Other titles published in this Series:

Performance Assessment of Control Loops: Theory and Applications
Biao Huang & Sirish L. Shah

Advances in PID Control
Tan Kok Kiong, Wang Qing-Guo & Hang Chang Chieh with Tore J. Hägglund

Advanced Control with Recurrent High-order Neural Networks: Theory and Industrial Applications
George A. Rovithakis & Manolis A. Christodoulou

Structure and Synthesis of PID Controllers
Aniruddha Datta, Ming-Tzu Ho and Shankar P. Bhattacharyya

Data-driven Techniques for Fault Detection and Diagnosis in Chemical Processes
Evan L. Russell, Leo H. Chiang and Richard D. Braatz

Bounded Dynamic Stochastic Systems: Modelling and Control
Hong Wang

Non-linear Model-based Process Control
Rashid M. Ansari and Moses O. Tadé

Identification and Control of Sheet and Film Processes
Andrew P. Featherstone, Jeremy G. VanAntwerp and Richard D. Braatz

Precision Motion Control
Tan Kok Kiong, Lee Tong Heng, Dou Huifang and Huang Sunan

Nonlinear Identification and Control: A Neural Network Approach
Guoping Liu

Digital Controller Implementation and Fragility: A Modern Perspective
Robert S.H. Istepanian and James F. Whidborne

Optimisation of Industrial Processes at Supervisory Level
Doris Sáez, Aldo Cipriano and Andrzej W. Ordys

Applied Predictive Control
Huang Sunan, Tan Kok Kiong and Lee Tong Heng

Hard Disk Drive Servo Systems
Ben M. Chen, Tong H. Lee and Venkatakrishnan Venkataramanan

Robust Control of Diesel Ship Propulsion
Nikolaos Xiros

Hydraulic Servo-systems
Moheiddine Jelali and Andreas Kroll

Model-based Fault Diagnosis in Dynamic Systems Using Identification Techniques
Silvio Simani, Cesare Fantuzzi and Ron J. Patton

Strategies for Feedback Linearisation: A Dynamic Neural Network Approach
Freddy Garces, Victor M. Becerra, Chandrasekhar Kambhampati and Kevin Warwick

Robust Autonomous Guidance: An Internal Model Approach
Alberto Isidori, Lorenzo Marconi and Andrea Serrani

Gennady G. Kulikov and Haydn A. Thompson
(Eds.)

Dynamic Modelling of Gas Turbines

Identification, Simulation, Condition Monitoring and Optimal Control

With 184 Figures

Springer

Gennady G. Kulikov, Ph.D, D.Sc
Department of Automated Control Systems
Ufa State Aviation Technical University
Russia

Haydn A. Thompson, Ph.D, CEng
Department of Automatic Control and Systems Engineering
The University of Sheffield
Sheffield, UK

British Library Cataloguing in Publication Data
Dynamic modelling of gas turbines : identification,
simulation, condition monitoring and optimal control. -
(Advances in industrial control)
1. Gas turbines - Mathematical models 2. Gas turbines -
Dynamics 3. Gas turbines - Automatic control
I. Kulikov, Gennady G. II. Thompson, Haydn A.
621.4′33′015118

Library of Congress Cataloging-in-Publication Data
Dynamic modelling of gas turbines : identification, simulation, condition monitoring,
and optimal control / Gennady G. Kulikov and Haydn A. Thompson, eds.
p. cm. — (Advances in industrial control)
Includes bibliographical references and index.

1. Airplanes—Turbojet engines—Mathematical models. I. Kulikov, Gennady G.
(Gennady Grigoryevich) II. Thompson, Haydn A., 1965—III. Series.

TL709.3.T83D96 2004
629.134′3533′0151—dc22 2003067315

ISBN 978-1-84996-914-7
Springer-Verlag is part of Springer Science+Business Media
Springeronline.com

69/3830-543210 Printed on acid-free paper

Advances in Industrial Control

Professor T.H. Lee
Department of Electrical Engineering
National University of Singapore
4 Engineering Drive 3
Singapore 117576

Professor Emeritus O.P. Malik
Department of Electrical and Computer Engineering
University of Calgary
2500, University Drive, NW
Calgary
Alberta
T2N 1N4
Canada

Professor K.-F. Man
Electronic Engineering Department
City University of Hong Kong
Tat Chee Avenue
Kowloon
Hong Kong

Professor G. Olsson
Department of Industrial Electrical Engineering and Automation
Lund Institute of Technology
Box 118
S-221 00 Lund
Sweden

Professor A. Ray
Pennsylvania State University
Department of Mechanical Engineering
0329 Reber Building
University Park
PA 16802
USA

Professor D.E. Seborg
Chemical Engineering
3335 Engineering II
University of California Santa Barbara
Santa Barbara
CA 93106
USA

Doctor I. Yamamoto
Technical Headquarters
Nagasaki Research & Development Center
Mitsubishi Heavy Industries Ltd
5-717-1, Fukahori-Machi
Nagasaki 851-0392
Japan

Series Editors' Foreword

The series *Advances in Industrial Control* aims to report and encourage technology transfer in control engineering. The rapid development of control technology has an impact on all areas of the control discipline. New theory, new controllers, actuators, sensors, new industrial processes, computer methods, new applications, new philosophies . . . , new challenges. Much of this development work resides in industrial reports, feasibility study papers and the reports of advanced collaborative projects. The series offers an opportunity for researchers to present an extended exposition of such new work in all aspects of industrial control for wider and rapid dissemination.

Modelling and system identification are used in several different ways in industrial control. Modelling is an inexpensive way of studying the behaviour of complex industrial systems whilst system identification may be used to improve the accuracy of the model or some parameters in a model for a particular installation of an industrial system. Models and estimates of model fidelity or uncertainty can form the basis of advanced robust control designs. Meanwhile system identification experience with a particular system or class of systems can be used in the effective design of adaptive control systems. The identification of black-box and grey-box models is a development which tries to limit the involvement of physical and chemical process description and use convenient mathematical forms to capture the behaviour of an industrial system. The growth in the theory and methods of modelling and system identification has been considerable. The classic *System Identification* by Pieter Eykhoff was published in 1974 whilst thirteen years later the book *System Identification: Theory for the User* by Lennart Ljung (1987) emphasised the use of these techniques by the control engineer.

This long period of theoretical gestation has made modelling techniques and system identification a fundamental weapon in the control engineer's armoury. The area now forms a strong discipline with a high-level supporting conference. As a specific example of how the various modelling and system identification techniques may be used for a particular class of industrial processes we welcome this monograph *Dynamic Modelling of Gas Turbines* edited by Professors G.G. Kulikov and H.A. Thompson to the *Advances in Industrial Control* series. Nice authors from Russia and the United Kingdom have contributed chapters which employ the full diversity of modelling and identification methods to the gas turbine modelling problem. Much of the work is based on industrial data from various experimental gas turbine test rigs.

The monograph will be of significant interest to the gas turbine specialist who should find it to be a useful source and reference book. The wider control community will find the applications of the techniques interesting and potentially inspiring for other possible applications. The authors are to be congratulated on attaining such a high standard in technical application and for their uniformity and clarity of presentation.

M.J. Grimble and M.A. Johnson
Industrial Control Centre
Glasgow, Scotland, U.K.

Preface

This book presents a survey of modern methods for dynamic modelling of gas turbine engines. The manuscript was prepared using the results of research and unique practical experience of digital controller design for aero gas turbine engines. Experience of different research teams from Russia and the United Kingdom in the field of modelling of aero engines is summarised.

Dynamic models are traditionally used for describing behaviour of various systems. Dynamic models allow the design, analysis and identification of systems. Moreover, dynamic models can also be used for condition monitoring of complex systems providing information redundancy. The book describes various approaches to building and applying dynamic models. The approaches are compared and results of experiments on real engines are supplied.

The research team of Prof. G. Kulikov from Ufa State Aviation Technical University (Department of Automated Control and Management Systems) has a long history of participation in aero engine control design projects with Russian and Ukrainian industry. Chapters 1-4, 7, 10-13, 15 were contributed by Prof. G. Kulikov, Prof. V. Arkov and Dr. T. Breikin. Chapter 14 was contributed by Dr. O. Lyantsev.

The research team of Prof. P. Fleming works in close collaboration with British industry in the framework of the Rolls-Royce University Technology Centre for Control and Systems Engineering (Automatic Control and Systems Engineering Department, the University of Sheffield, UK). The introduction was contributed by Prof. P. Fleming and Prof. H. Thompson. Chapter 9 was contributed by Prof. P. Fleming, Prof. H. Thompson and Dr. K. Rodriguez-Vazquez.

The team of Dr. D. Rees (University of Glamorgan, UK) participates in research projects investigating modelling of Rolls-Royce aero engines. Chapters 5, 6 and 8 were contributed by Dr. D. Rees and Dr. N. Chiras.

Recently, a joint research programme was performed by all the mentioned teams funded by the British government enabling extended experimentation to be carried out at the Rolls-Royce test facility at DERA. Various identification techniques were demonstrated and compared during this project, and results are also featured in the book.

The monograph should be useful for control engineers, scientists and students.

Ufa, Russian Federation, August 2003 — *Gennady Kulikov*, *Valentin Arkov*, *Oleg Lyantsev*

Sheffield, UK, August 2003 — *Haydn Thompson*, *Peter Fleming*, *Timofei Breikin*

Glamorgan, UK, August 2003 — *David Rees*, *Neophytos Chiras*

Mexico city, Mexico, August 2003 — *Katya Rodriguez-Vazquez*

Introduction

In the early 1930s, jet engine technology was simultaneously developed in Britain by a team headed by Frank Whittle and in Germany by a team led by Hans von Ohain and Max Hahn. In the intervening years, gas turbine engines have revolutionised air transport, and aero derivative engines are now widely used in marine and industrial applications. In the early days, much experimentation was done using heuristic techniques to develop control systems and finetune performance. Modelling has always been a key area of investigation, and over the years considerable effort has been expended in gaining a greater understanding of engine dynamics and in representing these in increasingly complex models. With the escalating costs of engine programmes with high investment, and consequent high commercial risk, modelling is being used more and more in a variety of ways. Good high-fidelity models that accurately represent the engine are essential in development to analyse and predict engine performance. They are also crucial in control law development to optimise key commercial parameters such as specific fuel consumption and to meet increasingly stringent engine handling requirements driven by customers, certification bodies and environmental legislation.

In the 1970s and 1980s, with the advent of processing technology, hydraulic systems were replaced with digital systems. Since then, the level of processing power available has increased dramatically, allowing much greater scope for advanced engine control techniques to be used, and additionally a much greater level of health monitoring functionality. Indeed, health monitoring is seen as a key commercial driver in aftermarket business for the future.

The aim of this book is to provide an introduction to methods for dynamic modelling and systems identification of aero and power gas turbine engines. The book is organised into three parts. In the first part, Chapters 2-4, the traditional approaches to gas turbine engine modelling are discussed over the life cycle of a control system. In the second section, Chapters 5-8, an overview of systems identification is given and a number of common system identification techniques are considered. The final section of the book discusses new perspectives in modelling, identification, condition monitoring and control. This latter section contains practical applications and results of recent gas turbine engine research in the United Kingdom and Russia performed on test beds and in flight demonstration on both turbofan and prop fan engines. A more detailed overview of content is given in the following sections.

Chapter 1. Introduction to Gas Turbine Engine Control. In Chapter 1, an introduction to fundamentals of gas turbine operation is given followed by an overview of control systems design for aero power plants. Major types of gas turbine engines are characterised. In particular, the differences between turbo prop and turbo fan engines are highlighted. General principles for engine control are considered according to the type of engine operation: start-up, transient, maximum thrust, normal steady-state operation, minimum thrust and operation with afterburning. A typical set of control laws is presented for a bypass twin-shaft turbo-jet, and optimization criteria are discussed for integrated engine control. On-line optimization of the controller performance is given, maintaining the demanded thrust and minimum specific fuel consumption at cruise flight.

Part 1. Gas Turbine Models.

Chapter 2. Models and the Control System Design Cycle. In Chapter 2, techniques for mathematical modelling of gas turbine engines are presented considering the different stages in the life cycle from design, to demonstration, to commercial control system development, and finally through to in-service use. This allows an appreciation of the relationships between the models and their derivation for specific control engineering needs. Through the life cycle stages the fidelity of the engine model is refined, as more information becomes available. Different accuracies and degrees of detail for the mathematical models are required at various stages of the life cycle of a controller. The degree of approximation to individual engine characteristics also differs.

Apart from control design applications, mathematical models of gas turbine engines are extensively used in testing hardware, software, control laws and algorithms of the controller. This drives further investigation into identification and modelling of gas turbines to allow cost-effective testing based on computer simulation.

Chapter 3. Off-Line Models. In Chapter 3, existing methods for mathematical modelling of aero engines at the design phase are considered. During this phase complex off-line models are used that allow general engine performance to be analysed and control laws to be determined. Typically, during the controller design stage, a number of dynamic models are derived from a detailed nonlinear performance-based model comprised of dynamic and static models. Such models are based on gas dynamics laws and typically include over 100 algebraic and transcendent equations. The engine characteristics are partly defined in the form of empirical formulae, graphs and tables. The performance-based detailed nonlinear dynamic model is similar to the static model, but the equations are written in the dynamic form. These show the balance between characteristics of engine elements in the form of continuous flows and energy conservation equations. An analytical relationship exists between the dynamic and static parts of the model. This allows dynamic models to be obtained quickly using additional information about the engine geometry and mass. An example of detailed modelling of a single-shaft gas turbine engine is considered.

The dynamic characteristic is considered utilising a graphical interface to the nonlinear dynamic model. In addition to the static line, it also includes a set of constant acceleration lines. Methods for creation of the dynamic characteristic from static and dynamic models are explained. This type of model visually represents the main shape of the engine nonlinearity. Theory basics are explained for reducing engine parameters to the standard sea-level atmosphere conditions including reduction formulae and interpolation techniques.

Linear dynamic modelling techniques described in Chapter 5 are well-established in control engineering compared to nonlinear modelling. Control systems are usually designed and assessed using linear models, which are obtained from performance-based models via linearisation or identification methods. These describe engine dynamics around a steady-state operating point. A simple form of linear modelling accounts for the inertia of the turbocompressor rotors accumulating mechanical energy. Gas flow dynamics can also be introduced in linear modelling via conditional dynamic additions accounting for air, gas and heat accumulation in the engine volumes. However, when exploring control of transient operation, the models should incorporate nonlinear properties of the engine.

Chapter 4. On-Line Models. In Chapter 4, a real-time piecewise linear dynamic model (RPLDM) is considered combining the accuracy of a nonlinear model with the computing speed of a linear model. This approach extends the validity of linear models to the full operating range of the engine. The model consists of a set of linear models connected to the nonlinear static line. During simulation, the linear model parameters are interpolated according to the nearest point of the static line. Example diagrams of algorithms and SIMULINK structures are given demonstrating the application of the method. Practical experience of the model implementation in digital control design is discussed. RPLDM provides an acceptable level of accuracy and can be used in real time due to its simplicity compared with a detailed nonlinear dynamic model. This enables control algorithms to be effectively tested on models and the controllers to be tested on electronic test beds. RPLDM models also have the potential to be used within on-board engine condition monitoring systems.

Part 2. Gas Turbine System Identification.

Chapter 5. Linear System Identification. In Chapter 5, an introduction is provided to the fundamental concepts relating to the modelling and estimation of linear time-invariant systems in the time- and frequency-domain. Estimators are described for parametric time-domain modelling and for both nonparametric and parametric frequency-domain modelling. A comparison is made between discrete-time and continuous-time models, and their applications are discussed. A discussion is presented of the key issues in experiment design. Particular attention is paid to the selection of the input excitation and the use of periodic signals is advocated. The properties of multisine and maximum length binary sequences are discussed.

Chapter 6. Linear Gas Turbine Modelling. Linear gas turbine engine dynamic modelling techniques are described in Chapter 6. Control systems are usually designed and assessed using linear models, which are obtained from performance-based models via linearisation or identification methods. These describe engine dynamics around a steady-state operating point. A simple form of linear modelling accounts for the inertia of the turbocompressor rotors accumulating mechanical energy. Gas flow dynamics can also be introduced in linear modelling via conditional dynamic additions accounting for air, gas and heat accumulation in the engine volumes.

Chapter 7. Closed-Loop Control and System Identification. It is often practically impossible to carry out identification tests with open-loop control. Such tests often are not allowed from a safety viewpoint. Moreover, adaptive control systems require closed-loop identification. Such identification tests allow the plant model to be obtained under some conditions referred to as *identifiability conditions*. In Chapter 7 closed-loop identifiability is considered for gas turbine engines utilising various controllers in the real operating environment. Existing concepts of identifiability are analysed. Natural exciting factors are discussed which allow the plant model to be directly identified. A method for verification of identifiability conditions is proposed which does not require direct measurement of the excitations.

Chapter 8. Nonlinear Gas Turbine Modelling. In Chapter 8, several nonlinear model representations are presented along with a general methodology for nonlinear system modelling. Nonlinear dynamic modelling techniques are less established in control engineering compared to linear modelling. However, when exploring control of transient operation, the models should incorporate nonlinear properties of the engine and in order to model the global dynamics of the gas turbine a nonlinear model is required. In this chapter application of polynomial NARMAX models and neural networks are described in more detail and estimation of nonlinear models for the gas turbine engine is considered.

Part 3. New Perspectives in Modelling, Identification, Condition Monitoring and Control.

Chapter 9. Nonlinear Model Structure Selection Using Evolutionary Optimisation Methods. In Chapter 9, an evolutionary computing approach is considered to aid selection of an appropriate structure for gas turbine engine model. An example of a jet engine nonlinear model identification is presented to demonstrate this technique. Experimental data recorded from the real engine test bed are used for identification and comparison of the models with the real engine dynamics. In the example, model structure selection is realised through genetic programming, in a multiobjective fashion to generate global models, which describe the nonlinear dynamic behaviour of the gas turbine engine. The introduction of the validation stage of system identification into the multiobjective tool is also explored in this chapter, in order to direct the identification process to a set of global models of the system.

Chapter 10. System Identification Using Frequency Response Techniques with Optimal Spectral Resolution. In Chapter 10, the problem of nonparametric identification of dynamic models in the form of frequency response is considered. Frequency response identification has been widely discussed; however, the choice of spectral resolution is not well founded yet. Combining multiple spectral estimation and *a priori* knowledge allows an improvement in estimate accuracy. Correlation window analysis demonstrates that the optimal resolution varies for different frequencies. A similarity with wavelet analysis is considered where the optimal window is chosen prior to the estimation. A formalised procedure is given for the practical use of the method in research and control engineering.

Chapter 11. Turbo Prop Fan Engine Identification: Practical Issues. In Chapter 11, an example of closed-loop identification is given of a turbo prop fan engine. The identification techniques are applied to a real-life power plant. Experimental data was gathered at ground-based engine test bed using the controller as a data acquisition system. Practical issues of systems identification are demonstrated including the engine model decomposition, the choice of the optimal spectral resolution, investigation into closed-loop identifiability and validation of the performance-based model. Useful outcomes from dynamic model identification are demonstrated via improved control accuracy and introduction of test signal generators in a systems test facility.

Chapter 12. Stochastic Gas Turbine Engine Models. A gas turbine engine is usually considered for control engineering purposes as a deterministic plant. Its dynamics are described by means of linear or nonlinear differential equations. However, stochastic dynamics are observed during transient and steady-state conditions. Such behaviour is derived from internal and external excitations in the engine and controller. Accounting for stochastic processes could allow analysis of control accuracy when testing the controller. Also, this might provide additional information for condition monitoring of the system. Considering deterministic and stochastic dynamics as a unified model makes it possible to design simple reliable algorithms for optimal control.

In Chapter 12, the basics of dynamic Markov modelling are considered for describing engine dynamics. This class of nonparametric model in the time domain describes nonlinear stochastic plant dynamics in the form of multi-dimensional distribution density. Markov modelling in a natural way incorporates stochastic dynamics of the process and allows simulation and analysis of control processes. Basic descriptive properties of dynamic Markov models are derived. Transition to conventional deterministic models is then given.

A novel modelling technique for automatic control purposes is discussed. A fuzzy Markov system is proposed to describe both determined and random behaviour of complex dynamic plants. The main advantage is its high computational speed. Another benefit of this method is its flexibility and applicability to both linear and nonlinear systems. A controlled Markov chain represents a fuzzy system with a rectangular membership function. Its output is the probability distribution, not a variable value. This approach represents an attempt to overcome the primary difference between nonrandomness of fuzzy sets and Markov chain theory, which deals with random phenomena. This provides new opportunities for further research in conventional fuzzy control systems.

Chapter 13. Markov Modelling Of Turbo Prop Fan. Application areas for Markov modelling of gas turbine engines include identification, control system design, optimal control and condition monitoring. Application examples are demonstrated in Chapter 13 including Markov model identification and hardware-in-the-loop simulation at a system test facility for electronic engine controllers. Discussion of effectiveness of the Markov modelling technique for simulation of the identified models at the HIL test-bed facility is provided. Application Markov modelling for condition monitoring of the aero engine and its controller is discussed in this Chapter.

Chapter 14. Optimal Control of Gas Turbine Engines Using Mathematical Programming. The aero engine performance can be improved through the controller optimization. This can be split into determination of optimum control laws and optimum realisation of the programs. The engines' characteristics differ considerably over the fleet and also change during in-service use. This leads to the need for a real-time optimizing control. In Chapter 14, numerical algorithms for multivariable gas turbine engine control are considered taking into account individual characteristics of the plant. The problem of optimal gas turbine engine control is reduced to a quadratic programming problem solved with a real-time algorithm. An example is given, which proves the efficiency of the method. It optimizes the speed of response for a real power plant while minimizing specific fuel consumption at aircraft cruise mode.

Chapter 15. Dynamic Model Identification of a Turbo Jet Engine. In Chapter 15, an example of open-loop and closed-loop identification of dynamic models of a turbo fan engine with reheat is given. Experiments were performed with and without test signals. The dynamic gains derived from dynamic and static models are compared. After the frequency response was estimated, a parametric model was built in the form of a transfer function with common denominators for shaft speed models. An approximation formula is introduced for the optimal spectral resolution choice providing a smooth shape of estimates. Closed-loop identifiability is further discussed using data from an experiment with a shaft speed controller. The real-life example of a turbo jet engine demonstrates practical problems of implementation for various identification techniques.

Concluding Remarks. Readers who are new to the area of modelling and control are encouraged to read Chapters 2-4. Experienced practitioners of systems identification may wish to skip these sections although they do contain practical advice with respect to application to gas turbine engines. The latter stages of the book consider a number of different techniques presenting a control engineer with a suite of identification tools. The engineer can choose from these dependent on the situation faced. The techniques have been practically applied to real systems and the results of these investigations are presented.

Acknowledgments. The work on the manuscript was supported by the European Commission grant INTAS-2000-757. Prof. Kulikov and Prof. Fleming were scientific editors of the monograph. Prof. Arkov performed English translation of Chapters 1-4, 7, 10-13 and 15. Prof. Thompson and Prof. Arkov edited general style. Authors gratefully acknowledge Dr. D. Hill (Rolls-Royce plc.) for kindly supplying data from the Spey engine tests and assistance with organising the experimentation. Dr. Rees and Dr. Chiras would like to express their deepest gratitude to the late Dr. Ceri Evans, for his invaluable contribution to the work presented in Chapters 5, 6 and 8.

Contents

Notation

Variables

$\lvert . \rvert$	absolute value
$\lVert \cdot \rVert$	norm of vector
$\overline{F}_{\text{ef}}$	effective thrust factor
$\dot{n}$	shaft speed derivative
A	area
B	bandwidth
C_1, C_2, C_3	constants
C_n	normalised gas flow velocity in nozzle
c_p	specific constant-pressure thermal capacity
C_{sp}	specific fuel consumption
c_V	specific constant-volume thermal capacity
E	excess
E[.]	average
F	thrust
$F(s)$, $G(s)$, $H(s)$	transfer functions
$F(\omega)$, $G(\omega)$, $H(\omega)$	frequency response functions
F[.]	Fourier transform
g	gravity acceleration
H	flight height
H_u	low calorific value of fuel
J	inertia moment
J^*	total enthalpy
K_a	aeroplane lift-to-drag ratio
K_g	adiabatic coefficient
L	flight distance
m	by-pass ratio; mass
M	Mach number; angular shaft moment
N	power
n	shaft speed
p	pressure
P_{ijk}	elements of transition probability matrix
Prob{A}	probability of event A
$q(.)$	gas dynamic function
Q_{comb}	amount of heat delivered to air flow in combustor
R	gas constant
$R(t)$	correlation function
Re	Reynolds number
Re[.]	real part

s	Laplace complex variable
$S(\omega)$	spectral density
T	temperature
t	time
T, T_1, T_2, T_3	time constants
T_s	sampling time interval
T_0	sea-level temperature
T_r	record length
$u(t)$	correlation window function
$u(t), v(t), x(t), y(t)$	random processes in engine
$U(\omega)$	spectral window function
v	aeroplane flight speed
V	volume
var[.]	parameter variance
V_{PI}	output signal of PI controller
W	mass flow
$w(t)$	impulse response
W_1, W_2	initial and final aircraft weight
W_f	fuel flow
$W_{f.\mathrm{main}}$	fuel flow into main combustor
$W_{f.s}$	fuel consumption per second
X	drag force
Δ	deviation about mean value
Δr	blade tip clearance in compressor or turbine
$\Delta u(t)$	deviation from average or static point
$\Delta\omega$	filter pass band
Φ	transitional matrix
Θ	heating degree of propulsive mass
α	angular position
α_a	air excess coefficient
ε_b	bias (systematic error)
ε_r^2	variance (random error)
ε^2	mean square error
γ^2	coherence function
η	efficiency; operating parameter of RPLDM
η_0	total efficiency of turbine engine
η_g	fuel combustion efficiency
$\varphi(\omega)$	frequency response phase
λ_a	normalised velocity of the air flow
μ	mean value
π	pressure ratio
θ	propulsive mass temperature
σ	hydraulic resistance
τ_m	correlation window size
ω	angular frequency

∂	partial differential
$\angle$	angle

Matrices and vectors

A, **B**, **C**, **D**	matrices of linear dynamic model coefficients
F, **f**	nonlinear matrix operators
P	transition probability matrix
U	control vector
V	flight atmospheric conditions vector
X	state vector
Y	observed coordinates vector
Z	discrepancy vector
$\boldsymbol{\lambda}$	vector of final probabilities

Abbreviations

3D	three-dimensional
A/D	analogue-to-digital (converter)
ARMAX	auto regression with moving average and exogenous input
D/A	digital-to-analogue (converter)
DAS	digital acquisition system
DC	dynamic characteristic
DMC	dynamic modelling complex
FADEC	full authority digital engine control (system)
FFT	fast Fourier transform
FRF	frequency response function
GTE	gas turbine engine
HIL	hardware-in-the-loop
HP	high pressure
LDM	linear dynamic model
LP	low pressure
MOGA	multiobjective genetic algorithms
PC	personal computer
PI	proportional integrate (control)
PLA	pilot lever angle
RPLDM	real-time piecewise linear dynamic model
SFC	specific fuel consumption
SISO	single input and single output
SLS	sea-level static
STD	standard deviation
STF	systems test facility
TF	transfer function
VSV	variable stator vanes
VIGV	variable inlet guide vanes

Indices (subscripts and superscripts)

$\bar{}$	normalised
$\dot{}$	first derivative
*	stopped parameter; complex conjugate function
0	sea-level parameter
Σ	overall
^	estimate
$^{(2)}$	second derivative
a	air
ab	afterburner
c	compressor
comb	combustor
cooling	cooling air
cr	critical nozzle
dem	demanded value
den	denominator
dyn	dynamic addition
e	experimental
ef	effective
eng	engine
exp	expected
f	fan
g	gas
H	parameter at height H
han	Hanning (window)
I	inner duct parameter
II	bypass duct parameter
in	intake
main	main combustor
mech	mechanical
mix	mixing chamber
mv	metering valve
n	nozzle
opt	optimal parameter
phys	physical
pp	power plant
rec	rectangular window
red	reduced
sp	specific
st	static
T	matrix transpose
t	turbine

Chapter 1
Introduction to Gas Turbine Engine Control

1.1 Introduction

In this chapter, an introduction to gas turbine engine operation and control is given. The chapter starts by emphasising engine control in general through a brief survey of conventional control strategies. Engines need to attain high levels of overall efficiency. A number of high-level efficiency criteria are introduced and the use of optimisation within the design cycle is considered. The chapter concludes with a section considering examples of on-line optimisation from a number of different viewpoints to maximise: range at cruise condition, specific fuel consumption and thrust. The aim is to globally optimise the engines overall performance through selection of appropriate control strategies.

The main task of an aircraft power plant is to create a demanded value of effective thrust F_{ef}. The power plant structure includes the gas turbine engine, intake and exhaust nozzle. Effective thrust is a function of the flight parameters H, M and the operating conditions of the engine, air intake and nozzle.

Aero power plants are based on gas turbine engines, which represent a class of reaction engines developing its thrust via gas flow ejected [1, 2]. The value of the thrust developed is approximately defined by the formula:

$$F \approx W_a(C_n - v_f) \tag{1.1}$$

where F is the thrust;

W_a is the mass flow of the air thrown out;

C_n is the final air velocity (nozzle gas/air speed);

v_f is the start air speed with respect to the aircraft (flight speed).

The kinetic energy of the unit mass of air is increased by the work:

$$\Delta E = C_n^2/2 - v_f^2/2 \tag{1.2}$$

The additional energy is obtained by conversion of chemical fuel energy to thermal energy and then into mechanical work using the thermodynamic cycle. Any power plant performs two functions:

- converts the chemical energy of the fuel into mechanical energy (strictly speaking, an engine, or a thermal machine, refers to the part of the power plant performing this function);
- uses the obtained mechanical energy to accelerate the air flow and does the useful traction work (this function is carried out by the mover).

Further in the text, the term *engine* will refer to the complex of a thermal machine and a mover. Aero engines with no bypass duct are direct-reaction engines, where the thermal machine and mover are combined in the same device. Turboshaft helicopter engines are indirect-reaction engines, where different devices execute the functions of the thermal machine and mover; here the mover is the rotor. Bypass turbo jet, turbo prop and turbo prop fan engines, where the joined mass of air participates in the creation of the thrust, can be referred to as mixed-reaction engines.

A gas turbine engine of any type contains two major elements:

- the compressor squeezing the air;
- the turbine driving the compressor rotor.

The power plant control system objectives depend on the flight mode of the aircraft. In particular, take-off requires maximum thrust, cruise flight control must maintain maximum power plant efficiency, or minimum specific fuel consumption. Design of control systems for aero engines requires gas turbine theory, control theory and differential calculus methods [3, 4].

1.2 Calculation of Effective Thrust

For aircraft performance, an essential parameter is the effective thrust provided by the power plant [5]. In this section, derivation of the effective thrust is given for a typical military aero engine. This is used in later sections to demonstrate control system development.

The effective thrust of a power plant is less than the actual thrust created by the engine, by the value of external losses ΣX_i:

$$F_{\text{ef}} = F_{\text{eng}} - \sum X_i = F_{\text{eng}} \overline{F}_{\text{ef}} = f(M, H, \alpha_{\text{PLA}}) \tag{1.3}$$

The value of the effective thrust factor $\overline{F}_{\text{ef}} = F_{\text{ef}} / F_{\text{eng}}$ depends on Mach number M and the operating point of the engine. Considering a military engine, cruise is $M = 0.8,...,0.85$ and the value $\overline{F}_{\text{ef}}$ is 0.92,...,0.97. In horizontal flight without acceleration the required thrust of a flight vehicle approximately equals the drag force X. The required thrust depends on flight conditions and the lift-to-drag ratio of the aircraft. The output parameters of the engine and the power plant include thrust F_{eng}, specific thrust F_{sp}, overall efficiency η_0 and specific fuel consumption C_{sp}. These generally depend on the main parameters of the working process of the thermal machine and propulsor. Other important factors are the following:

- the heating degree of the propulsive mass in the main and afterburning combustion chambers $\Theta_{\text{main}} = T_g^* / T_H$, $\Theta_{\text{ab}} = T_{\text{ab}}^* / T_{\text{in.ab}}^*$,
- gas temperature before the turbine T_g^*,
- compressor pressure ratio π_c^*,
- engine pressure ratio π_{eng}^*,
- air flow W_a,
- bypass ratio $m = W_{a\text{II}} / W_{a\text{I}}$,
- fan pressure ratio $\pi_f^* = p_f^* / p_{\text{in}}^*$.

Also, there are additional parameters of the working process which need to be considered such as efficiency of units of the power plant (such as intake, compressor, combustion chambers, turbines, and exhaust nozzle). Therefore, control of the output parameters F_{eng} and C_{sp} affects parameters of the working process, which act as controlled parameters. The effect on controlled parameters is achieved by changing control factors. For independent change of several controlled parameters it is necessary to have the same number of control factors. The controlled parameters are connected implicitly to the control factors by a control program:

$$\text{Program} = f(M, H, \alpha_{\text{PLA}}) \quad \text{where} \quad i = 1, \ldots, n \tag{1.4}$$

1.3 Military Engine Example

Continuously increasing demands for flight vehicles require improvements in engine performance. This drives reduction of specific fuel consumption, increase of specific thrust and reduction of specific weight. These requirements can be fulfilled by increasing or optimising the absolute values of the working process parameters (π_c^*, T_g^*, *etc.*) and by using more complex geometry.

A generalised schematic of a hypothetical reheat bypass turbojet engine is shown in Figure 1.1. The schematic includes all the control factors mentioned above. The main constituent units – compressor, combustor and turbine stages – are also shown.

The main control factors include fuel flow into the main combustion chamber W_f and afterburner $W_{f.\text{ab}}$, critical nozzle area A_{cr} and output nozzle area A_n. The characteristics of separate units of the engine must be optimised at different operating points in a broad band of altitudes and flight speeds. To perform the optimisation, the following control factors are needed: variable stator vanes (VSV) and the air bypass. Variable stator vanes create optimum conditions for the air flow in the LP and HP compressors ($\alpha_{\text{VSV.LP}}$ and $\alpha_{\text{VSV.HP}}$). Bypass air from the first stages of the compressor provides necessary reserves of gas dynamic stability for the fan and compressor. Control of radial clearance in the final stages of the compressor (Δr_c) and in the high-pressure turbine ($\Delta r_{t.HP}$) ensures the maximum efficiency of the turbocompressor. Control of radial clearances in the fan turbine (Δr_{ft}) together with variable stator vanes in the turbine ($\alpha_{VSV.t}$) maintains the maximum efficiency of the turbofan stage. In a bypass turbo jet, some other control factors can be used as well; see Figure 1.1. Exerting an integrated influence on the indicated factors provides control of engine parameters in a broad band of operating conditions. The number of control elements in modern engines can reach 15 or more.

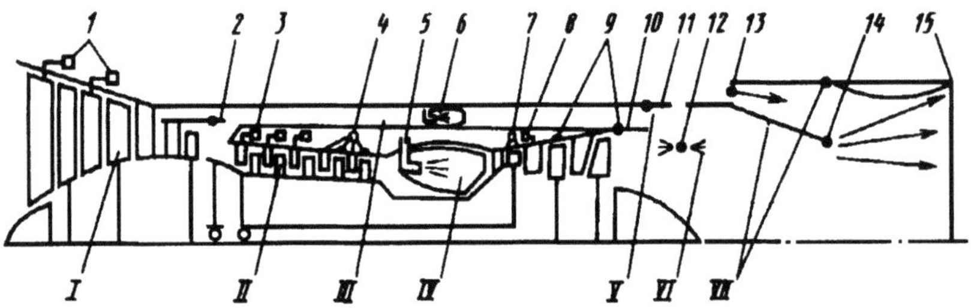

Figure.1.1. Generalised schematic of reheat bypass twin-shaft turbo jet with variable working process

The possible control factors shown in Figure 1.1 include: 1 – variable stator vanes of fan ($\alpha_{VSV.f}$); 2 – air bypass from first stages of compressor into bypass duct (m, π_f^*); 3 – variable stator vanes of compressor ($\alpha_{VSV.c}$); 4 – variable radial clearance in final stages of compressor (Δr_c); 5 – fuel supply in combustion chamber (W_f); 6 – fuel supply in combustion chamber of bypass duct ($W_{f.II}$); 7 – variable radial clearance in HP turbine ($\Delta r_{t.HP}$); 8 – variable stator vanes of fan turbine ($\alpha_{VSV.ft}$); 9 – variable radial clearance in fan turbine (Δr_{ft}); 10 – mixing area (A_{mix}); 11 – system of reverse thrust (F_{rev}); 12 – fuel supply in afterburner ($W_{f.ab}$); 13 – air supply in ejector nozzle (n_{ej}); 14 – critical nozzle area (A_{cr}); 15 – output nozzle area (A_n).

The units of the flowing part of the engine are as follows: I – fan (LP compressor); II – high-pressure compressor; III – bypass duct; IV – main combustion chamber; V – mixing chamber; VI – afterburner; VII – variable jet nozzle.

The programs for turbine engine control are determined with respect to the aircraft requirements. For example, a controller may include programs for maintenance of maximum maneuverability of the aircraft, for maximum efficiency ($C_{sp} \rightarrow \min$) and specific programs for take-off and landing. Each program is determined via optimisation of some criterion or a group of criteria.

Having determined the aircraft requirements, the control programs are transformed into a set of programs for direct control of the turbine engine. The existing principles of control program selection are described in [6–9]. Automatic control systems should ensure the following characteristics:

- engine operation, optimal by thrust and efficiency at demanded conditions;

- fulfilment of requirements for transient operation and control accuracy;
- preservation of stability of working processes and structural strengths of the engine by warning of intolerable mechanical and thermal loads on components during steady-state and transient operation.

1.4 Engine Control Programs

Engine control programs are divided according to the type of engine operation: start-up, transient operation, maximum thrust, normal steady-state operation, minimum thrust and operation with afterburning.

Consider a twin-shaft afterburner turbo jet with six control factors: W_f, A_n, A_{cr}, $W_{f.ab}$, $\alpha_{VSV.HP}$, $\alpha_{VSV.LP}$; see Figure 1.2. Control at start-up and transient operation (acceleration and deceleration) is set by the following relation:

$$W_f / p_c^* = f(n_c, T_{in}^*) \tag{1.5}$$

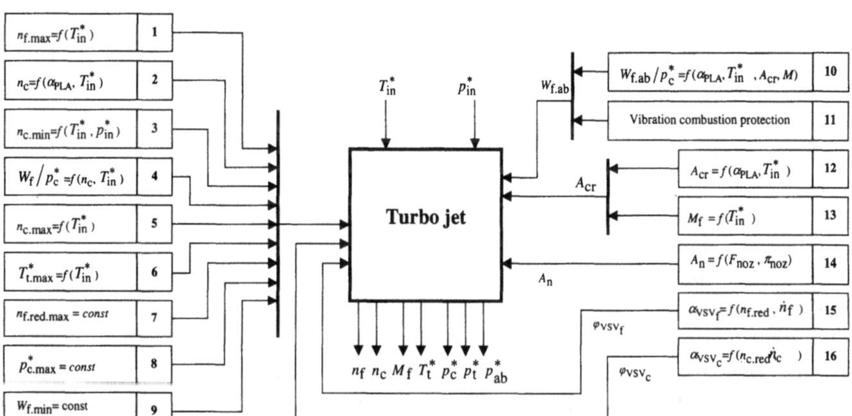

Figure.1.2. Combination of control programs for reheat bypass twin-shaft turbo jet: 1, 2, 3 – shaft speed control at steady-state, minimum and maximum thrust; 4 – control of start-up, acceleration and deceleration; 5–9 – limiters of critical parameters; 10, 11 – fuel flow control in afterburner; 12–14 – control of jet nozzle; 15, 16 – control of compressor

At maximum thrust and steady-state operation, the following program is used, controlling the engine via the fuel flow:

$$n_c = f(\alpha_{PLA}, T_{in}^*) \tag{1.6}$$

Maximum values of some parameters are limited:

- fan shaft speed $n_{f.max} = f(T_{in}^*)$;

- compressor shaft speed $n_{c.\max} = f(T^*_{\text{in}})$;
- turbine temperature $T^*_{t.\max} = f(T^*_{\text{in}})$;
- reduced fan shaft speed $n_{f.\text{red}.\max} = \text{const}$;
- compressor pressure $p^*_{c.\max} = \text{const}$.

Control of the nozzle area:

$$A_{\text{cr}} = f(\alpha_{\text{PLA}}, T^*_{\text{in}}) \tag{1.7}$$
$$A_n = f(A_{\text{cr}}, \pi_n) \tag{1.8}$$

Variable stator vanes are controlled accordingly to:

$$\alpha_{\text{VSV.LP}} = f(n_{\text{LP.red}}, \dot{n}_{\text{LP}}) \tag{1.9}$$
$$\alpha_{\text{VSV.HP}} = f(n_{\text{HP.red}}, \dot{n}_{\text{HP}}) \tag{1.10}$$

Engine control when afterburning is performed as follows:

$$W_{f.\text{ab}} / p^*_c = f(\alpha_{\text{PLA}}, T^*_{\text{in}}, A_{\text{cr}}, M) \tag{1.11}$$

Engine control at minimum thrust is the following:

$$n_c = f(T^*_{\text{in}}, p^*_{\text{in}}) \tag{1.12}$$

Table 1.1 shows control factors and types of automatic control systems used on turbojets produced by the companies Pratt-Whitney (PW) and General Electric (GE), in the United States.

Table 1.2 summarises control programs for a typical turbo jet with afterburning. The synthesis of these programs is performed at the engine design stage using average or estimated characteristics of the engine units in order to fulfil requirements. During design of the automatic control system, the control programs are realised in the engine control unit with maximum accuracy, response speed, etc.

Complication of engine control systems and increase of aircraft requirements have resulted in the necessity to use on-board computers in automatic control systems in addition to, or instead of hydraulic and pneumatic parts. Note that most actuators are still hydraulic and pneumatic devices.

Computational capabilities of modern on-board computers enable optimisation of control programs to be performed, accounting for flight conditions and individual characteristics of the engine. Simultaneously with current control, optimal automatic adjustment of control programs can also be realised. This requires new control methods to be developed.

Table 1.1. Control factors and automatic control systems of aero engines: used (•), not used (○), second-generation supersonic passenger aircraft (SuperPA-II), subsonic passenger aircraft (SubPA), hydromechanical (H/m), electronic (el.)

Engine (company)	Aircraft	Control factors															Controller
		$a_{VSV.f}$	$\dot{m}$	a_{VSVc}	Δr_c	W_f	$W_{f.II}$	$W_{f.mix}$	Δr_{tLP}	$a_{VSV.ft}$	Δr_{tHP}	A_{mix}	F_{rev}	n_{ej}	A_{cr}	A_n	
F101DFE (GE)	Fighter F14, F16	●	○	●	○	●	○	●	●	○	○	○	○	○	●	●	H/m with el. limiter
F101-GE-102 (GE)	Bomber B1B	○	○	●	○	●	○	●	●	○	○	○	○	○	●	●	H/m with el. limiter
F404-GE-400 (GE)	Fighter F18, F5G	○	○	●	○	●	○	●	○	○	○	○	○	○	●	●	El. and H/m (DEEC)
PW1120 (PW)	Fighter F4	○	○	●	○	●	○	●	○	○	○	○	○	○	●	●	DEEC
PW 2037 (PW)	B757, B52G, C17	○	○	●	●	●	○	○	●	○	●	○	○	○	○	○	DEEC
GE21/J11B3 (GE)	Super PA-II	●	●	●	○	●	○	●	○	●	○	●	○	○	●	●	FADEC
VSCE-502B (PW)	Super PA-II	●	●	●	●	●	●	○	○	○	○	○	●	●	○	○	FADEC
E^3 (GE)	SubPA	○	●	●	●	●	○	○	●	○	●	○	○	○	○	○	FADEC
PW4000 (PW)	SubPA B757, A300, A310	○	●	●	●	●	○	○	●	○	○	●	○	○	○	○	FADEC

Table 1.2. Control laws of bypass twin-shaft turbo jet with after burner

Control law	Formulae	Accuracy
Acceleration fuel feed limit	$\frac{\dot{n}_c}{p_c} = f_1(n_{\text{red}}) \cdot f_2(T^*_{\text{in}})$ $f_1(n_{c.\text{red}}) = C_0(n_{c.\text{red}})^2 + C_1(n_{c.\text{red}}) + C_2$ $f_2(n_{c.\text{red}}) = C_3(n_{c.\text{red}})^2 + C_4(n_{c.\text{red}}) + C_5$	± 0.5 %
HP shaft speed limit	$n_c = f(T^*_{\text{in}}, p_H)$	± 0.1 %
LP shaft speed limit	$n_f = f(T^*_{\text{in}})$	± 0.1 %
Gas temperature limit	$T^*_t = f(T^*_{\text{in}}, p_H)$	± 2.0 K
Cruise LP shaft speed control	$\begin{cases} n_{f.\text{red}} = \min[n^{(1)}_{f.\text{red}}, n^{(2)}_{f.\text{red}}] \\ n^{(1)}_{f.\text{red}} = a_1 n_{c.\text{red}} + b_1 + c_1(T^*_{\text{in}} - T^*_5) \\ n^{(2)}_{f.\text{red}} = a_2 n_{c.\text{red}} + b_2 + c_2(T^*_{\text{in}} - T^*_5) \end{cases}$	± 0.1 %
Max and AB LP control	$n_f = f(T^*_{\text{in}}, \alpha_{\text{PLA}})$; $W_{f.\text{AB}} = f(p^*_c)$	± 0.1 %
Select LP law for nozzle area control	$\min\{n_{f.\text{red}} = f(n_{c.\text{red}}), n_{f.\text{red}} = f(T^*_{\text{in}})\}$ $A_{\text{cr}} = f(\pi_n)$	N/A
Inlet guidance vanes (IGV)	$\alpha_{\text{IGV}} = f(n_{f.\text{red}})$	± 0.2 %
Air inlet	$\alpha_{\text{wedge}} = f(n_{f.\text{red}})$	± 2.5 %

1.5 On-line Optimisation

In this section, on-line optimisation of the controller is considered for a turbine engine at cruise. Such cruise flight takes place under the following conditions:

- maximum range flight with static ceilings at $H \geq 11$ km ($M =$ const, $H\uparrow$);
- flight with maximum range at altitudes $0 < H \leq 11$ km at a fixed operating point of the engine and changes in the flight altitude and speed with fuel use;
- flight at H = const, M = const; the engine operating point changing with fuel use.

Some other general conditions for cruise flight can be defined – for example, minimum flight time and maximum flight range corresponding to minimum fuel consumption per kilometre.

1.5.1. Example of Maximum Distance Calculation at Cruise Flight

In this example, maximum flight range L is chosen to be the objective function when solving the problem of optimum control of a power plant at cruise flight with H = const, M = const:

$$L \to \max \tag{1.13}$$

The aircraft flight range on a straight steady-steady flight is calculated by the formulae:

$$L = \int_{W_2}^{W_1} \frac{K_a}{g} \overline{F}_{\text{ef}} \eta_0 H_u \frac{dW}{W} = L_{\text{sp}} \ln \frac{W_1}{W_2} \quad \text{or} \tag{1.14}$$

$$L = \int_{W_2}^{W_1} \frac{K_a v}{g C_{\text{sp}}} \overline{F}_{\text{ef}} \frac{dW}{W} = L_{\text{sp}} \ln \frac{W_1}{W_2}\ , \tag{1.15}$$

where $K_a = C_y / C_x$ is the lift-to-drag ratio of the aircraft at cruise flight;
H_u is the low calorific value of fuel;
g (m/s^2) is the acceleration of gravity;
$\eta_{0\text{pp}} = \eta_0$ is total efficiency of the power plant;
η_0 is total efficiency of the turbine engine;
v (m/s) is the aircraft flight speed;
$\underline{C}_{\text{sp.ef}}$ is effective specific fuel consumption of the power plant;
$\overline{F}_{\text{ef}} = F_{\text{ef}} / F_{\text{eng}}$ is the effective thrust factor accounting for relative thrust reduction because of the external drag of the power plant;
W_1 and W_2 are the initial and final values of the aircraft weight on a cruise flight;

$$L_{\text{sp}} = \frac{K_a}{g} \overline{F}_{\text{ef}} \eta_0 H_u = \frac{K_a v}{g\, C_{\text{sp}}} \overline{F}_{\text{ef}} \tag{1.16}$$

is the conditional distance of the flight.

The theoretical distance of a cruise flight $L_{\text{theor}}(M)$ can be obtained at v = const, K_a = const, $C_{\text{sp.ef}} = C_{\text{sp}} / \overline{F}_{\text{ef}}$ = const, $\eta_{0\text{pp}} = \eta_0 \overline{F}_{\text{ef}}$ = const, H_u = const from Equation (14) and (15):

$$L_{\text{theor}} = \frac{K_a}{g} \eta_0 \overline{F}_{\text{ef}} H_u \ln \frac{1}{1 + m_f}\ ; \tag{1.17}$$

$$L_{\text{theor}} = \frac{K_a}{g} \frac{V \overline{F}_{\text{ef}}}{C_{\text{spec}}} H_u \ln \frac{1}{1 + m_f}\ , \tag{1.18}$$

where $m_f = \Delta W_f / W_1$ is the ratio of the weight of the fuel spent to the initial weight of the aircraft in the beginning of the cruise segment;

$C_{sp}(\mathrm{kg}/N \cdot s)$ is the specific fuel consumption of the engine.

1.5.2. Power Plant Efficiency

The total efficiency of the power plant $\eta_{0\mathrm{pp}}$ represents the ratio of the useful power effect (useful work of moving the flight vehicle N_w) to the overall power spent in a unit of time (energy of fuel $N_{\mathrm{spent}} = N_{f.\mathrm{chem}} + N_{f.\mathrm{kin}}$):

$$\eta_{0\mathrm{pp}} = \frac{N_w}{N_{\mathrm{spent}}} = \frac{F_{\mathrm{ef}} v}{N_{f.\mathrm{chem}} + N_{f.\mathrm{kin}}} = \frac{F\overline{F}_{\mathrm{ef}} v}{W_{f.s}\left(H_u + v^2/2\right)} \approx \frac{F\overline{F}_{\mathrm{ef}} v}{W_{f.s} H_u}, \tag{1.19}$$

where $F_{\mathrm{eng}} = f(H)$ is the internal engine thrust;

$W_{f.s}$ is fuel consumption per second.

Using the relation $C_{\mathrm{sp.ef}} = W_{f.s} / (F_{\mathrm{eng}} \overline{F}_{\mathrm{ef}})$, the total power plant efficiency is obtained:

$$\eta_{0\mathrm{pp}} = \frac{v}{C_{\mathrm{sp.ef}} H_u}. \tag{1.20}$$

Analysis of Equation (1.11), (1.14) and (1.15) shows that the cruise flight range is maximum with the maximum value η_0 or with minimum specific fuel consumption $C_{\mathrm{sp.ef}}$, given all other factors being equal. Thus, the maintenance of $\eta_{0\mathrm{pp}} \to \max$ or $C_{\mathrm{sp.ef}} \to \min$ can be adopted as the objective function for the engine control during cruise flight:

$$C_{\mathrm{sp.ef}} = C_{\mathrm{sp}} / \overline{F}_{\mathrm{ef}} \to \min. \tag{1.21}$$

Note that the specific fuel consumption (being equal to the ratio of fuel consumption per hour to the engine thrust) does not fully represent total efficiency of a turbine engine within a power plant, because the same thrust produces different work at different flight speeds.

Consider Equation (1.21). Specific fuel consumption is connected to the total engine efficiency η_0 by the relation

$$C_{\mathrm{sp}} = \frac{v}{H_u \eta_0} = \frac{W_f}{F}.$$

In order to define the value of C_{sp} it is necessary to measure the thrust F and the fuel flow W_f. However, the total efficiency can be calculated based on internal parameters of the engine without measuring W_f and F. Both η_0 and F depend on the following parameters of the working process:

- engine pressure ratio π^*_{eng},
- temperature of the propulsive mass $\theta = T^*_g / T_H$,
- by-pass ratio m,
- efficiency of the engine units etc.

Therefore, when controlling demanded values of the parameters of the working process, the required thrust and effectiveness characteristics of the engine are maintained indirectly.

1.5.3 Limitations of Current Controllers

Presently, control programs for engines are developed based on the average characteristics of the engine and the flight speeds and altitudes (or parameters describing v and H). Demanded values are set as a function of the pilot's lever angle (PLA). For example, when not afterburning, the controlled engine parameters are shaft speeds n or compressor pressure ratio π^*_c and gas temperature before the turbine T^*_g. The following control factors are used: fuel flow into the main combustion chamber W_f and nozzle area A_n. Here the demanded values of the controlled parameters are selected to ensure optimum technical and efficiency characteristics on given flight path segments of the aircraft in expected atmospheric conditions (at a standard sectional view of the atmosphere). In this case, the control programs are not optimal if the engine is operating in non-standard atmospheric conditions.

Such a control approach has a number of essential disadvantages originating from the restricted functionality of hydraulic and pneumatic control systems:

- the control programs do not account for deviations between the characteristics of engine units and their degradation due to wearing and ageing, which reduce engine efficiency. Additionally, different combinations of control parameters can be used to achieve the demanded thrust;
- the control programs do not take into account possible deviations of the air temperature compared with the standard relation $T_H = f(H)$. These deviations can reach values $\Delta T = \pm 20^\circ\text{C}$ with the probability $p = 0.993$. This can affect internal thrust of the engine F_{eng}, the effective thrust F_{ef} of the power plant, the flight lift-to-drag ratio of an aircraft K_a and the required thrust of the power plant. In such circumstances, the demanded values of controlled parameters calculated within the controller based on the pilot's lever angle are not optimum;
- the influence of additional control factors ($\alpha_{VSV.t}$, $\alpha_{VSV.c}$, $W_{cooling}$ etc.) on main parameters of the working process is not considered;
- a control program, optimum for the expected flight conditions H and M, is not optimum at deviations from such conditions, i.e. at $H \neq H_{exp}$, $M \neq M_{exp}$.

As a result, the optimum values of controlled parameters demanded by PLA deviate from actual.

The factors mentioned above require correction of the "nominal" conventional control program for actual engine characteristics and flight conditions. Computer-based electronic control systems open possibilities for adaptive correction of control programs with feedback on global optimisation criteria.

In particular for cruise flight, such a correction program could be activated automatically or by a pilot's command. This would enable the control system to maintain $\eta_{0\mathrm{pp}} \to \max$ with constant H and M. The value $\eta_{0\mathrm{pp}}$ can be adjusted at $M = \mathrm{const}$ by changing W_f and A_n and evaluated by internal engine parameters measured with appropriate sensors.

1.5.4. Design of Optimising Control

The advent of digital controllers allows the incorporation of more complex control strategies. These can be used to remove the existing limitations found in current gas turbine engine controllers. In this section an optimising control strategy is outlined for a typical military gas turbine engine. Control for $\eta_{0\mathrm{pp}}$ can be obtained in the following way:

$$\eta_{0\mathrm{pp}} = \eta_0 \overline{F}_{\mathrm{ef}} = \frac{F\overline{F}_{\mathrm{ef}} v}{N_{\mathrm{spent}}} \tag{1.22}$$

$$F_{\mathrm{ef}} = W_a F_{\mathrm{sp}} \approx W_a (C_n - v) \tag{1.23}$$

$$N_{\mathrm{spent}} = Q_{\mathrm{comb}} / \eta_g = \left(J_n^* - J_H^*\right) / \eta_g \approx W_a \left(C_{p_g} T_n^* - C_{p_a} T_H^* \right) / \eta_g \tag{1.24}$$

$$C_n = \varphi_n \sqrt{2 C_{p_g} T_n^* \left[1 - (p_H / p_n^*)^{\frac{K_g - 1}{K_g}} \right]} \tag{1.25}$$

where $T_n^* = T_t^*$ for a turbo jet and $T_n^* = T_{\mathrm{mix}}^*$ for a bypass engine. In view of Equation (1.22)-(1.25) the following is obtained:

$$\eta_{0\mathrm{pp}} = \frac{\left\{ \varphi_n \sqrt{2 C_{p_g} T_n^* \left[1 - \left(p_H / p_n^* \right)^{(K_g - 1)/K_g} \right]} - v \right\} v \overline{F}_{\mathrm{ef}} \eta_g}{C_{p_g} T_n^* - C_{p_a} T_H^*} \tag{1.26}$$

In Equation (1.14), the values $\varphi_n, C_{p_g}, K_g, \overline{F}_{\mathrm{ef}}, \eta_g, C_{p_a}$ can be set constant. In this case, the influence of possible inaccuracies will be eliminated. The optimisation algorithm will include consecutive comparison of the values $\eta_{0\mathrm{pp}}$ with changing control factors W_{f} and A_{n} in order to maintain $\eta_{0\mathrm{pp}} \to \max$. The value $\eta_{0\mathrm{pp}}$ is calculated via Equation (1.24) using temperature and pressure measurements. Random measurement errors can be reduced at the expense of averaging of repeated sensor reads at each step of control.

Consider a possible approach to the introduced correction by maximising the total efficiency for a bypass turbo jet with mixing gas flows. Assume that the nominal engine control system maintains a cruise flight with v=const and H=const. The corresponding effective thrust is defined by the formula:

$$F_{\text{ef}} = W_{a\Sigma}\left\{\varphi_{\text{n}}\sqrt{2C_{p_g}T_n^*\left[1-\left(p_H/p_n^*\right)^{(K_g-1)/K_g}\right]}-v\right\}\overline{F}_{\text{ef}}\,; \tag{1.27}$$

where $W_{a\Sigma} = m_a p_{\text{in}}^* A_{\text{in}} q(\lambda_a)\big/\sqrt{T_{\text{in}}^*}$ (1.28)

The temperature $T_{\text{in}}^* = T_H^*$ and pressure p_{in}^* are measured in the inlet device before the fan. p_n^* and T_n^* are measured or evaluated. All coefficients are considered known. The effective engine thrust can be estimated with the measured engine pressure ratio $\pi_{\text{eng}}^* = p_n^*/p_H^*$ and critical nozzle area A_{cr}. The full efficiency of the power plant is calculated by Equation (1.24).

The effective thrust factor can be considered constant $\overline{F}_{\text{ef}} = \text{const}$, if the operation of the air intake and output equipment is automatically co-ordinated with the engine operation. This results in $F_{\text{ef}} = \text{const}$ under the condition of $F_{\text{eng}} = \text{const}$ and $v = \text{const}$. In these circumstances, the overall efficiency can be consecutively improved with $F_{\text{eng}} = \text{const}$; the system of automatic correction can optimise the engine operation by changing control factors. Finally, the engine would reach the point of $\eta_0 = \max$, thus ensuring the maximum flight length at $v = \text{const}$, $H = \text{const}$. The example considered above shows how the maximum overall engine efficiency could be used as the criterion for optimisation.

1.6 Concluding Remarks

In this chapter, the problems of selection and coordination of control programs for turbine engines have been discussed. Conventional approaches to engine control design consist of optimisation of control programs for a generalised engine and average flight conditions. Therefore, the best values of controlled parameters deviate from that expected because of characteristic deviation of engine units, gradual wearing of the engine and changes in the engine operation conditions.

Modern aero engines possess a large number of control factors. This creates necessary conditions for optimising the operation of the power plant and aircraft as a whole using an on-board computerised controller.

The potential to incorporate optimising control in modern digital controllers offers great opportunities for minimisation of fuel consumption, leading to greater range and reduced emissions. In addition, the control system can be optimised to reduce stress on engine components, leading to reduced maintenance costs and greater longevity.

PART I. GAS TURBINE MODELS

Chapter 2
Models and the Control System Design Cycle

2.1 Introduction

In this chapter techniques for mathematical modelling of gas turbine engines are presented considering the different stages in the life cycle from design, to demonstration, to commercial control system development, and finally through to in-service use. Through these stages the fidelity of the engine model is refined, as more information becomes available. The chapter thus considers evolution of models from the initial stages of design, where the general characteristics of the engine are known, to the latter stages, where a controller is tailored to the specific characteristics of the engine. The models required over the life cycle are discussed in greater detail in subsequent chapters. The key modelling stages are as follows:

- nonlinear performance–based modelling;
- dynamic characteristic modelling;
- linear dynamic modelling;
- piecewise linear dynamic modelling for real-time development.

In this chapter, a brief introduction is given to each of these techniques to highlight the use of the models at the different life cycle stages. It should be noted that the controller is an essential part of a GTE (as well as of the whole aircraft power unit). In practice, the engine and controller are designed simultaneously and are cooperatively tested. Therefore, the engine and controller are an inseparable integrated unit, consisting of transducers, converters, digital control electronics and actuators. The system has to be modelled as a whole considering the transducers and actuators to be part of the plant. Within this framework the control algorithms can be adjusted to provide the desired levels of performance.

2.2 Mathematical Models and Controller Life Cycle

The life cycle of the engine and its controller can be split into the following stages: scientific research, engineering design, demonstration, mass production, and the usage stage.

The *controller design stage* requires construction of a system with standard modules (transducers, actuators, processors, and system software), control algorithms, software development, controller synthesis and operational development. At this stage, mathematical models of the plant are widely used. After controller performance has been tested with test beds and engines, flight tests are performed (Figure 2.1). Joint GTE and controller tests enable experimental data to be gathered for system identification and controller operational development. Hence, design models are refined to reflect the individual characteristics of a particular engine, and the control algorithms, which are originally developed using a general engine model, are improved using more precise individual models.

Different accuracies and degrees of detail for the mathematical models are required at various stages of the life cycle of a controller (see Table 2.1). The degree of approximation to individual engine characteristics also differs.

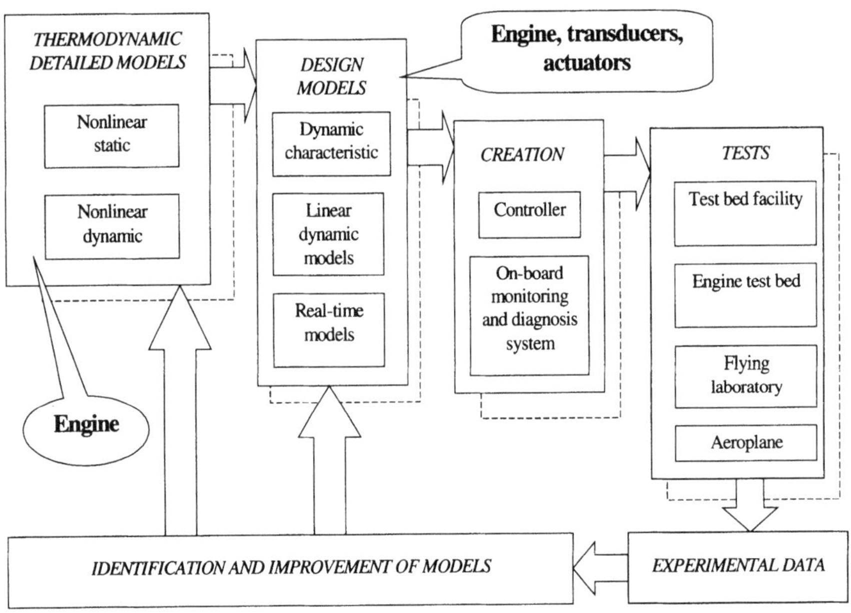

Figure.2.1. Design of controllers for gas turbine engines

Table 2.1. Engine modelling through controller life cycle

Stage	Models	Purpose
Design	General deterministic (simplified thermodynamic)	Control algorithms
Demonstration, tuning, production	Improved general + stochastic	Quality control
In-service use	Individual models	Monitoring and diagnosis, adaptation and optimisation

2.2.1 Design Stage Models

At the design stage, only general characteristics and parameters of the plant are necessary and the mathematical models are greatly simplified compared with the detailed nonlinear models used in engine design [1]. The design model of an engine is obtained by means of simplification of thermodynamic models. During the controller design stage, it is necessary to take into account that engine characteristics vary within broad limits due to flight and atmospheric conditions as well as due to gradual aging and deterioration of engine elements. Some change in characteristics also occurs after repair and replacement of elements of the engine and controller. Hence, the design model should reflect general characteristics of a whole class of engines. As a result, a controller is designed for an average plant. This is followed by individual tuning after installation of the controller on a particular power plant.

2.2.2 Experimental Demonstration and Production Stage Models

During the experimental demonstration and production stage, mathematical models are used for monitoring the fulfilment of requirements with respect to quality of control. Test bed facilities are designed for testing manufactured digital controllers. The term *test bed* refers to testing the controller, with the engine imitated by mathematical models. Such substitution provides a decrease in the amount of experimental research required on real engines and a considerable decrease in the research cost. Existing test beds for electronic controllers contain an electronic model of the controlled plant, imitators of transducers and actuators, and a logging device for recording and analysing experimental data. The most important components of such test beds are real-time dynamic models. Usually, these are deterministic linear dynamic models combined with nonlinear static models.

2.2.3 In-Service Use Models

In-service models can be used for health monitoring and on-line optimisation. These need to take into account the characteristics of the specific engine and the current operating conditions. A promising approach in this area is the use of identification techniques to produce accurate estimated models both off-line and on-line. A schematic of how this could be used in practice for condition monitoring is shown in Figure 2.2. Here, identification techniques are shown being used to aid in preflight check and scheduled maintenance to monitor the condition of the engine. The main types of engine modelling for control engineering purposes are discussed more thoroughly in the sections below; new trends in the development and application of such models are also described.

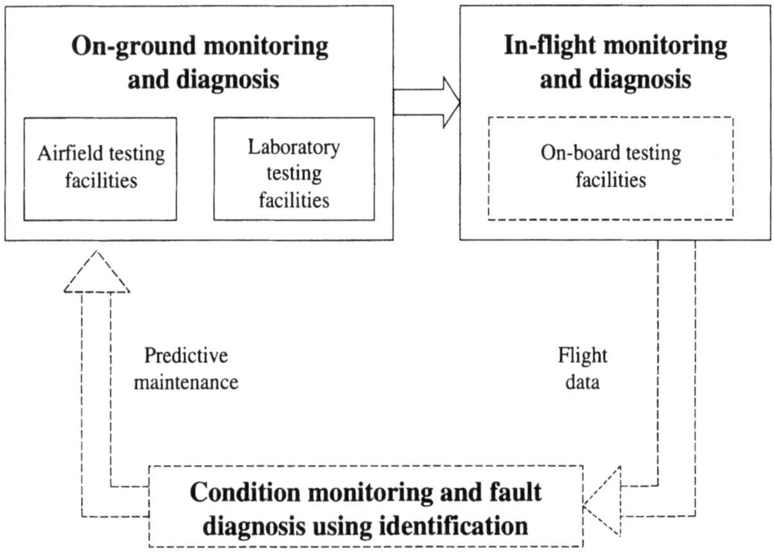

Figure.2.2. Use of identified models in scheduled maintenance

2.3 Dynamic Modelling Techniques

During the controller design stage, a number of dynamic models are derived from a detailed nonlinear performance-based model comprised of dynamic and static models. The aim of the nonlinear static model is to investigate engine characteristics during its creation. The model includes, for instance, the altitude-velocity and fuel-scheduling characteristics. Such models are based on gas dynamics laws and typically include over 100 algebraic and transcendent equations [2-4]. The characteristics are partly defined in the form of empirical formulae, graphs and tables. The static operating line is then calculated by means of iterative numeric methods.

The detailed nonlinear dynamic model is similar to the static model, but its equations are written in dynamic form. These show the balance between characteristics of GTE elements in the form of continuous flows and energy conservation equations. In algebraic form, they define a static model; with differential notation defining a dynamic model. The number of determining parameters in this model depends on the engine scheme and is defined by the number of energy accumulators. The purpose of the detailed nonlinear dynamic model is to analyse engine transients. Although the above-mentioned model can be used for controller performance analysis, a large amount of calculation is involved and so generally simplified models are used. However, as processing power increases, it is becoming more feasible to use this model directly.

2.4 Dynamic Modelling for Control Systems Development

In this section, dynamic modelling techniques are introduced for controller design purposes including dynamic characteristic models (DCM), linear dynamic models (LDM) and real-time piecewise linear dynamic models (RPLDM). The starting point in any gas turbine engine controller development is the performance-based model. This model is a highly complex thermodynamic model of the engine. A full description of this type of model is given in Chapter 3. The detailed performance-based nonlinear model can be simplified to create a set of dynamic models. An analytical relationship exists between the model's dynamic and static parts; this relationship allows dynamic models of the controlled plant to be obtained quickly. As an example considering a single-shaft gas turbine engine, the detailed nonlinear static model is represented by the following system:

$$\begin{cases} \mathbf{f}_x(\mathbf{X},\mathbf{U},\mathbf{V}) = \mathbf{0} \\ \mathbf{Y} = \mathbf{f}_y(\mathbf{X},\mathbf{U},\mathbf{V}) \end{cases} \tag{2.1}$$

where $\mathbf{X} = [n, p_g^*, T_g^*, T_n^*, p_n^*]^T$ is the state vector;
$\mathbf{U} = [W_f, A_n]^T$ is the control vector;
$\mathbf{V} = [M, H, p_H, T_H]^T$ is the flight atmospheric conditions vector;
$\mathbf{Y} = [n_{\text{red}}, p_c^*, \pi_c^*, \ldots]^T$ is the observed coordinates vector.

Taking the detailed dynamic model from the performance-based model, Equation (2.1) can be transformed into a system of differential equations with the same vectors **X**, **U**, **V** and **Y**:

$$\begin{cases} \dot{\mathbf{X}} = \mathbf{F}_x(\mathbf{X},\mathbf{U},\mathbf{V}) \\ \mathbf{Y} = \mathbf{F}_y(\mathbf{X},\mathbf{U},\mathbf{V}) \end{cases} \tag{2.2}$$

In steady-state conditions, all derivatives are equal to zero, and Equation (2.1) transforms into the system Equation (2.2). Consider a single-shaft engine static model and dynamic model. For this the compressor–combustor continuous flow equation is as follows:

$$W_g - W_a - W_f = \Delta W_a^{\text{dyn}} = Z_1 \Rightarrow 0 \tag{2.3}$$

where Z_1 is a discrepancy or an equality violation factor. By performing a differentiation on the gas condition equation for the combustion chamber with:

$$\frac{pV}{T} = mR \tag{2.4}$$

the following equation is obtained:

$$\dot{p}_g^* = \frac{RT_g^*}{V_{\text{comb}}} \Delta W_a^{\text{dyn}} + \frac{p_g^*}{T_g^*} \dot{T}_g^* \tag{2.5}$$

Equation (2.5) provides a link between the pressure and temperature derivatives and the dynamic addition ΔW_{dyn}. Equations like (2.5) linking (2.1) and (2.2) are the basis for the methodology of dynamic modelling using a detailed static model and additional information about the engine construction. For instance, Equation (2.5) includes the combustion chamber volume V_{comb}.

The linear dynamic model can be obtained from the detailed nonlinear model using identification or linearisation, provided the model structure is determined *a priori* from thermodynamic model analysis. Usually, the linearity zone for such parameters as shaft speed, temperature and pressure is approximately 3-5%. In order to determine the linear dynamic model coefficients, a transient with small amplitude is used around the steady-state conditions. Linear models describe engine dynamic properties in the form of linear differential equations and transfer functions. These models are very visual and convenient for stability and control quality analysis. However, the accuracy of results decreases due to approximation.

2.5 Dynamic Characteristic

The dynamic characteristic (DC) is a graphical image of the nonlinear dynamic model [2]. In addition to the static line $\dot{n} = 0$, the DC of a single-shaft engine also includes a set of lines of constant accelerations $\dot{n}$ =const. In order to derive a DC from the static model, a series of operating points is calculated. The operating conditions are formed with varying discrepancies, which are then converted into corresponding accelerations as described above. A full explanation of the dynamic characteristic models is given in Chapter 3. When creating the DC, solutions are selected in which all the discrepancies are equal to zero except the discrepancy determining the mechanical energy balance between the turbine and compressor. The DC derivation from the detailed dynamic model is performed by means of setting up the control laws $\dot{n} = \text{const}$, $p_c = \text{const}$, $T_t = \text{const}$ and calculating the corresponding lines in the engine parameters plane with the axes: "shaft speed *versus* fuel flow."

The DC is widely used because of the visual comprehensibility of engine dynamic properties and because it retains the main shape of the engine nonlinearity [3]. However, the DC of engines with a lot of variable geometry is multi-dimensional. Significant assumptions and simplifications are introduced to represent the dynamic characteristic on the plane or in 3D space, which leads to the loss in their visual comprehensibility.

2.6 Adaptation of Controller Structure and Parameters at the Design Stage

The engine operation and flight conditions are usually known at the design stage. This enables the controller's properties to be altered within the aircraft flight envelope. The GTE characteristics for any flight conditions can be accurately predicted using a detailed nonlinear model. However, this would require a large amount of calculation. Therefore, the engine model is usually simplified to produce the DC combined with the formulae for the reduction to the sea-level conditions with linear interpolation [4].

The flight envelope is divided into three subareas, different in terms of gas-dynamic similarity (Figure 2.3). In area **I**, the engine is described by a single DC in reduced coordinates, and fairly accurate results are produced by the reduction formulae:

$$n_{\text{red}} = n_{\text{phys}}\sqrt{\frac{288.15}{T^*_{\text{in}}}}, \quad \dot{n}_{\text{red}} = \dot{n}_{\text{phys}}\frac{101325}{p^*_{\text{in}}}, \ etc. \qquad (2.6)$$

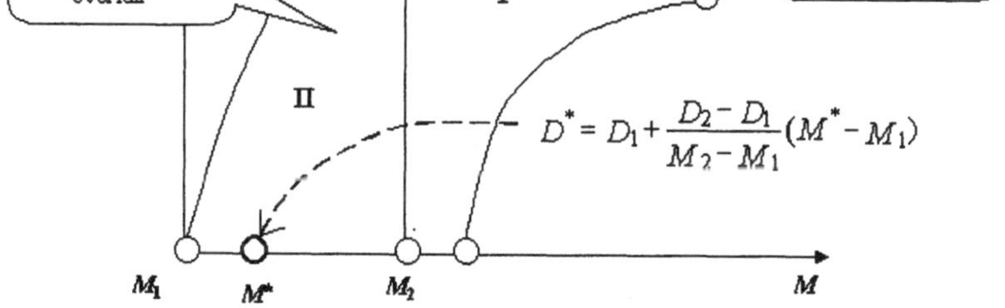

Figure 2.3. Adaptation of controller to flight conditions

In order to use the relationships indicated in Equation (2.6), the gas dynamic similarity conditions should be fulfilled. This includes the constancy of the engine geometry and the air properties such as humidity and *Re* number.

In other areas, these relationships are correct only at a qualitative level. In area II, the pressure overfall at the reactive nozzle becomes significant, and consequently, the DC is "stratified," and a separate model is created for each operating point. In area III, the effect of the *Re* number change is additionally taken into account.

In such a case, correcting coefficients are usually introduced into the thermodynamic model computations. In practice, the dynamic characteristics for particular flight conditions are determined by means of linear interpolation between the operating points.

A control program in reduced parameters:

$$n_{\text{red}} = f(\alpha_{\text{PLA}}) \tag{2.7}$$

is satisfactory if the engine is described by a single dynamic model. The program in Equation (2.7) sets up the demanded reduced shaft speed as a function of the angular position of PLA. Such control holds the engine within the gas dynamic stability region.

During transients, the $\dot{n}$ acceleration control program is used, where the shaft speed derivative is reduced using the engine inlet pressure. The various reduced parameter equations can also be used to control the transient operation. The temperature and pressure constraint programs are additionally introduced over the flight envelope to ensure gas dynamic stability. Thus, the controller acquires the ability to adapt in an open-loop scheme using *a priori* knowledge of the controlled plant properties. Note that the programs adapt to the flight conditions dependent upon p_{in}^* and T_{in}^*.

2.7 Linear Dynamic Models

This type of model is widely used in design and analysis of control systems. Linear dynamic modelling techniques are well established compared to nonlinear modelling. Control programs for steady-state conditions are usually designed and investigated using linear models.

Linear models are obtained from performance-based models via linearisation or identification methods. Also, they can be estimated from experimental data. Linear dynamic models are usually used in the form of state-space equations, transfer functions or Bode diagrams. These describe engine dynamics around a steady-state operating point. In Chapter 3, this modelling technique is discussed in more detail.

However, when exploring control programs for transient operation, linear models should incorporate nonlinear properties of the engine. The next section outlines a solution for this problem.

2.8 Real-time Piecewise Linear Dynamic Models

The real-time piecewise linear dynamic model (RPLDM) combines the engine's nonlinearity and the LDM linearity. The source data for building the RPLDM are the nonlinear static lines and the LDM coefficients. The static characteristics are approximated by piecewise linear relationships determined at typically 7–10 operating points. The static line parameters between these points are then determined by means of interpolation. For the LDM coefficients, interpolation is used in the same manner. This is described in detail in Chapter 4. At the same time, it is assumed that the linear model parameters during transients are equal to the LDM coefficients for the closest static line point. The final calculated relationships Equation (2.8) establish the link between the **A**, **B**, **C** and **D** matrices of the LDM coefficients and the static line coordinates $\mathbf{X}_{st}$, $\mathbf{U}_{st}$ and $\mathbf{Y}_{st}$ by means of the operating parameter η :

$$\begin{aligned} \dot{\mathbf{X}}(t) &= \mathbf{A}(\eta)\big(\mathbf{X}(t)-\mathbf{X}_{st}(t)\big)+\mathbf{B}(\eta)\big(\mathbf{U}(t)-\mathbf{U}_{st}(t)\big) \\ \mathbf{Y}(t) &= \mathbf{C}(\eta)\big(\mathbf{X}(t)-\mathbf{X}_{st}(t)\big)+\mathbf{D}(\eta)\big(\mathbf{U}(t)-\mathbf{U}_{st}(t)\big)+\mathbf{Y}_{st} \end{aligned} \tag{2.8}$$

$$\eta = \sum_{i=1}^{n} z_i x_i(t) \tag{2.9}$$

RPLDM provides an acceptable level of accuracy and can be used in real time due to its simplicity as compared with a detailed nonlinear dynamic model. This enables control algorithms to be effectively tested on models and the controllers to be tested on electronic test beds. RPLDM models also have the potential to be used within on-board engine condition monitoring systems.

2.9 Control System Testing

Once a controller has reached this stage of design it is necessary to perform testing. There are several stages of testing which begins with a System Test Facility (STF) leading to an engine test bed and finally on the aircraft itself.

During initial testing it is common to use a STF. In a STF the control hardware and associated control laws can be exercised against a high fidelity model of the engine run in a computer. The control hardware may also include key actuation elements of the system for probing. This is commonly called hardware-in-the-loop or HIL.

In the second stage of testing the controller is exercised on the real engine on an engine test bed. This testing brings the operational environment of the controller closer to real-life operation. It is also common to operate an engine at an altitude test facility to explore the effects of the flight envelope.

The final stage of testing is on-wing testing where the engine is operated on a real aircraft and tests are performed over the flight envelope.

The different levels of testing become increasingly expensive as they approach real life. It is thus preferred to perform as many tests as possible at the STF stage. This is driving the development of more accurate identification and modelling techniques.

2.10 Concluding Remarks

This chapter has given a brief overview of the use of engine models at different stages in the controller development cycle. In Chapter 3, the derivation of off-line models is discussed in more detail. In Chapter 4, on-line dynamic models are discussed. This allows an appreciation of the relationships between the models and their derivation for specific control engineering needs.

The overview of mathematical models of gas turbine engines at life cycle stages reveals a certain contradiction between individual and general engine characteristics. The more accurate the mathematical models are, the closer they approach the particular plants, whereas their general properties are not taken into account. A general model for a whole class of engines is created in the process of the controller design. At the same time, characteristics of each particular engine are different from the general predicted model. A "draft" model of an engine can be more accurately defined by analysing characteristics of several engines. At the controller design stage, the general models enable the system characteristics to be improved. A controller created for general GTE is adapted individually for each particular engine. Every engine has its own individual properties and control characteristics. This is why, at the usage stage, the requirements qualitatively change, and every individual model provides useful information, especially if it is used for condition monitoring and fault diagnosis on the ground and in the air.

Mathematical models of gas turbine engines are extensively used in testing hardware, software, control laws and algorithms of the controller. This drives further investigation into identification and modelling of gas turbines to allow cost-effective testing based on computer simulation.

Chapter 3
Off-line Models

3.1 Introduction

The level of complexity of aero gas turbine engine modelling in terms of dynamic and static accuracy can vary dramatically depending on the model's intended usage: research, analysis or synthesis of control systems. In this chapter the development of models for off-line use is discussed. Typically these models would be used in simulation exercises during engine design and development of the control system to establish performance criteria. For example, the stability study of a system for a given operating point can be conducted using classical control theory methods [1, 2] on a linear model of a turbine engine. However, the same problem can be solved with the use of a detailed thermodynamic model in the time domain.

Using a nonlinear model has the advantage that it deals with the errors of numerical integration methods; however, this is at the expense of computation which can increase tenfold compared to linear analysis. Using the simpler linear model allows more rapid development cycles, but the estimation of errors represents an additional problem. In this chapter the development of detailed performance models of a gas turbine engine will be first introduced followed by the derivation of simpler models that give the dynamic characteristics of the engine. From these, linear dynamic models suitable for assessment of stability and robustness of control at set operating points will be derived. This analysis is essential to prove to the certification authorities that the control strategy meets the requirements for performance and safety.

Generalisation of existing experience in the development and usage of mathematical models of aero engines enables the following requirements to be formed:

- aggregation (representation of a model as a set of standard modules);
- accuracy (model complexity should agree with the confidence and amount of the source data);
- frequency band limitation in simulation of the engine dynamics;
- possibility to identify the model characteristics from engine and flight test data;
- capability for changes in the scheme of the modelled engine;
- potential for updating and extension of the source information.

The listed requirements enable effective methods for mathematical modelling of aero engines for control development to be devised.

Optimisation of general performance of an engine during control design is usually performed in two stages. Firstly, control programs are determined based upon a detailed static model. Secondly, the structure and parameters of the control system are developed in order to fulfil demanded accuracy under various parametric and external disturbances. Therefore, control systems design requires deep inside knowledge of the general principles of engine modelling and relations between static and dynamic models.

3.2 Detailed Nonlinear Static Modelling

Consider a general form for expressing detailed static and dynamic models on an example of a single-shaft turbine engine [3]. Under certain assumptions, nonlinear equations describing operation of engine units take following form:

$$
\begin{aligned}
&\left.\begin{aligned}
&1.\ T_H = 288 - 0.0065H \\
&2.\ p_H = p_0 (T_H / T_0)^{5.25}
\end{aligned}\right\}, \quad \text{if } H < 11{,}000\ \text{m} \\
&\left.\begin{aligned}
&3.\ T_H = 216.5\text{K} \\
&4.\ p_H = p_{11{,}000} \exp\left(\frac{11{,}000 - H}{6{,}318}\right)
\end{aligned}\right\}, \quad \text{if } H > 11{,}000\ \text{m} \\
&5.\ T^*_{\text{in}} = T_H (1 + 0.2M^2) \\
&6.\ p^*_{\text{in}} = \sigma_{\text{in}} p_H (1 + 0.2M^2)^{3.5} \\
&7.\ n_{\text{red}} = n\sqrt{288/T^*_H} \\
&8.\ p^*_c = p^*_g / \sigma_{\text{comb}} \\
&9.\ \pi^*_c = p^*_c / p^*_{\text{in}} \\
&10.\ W_{a.\text{red}} = f(\pi^*_c, n_{\text{red}}) \\
&11.\ W_a = f(W_{a.\text{red}}, T^*_{\text{in}}, p^*_{\text{in}}) \\
&12.\ \eta^*_c = f(\pi^*_c, n_{\text{red}}) \\
&13.\ T^*_c = f(\eta^*_c, \pi^*_c, T^*_{\text{in}}) \\
&14.\ p^*_t = p^*_{\text{noz}} / \sigma_{\text{noz}} \\
&15.\ \pi^*_t = p^*_g / p^*_t \\
&16.\ n_{t.\text{red}} = n\sqrt{288/T^*_g} \\
&17.\ W_{g.\text{red}} = f(\pi^*_t, n_{t.\text{red}}) \\
&18.\ W_g = f(W_{g.\text{red}}, T^*_g, p^*_g)
\end{aligned}
\qquad (3.1)
$$

19. $\eta_t^* = f(\pi_t^*, n_{t.\text{red}})$

20. $T_t^* = f(T_g^*, \pi_t^*, \eta_t^*)$

21. $W_n = f(A_n, p_n^*, T_n^*)$

22. $N_t = f(W_g, T_g^*, T_t^*)$

23. $N_c = f(W_a, T_{\text{in}}^*, T_c^*)$

In addition to Equations (3.1), one should take into account the following balance relationships: air/gas flow balance, thermal balance in the combustion chamber and jet nozzle and power balance between the compressor and turbine:

$$
\begin{aligned}
& W_g - (W_a + W_f) = 0 \\
& c_{p_a} T_c^* W_a + H_u \eta_g W_f - c_{p_g} T_g^* W_g = 0 \\
& W_n - W_g = 0 \\
& c_{p_g} T_t^* W_g - c_{p_g} T_n^* W_n = 0 \\
& \eta_{\text{mech}} N_t - N_c = 0
\end{aligned}
\tag{3.2}
$$

The system of Equations (3.1) and (3.2) is complemented by the control program equations specifying the control factors W_f and A_n. For a single-shaft turbine engine such programs can be, for example, the following:

$$W_f = \text{const};\ A_n = \text{const} \ \text{ or} \tag{3.3}$$

$$n = f(\alpha_{\text{PLA}}, T_{\text{in}}^*) \text{ by affecting through } W_f \tag{3.4}$$

$$A_n = f(\alpha_{\text{PLA}})$$

The analysis of Equations (3.1) and (3.2) shows that the five variables – W_a, W_g, T_c^*, T_g^*, W_n – are unknown, thus determining the operation point of the engine at the demanded values W_f and A_n. Introduce the following form of the model:

$$\mathbf{f}_x(\mathbf{X}, \mathbf{U}, \mathbf{V}) = \mathbf{0} \tag{3.5}$$

$$\mathbf{Y} = \mathbf{f}_y(\mathbf{X}, \mathbf{U}, \mathbf{V}) \tag{3.6}$$

where $\mathbf{X} = [W_a, W_g, T_c^*, T_g^*, W_n]^T$ is the vector of state coordinates;

$\mathbf{U} = [W_f, A_n]^T$ is the vector of control coordinates;

$\mathbf{V} = [M, H, p_H, T_H]^T$ is the environment vector of flight conditions;

$\mathbf{Y} = [n_{\text{red}}, n, p_c^*, \pi_c^*, ...]^T$ is the vector of output observable coordinates;

Both $\mathbf{f}_x$ and $\mathbf{f}_y$ are nonlinear operators; note that the operator $\mathbf{f}_x$ implicitly includes $\mathbf{f}_y$.

The state vector **X** can be formed arbitrarily to some degree: if Equation (3.2) is transformed in view of Equation (3.1), then the vectors **X, Y** and the structure of $\mathbf{f}_x$ and $\mathbf{f}_y$ change accordingly.

The dimension of the state vector **X** is not changed, when considering "secondary" factors such as:

- the efficiency of fuel combustion $\eta_g = f(p_g^*, \alpha)$;
- nonlinearity of coefficients $K_1 = f(T_1^*, \alpha_1)$ and $c_{\mathrm{pi}} = f(T_1^*, \alpha_1)$;
- hydraulic resistance $\sigma = f(\lambda^2)$;
- cooling of the turbine;
- air bleed and mechanical power take-off from the engine.

This would affect only the structure of the operators $\mathbf{f}_x$ and $\mathbf{f}_y$.

3.3 Detailed Nonlinear Dynamic Modelling

In the detailed dynamic model, the system in Equation (3.1) is complemented by both differential equations describing the accumulation of the gas energy and mass within major volumes of the flowing part of the engine, and the differential equation of the shaft rotation.

$$
\begin{aligned}
\dot{p}_g^* &= \frac{p_g^*}{T_g^*}\dot{T}_g^* + \frac{RT_g^*}{V_{\mathrm{comb}}}\left(W_a + W_t - W_g\right) \\
\dot{T}_g^* &= \frac{1}{c_V m_c}\left[\left(c_{Pa}T_c^* W_a + H_u \eta_g W_f - c_{Pg}T_t^* W_g\right) - c_V T_g^*\left(W_a + W_t - W_g\right)\right] \\
\dot{p}_n^* &= \frac{p_n^*}{T_n^*}\dot{T}_n^* + \frac{RT_n^*}{V_n}\left(W_g - W_n\right) \\
\dot{T}_n^* &= \frac{1}{c_V m_n}\left[\left(c_{Pg}T_t^* W_g - c_{Pg}T_n^* W_n\right) - c_V T_n^*\left(W_g - W_n\right)\right] \\
\dot{n} &= \frac{\eta_{\mathrm{mech}} N_t - N_c}{Jn\left(\pi/30\right)^2}
\end{aligned}
\qquad (3.7)
$$

The dynamic model of a turbine engine shown in Equations (3.1) and (3.7) can be expressed in state space:

$$\dot{\mathbf{X}} = \mathbf{F}_x(\mathbf{X}, \mathbf{U}, \mathbf{V}) \qquad (3.8)$$

$$\mathbf{Y} = \mathbf{F}_y(\mathbf{X}, \mathbf{U}, \mathbf{V}) \qquad (3.9)$$

The operators $\mathbf{F}_x$ and $\mathbf{F}_y$ are reduced to the operators $\mathbf{f}_x$ and $\mathbf{f}_y$ at the steady-state conditions:

$$\mathbf{F}_x \equiv \mathbf{f}_x \equiv 0,\ \mathbf{F}_y \equiv \mathbf{f}_y \text{ at } \dot{\mathbf{X}} = \mathbf{0}$$

If $\dot{\mathbf{X}} \neq \mathbf{0}$ then $\mathbf{f}_x \neq \mathbf{0}$, *i.e.*, the solution of the system $\mathbf{f}_x \neq \mathbf{0}$ does not satisfy the static model. Such a solution corresponds to a dynamic point, not the steady-state conditions. The reverse problem is also true. If the solution for $\mathbf{f}_x(\ldots) \neq \mathbf{0}$ is known, then the derivative vector is easily calculated using Equation (3.7). Moreover, the following dynamic additions can be introduced into the "determining" Equation (3.2) of the static model:

$$\begin{aligned}
&W_g - (W_a + W_t) + \Delta W_a^{\text{dyn}} = 0 \\
&c_{p_a} T_c^* W_a + H_u \eta_g W_f - c_{p_g} T_g^* W_g + \Delta Q^{\text{dyn}} = 0 \\
&W_n - W_g + \Delta W_g^{\text{dyn}} = 0 \\
&c_{p_g} T_g^* W_g - c_{p_g} T_n^* W_n + \Delta Q_n^{\text{dyn}} = 0 \\
&\eta_{\text{mech}} N_t - N_c + \Delta N^{\text{dyn}} = 0
\end{aligned} \tag{3.10}$$

This allows interpretation at any intermediate point for solving the static model as a dynamic point in the state space. A particular dynamic condition of the engine is described by the derivative vector in Equation (3.8). The elements of the vector can be calculated based on the current values of **X** and the current dynamic additions in Equation (3.10).

This approach allows study of dynamic properties of a turbine engine using the static model and, in particular, estimation of parameters of linear dynamic models.

3.4 Relationship Between Static and Dynamic Models

The detailed static and dynamic models of a twin-shaft bypass turbo jet engine with an afterburner and a mixing chamber can be represented in the form:

$$\mathbf{f}_x(\mathbf{X},\mathbf{Y},\mathbf{V}) = \mathbf{0}\,;\ \mathbf{Y} = \mathbf{f}_y(\mathbf{X},\mathbf{Y},\mathbf{V}) \tag{3.11}$$

$$\dot{\mathbf{X}} = \mathbf{F}_x(\mathbf{X},\mathbf{Y},\mathbf{V}) = \mathbf{0}\,;\ \mathbf{Y} = \mathbf{F}_y(\mathbf{X},\mathbf{Y},\mathbf{V}) \tag{3.12}$$

where $\mathbf{X} = [n_f, n_c, p^*_{\text{comb}}, T^*_{\text{comb}}, p^*_t, T^*_t, T^*_{\text{II}}, p^*_{\text{II}}, p^*_{\text{ab}}, T^*_{\text{ab}}]^T$ is the state coordinates vector; $\mathbf{U} = [W_f, W_{f.\text{ab}}, A_n, \alpha_{\text{VSV}}]^T$ is the control coordinates vector; $\mathbf{V} = [M, H, p_H, T_H]^T$ is the environment vector of the flight conditions.

For comparison, selected equations of the static and dynamic models are summarised in Table 3.1. From the computational point of view, the detailed static Equation (3.11) and dynamic Equation (3.12) models differ by their methods for calculation. The static model requires straight iterative algorithms or enclosed iterative loops. The dynamic model employs integration methods such as Euler,

Runge-Kutta, *etc.* Note, however, that the time of calculation of one point of the static model is usually ten times less than that of the dynamic model.

Table 3.1. Comparison of static and dynamic models

Static model	Dynamic model
$W_g - W_a - W_f = \Delta W_a^{\text{dyn}} = Z_1 \Rightarrow 0$	$\dot{p}_g^* = \frac{RT_g^*}{V_{\text{cr}}} \Delta W_a^{\text{dyn}} + \frac{p_g^*}{T_g^*} \dot{T}_g^*$
$C_p T_g^* W_a + H_u \eta_g W_f -$ $- C_p T_g^* W_g = \Delta Q_c^{\text{dyn}} = Z_2 \Rightarrow 0$	$\dot{T}_g^* = \frac{1}{C_V m_{\text{comb}}} \left[C_p T_c^* W_a + H_u \eta_g W_f - \right.$ $\left. - C_p T_g^* W_g + \Delta Q_c^{\text{dyn}} - C_V T_g^* \Delta W_g^{\text{dyn}} \right]$
$W_n - W_g = \Delta W_g^{\text{dyn}} = Z_3 \Rightarrow 0$	$\dot{p}_n^* = \frac{p_n^*}{T_n^*} \dot{T}_n^* + \frac{RT_n^*}{V_n} \Delta W_g^{\text{dyn}}$
$C_p T_g^* W_g - C_p T_n^* W_n = \Delta Q_g = Z_4 \Rightarrow 0$	$\dot{T}_n^* = \frac{1}{C_V m_n} \Delta Q_g^{\text{dyn}} - C_V T_n^* \Delta W_g^{\text{dyn}}$
$N_t \eta_{\text{mech}} - N_c = \Delta N_{\text{dyn}} = Z_5 \Rightarrow 0$	$\dot{n} = \frac{\Delta N_{\text{dyn}}}{Jn(\pi/30)^2}$

In developing the detailed nonlinear model, an important problem is the identification of its parameters from experimental data. An optimisation method can be applied to solving this problem. As a result, the model should approach or fit the experimental data:

$$E = \sum_{k=1}^{m} q_k \lambda_k^2 \rightarrow \min \tag{3.13}$$

where $\lambda_k = f(\alpha_1, \alpha_2, \ldots, \alpha_n, t) = [y_e(t) - y_m(t)]$;
$y_e(t)$, $y_m(t)$ are the experimental (reference) and model signals;
$y_e(t) = [n, T_t^*, p_c^*, P, C_{\text{sp}}, W_a, \pi^*]_e$;
$y_m(t) = [n, T_t^*, p_c^*, P, C_{\text{sp}}, W_a, \pi^*]_m$;
$\alpha_{\min} \leq \alpha \leq \alpha_{\max}$ are the correction coefficients;
E is the total error on all parameters;
$q_k = (\partial R / \partial y)_k$ is the error weighting factor (influence for the kth parameter).

The minimum of the objective function in Equation (3.13) can be found, for example, by the least squares method using the condition:

$$\frac{\partial E}{\partial \alpha_i} = \frac{2}{T_s} \int_0^{T_s} \sum_{k=1}^{m} q_k \lambda_k (\alpha_1, \alpha_2, \ldots, \alpha_n) \frac{\partial \lambda_k}{\partial \alpha_i} dt = 0, \; i = 1, \ldots, n \tag{3.14}$$

3.5 Dynamic Characteristic Models

Performance-based models give great accuracy and are used extensively in the design process. Although with increasing processing capability it is possible to use performance models for control law development, control system engineers in the main do not require the level of detail which is provided by these models. Therefore, it is common practice to derive models suitable for control system design use that represent the dynamic characteristics of the engine.

The static and dynamic property information of a turbine engine is contained in detailed nonlinear performance-based models in a "hidden" form. Therefore, a visual representation of the engine properties is important for control systems design. A possible means for representation is *the dynamic characteristic* of a turbine engine [4].

The dynamic characteristic (DC) is a graphical means for representing the major nonlinear dynamic and static properties of a turbine engine for control systems design. DC can be obtained both from experimental data and by calculations on a detailed dynamic model. As a result, DC is both a quantitative and qualitative criterion of accuracy of detailed models. DC is usually built in a plane of reduced coordinates $W_{f.\text{red}}$ versus n_{red}. It consists of "grids" of lines of constant values of various parameters. These can be, for example, the lines of constant derivatives of the shaft speed $\dot{n}_{\text{red}} = c_i, i = \overline{1,n}$, the lines of constant turbine gas temperature T_t^*, compressor air pressure p_c^*, *etc.*

3.6 Deriving the Dynamic Characteristic Model

A method for building a DC is considered below using an example of a single-shaft turbine engine. The grid of constant accelerations $\dot{n}_{\text{red}} = \text{const}$ is gathered in the area of stable operation of the engine $n_{\text{red}}^{\min}, \ldots, n_{\text{red}}^{\max}$ under given external conditions. The controller is designed for $\dot{n}_{\text{red}} = \text{const}$ using the fuel flow W_f to control acceleration. Control programs for other control factors, *e.g.*, A_{cr} and air by-pass valves, are set constant or as a function of the shaft speed. Similarly, grids of the characteristics T_t^*=const, *etc.*, can be gathered by means of appropriate control programs. Also, the boundary of gas dynamic stability of the compressor can be determined experimentally. A calculated dynamic characteristic is obtained by executing "equivalent experiment" calculations on the nonlinear dynamic model. The DC is "stratified" for other control factors by constructing a series of characteristics against the appropriate factors.

The dynamic characteristic can also be obtained by the following calculations. Set a control program $n_{\text{dem}} = \text{const}$ with effect on W_{f}, while $A_n = \text{const}$. Parameters of the engine are calculated on the detailed static model consisting of the system of Equations (3.15) and (3.16) at $(K{\times}L)$ points with the following shaft speeds n_{dem} and dynamic additions Δ_{dyn}:

$$n^{\text{dem}} = \left\{ n_1^0, n_2^0, \ldots, n_k^0 \right\}$$
where $n_1^0 = n_{\min}, \ldots, n_k^0 = n_{\max}$
$$\Delta N^{\text{dyn}} = \left\{ \Delta N_1^{\text{dyn}}, \Delta N_2^{\text{dyn}}, \ldots, \Delta N_l^{\text{dyn}} \right\} \tag{3.15}$$
where $\Delta N_1^{\text{dyn}} = 0, \ldots, \Delta N_l^{\text{dyn}} = \Delta N_{\max}^{\text{dyn}}$
$$\Delta W_a^{\text{dyn}} = \Delta Q^{\text{dyn}} = \Delta W_n^{\text{dyn}} = \Delta W_n^{\text{dyn}} = 0$$

A table of values for drawing the DC is obtained by calculating derivatives from appropriate dynamic additions by the formula:

$$\dot{n}_k^l = \frac{\Delta N_{\text{dyn}}^l}{J n_k (\pi/30)^2}, \; k = \overline{1,K}, \; l = \overline{1,L} \tag{3.16}$$

The "grids" of constant accelerations $\dot{n} = c_i$, $i = \overline{1,N}$ and other parameters $p_c^* = \text{const}$, $T_g^* = \text{const}$ are calculated by interpolation in the coordinates $\{\dot{n}, W_f\}$, $\{T_t^*, W_f\}$, $\{p_t^*, W_f\}$ with the demanded values $n^{\text{dyn}} = n_k^0, k = \overline{1,K}$ and subsequent interpolation in the coordinates $\{n, W_f\}$.

With a sufficient degree of accuracy the dynamic characteristic "shaft speed n *versus* fuel flow W_f" can be approximated by the analytical relation [5]:

$$\frac{dn}{dt} = K(n)\left(W_f - W_f^{\text{st}}(n) \right) = K(n)\,\Delta W_f \tag{3.17}$$

where $K(n) = \dot{n}/\Delta W_f$ is the nonlinear impact coefficient defined from DC and the function $W_f(n)$ determines the static line of the steady-state conditions.

Parameters of a linear dynamic model are obtained by linearisation of that relationship around any steady-state point $\{W_f^0, n^0\}$:

$$T_{\text{eng}} \frac{d\Delta n}{dt} + \Delta n = K_{\text{eng}} \Delta W_f \tag{3.18}$$

where $K_{\text{eng}} = \dfrac{\partial n}{\partial W_f^{\text{st}}}; \; \tau_{\text{eng}} = \dfrac{K_{\text{eng}}}{K(n_0)}$

The parameters of the linear dynamic model concerning other coordinates of DC are obtained by linearisation of the characteristic $y = f(n, W_f)$, where $y = [p_c^*, T_t^*]^T$.

The dynamic characteristic at the segment of the autonomous power effect from a starter is built in the coordinates $\{M_{\text{st}}, n\}$. The grid of constant acceleration is formed according to the dynamic equation of the rotor of the turbocompressor:

$$\frac{dn}{dt} = \frac{M_{\rm st} - M_{\rm tc}^{\rm drag}(n)}{J\,\pi/30} \tag{3.19}$$

where $M_{\rm st}$ is the torque developed by the starter; $M_{\rm tc}^{\rm drag}(n)$ is the drag moment of the turbocompressor.

3.7 Graphical Representation

The dynamic characteristic of a single-shaft turbine engine for the ground and flight conditions is exemplified in Figure 3.1. The substantially wide environmental range of the operation conditions may violate similarity conditions when considering dynamic properties of a turbine engine.

3.8 Dynamic Characteristic in Reduced Parameters

The area of the flight conditions of the engine operation can be divided into three subareas according to the degree of violation of gas dynamic similarity (see Figure 2.3).

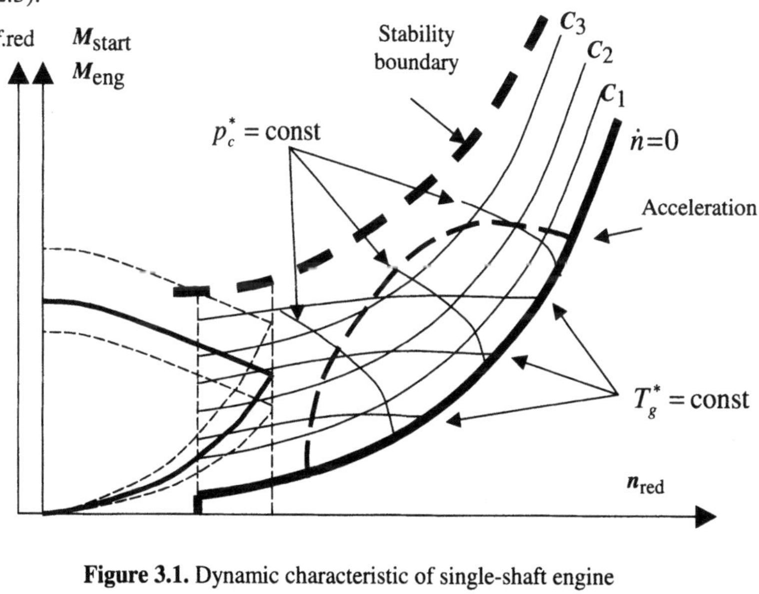

Figure 3.1. Dynamic characteristic of single-shaft engine

Subarea *I* is characterised by the presence of an over-critical pressure differential at the jet nozzle and by the absence of the influence of the *Re* number on the engine characteristics. This subarea starts with steady-state windmill. Here, the gas dynamic similarity conditions are fulfilled the most precisely; the static operating line of the turbocompressor in reduced parameters is stratified to a very small degree. This enables the engine dynamics in the whole subarea *I* to be described by a unified dynamic characteristic in reduced coordinates.

Subarea *II* is characterised by a subcritical pressure differential at the jet nozzle and by the absence of the influence of the *Re* number. In this subarea, the engine dynamics are described by a set of dynamic characteristics because of the stratification of the turbocompressor's static line by the flight speed.

In subarea *III*, the engine dynamics are dependent on the *Re* number; the pressure differential at the jet nozzle is over-critical. Hence, the dynamic model of the engine in this area should be described while accounting for the influence of the *Re* number to the characteristics of components of a turbine engine.

A gas turbine engine operates in continuous time in various conditions; and it is *a priori* known that smooth continuous models should describe physical processes within the engine. Hence, its static and dynamic parameters (including the dynamic response) are also smooth and continuous both by the environmental flight conditions, and by operating conditions. This conclusion covers all three subareas.

However, methods for accounting for environmental and operating conditions are different in the subareas. In subarea *I*, the reduction formulae are used to quantitatively describe the parameters' relations with operation conditions and with the flight altitude and speed (see Table 3.2).

Table 3.2. Formulae for reduction to sea-level conditions

Parameter	Reduction formula
Shaft speed	$n_{\text{red}} = n_{\text{phys}}\sqrt{288/T^*_{\text{in}}}$
Fuel feed	$W_{f.\text{red}} = W_{f.\text{phys}}\dfrac{1.033\sqrt{288}}{p^*_{\text{in}}\sqrt{T^*_{\text{in}}}}$
Shaft speed derivative	$\dot{n}_{\text{red}} = \dot{n}_{\text{phys}}\dfrac{1.033}{p^*_{\text{in}}}$
Compressor pressure	$p^*_{c.\text{red}} = p^*_{c.\text{phys}}\dfrac{1.033}{p^*_{\text{in}}}$
Gas temperature	$T^*_{g.\text{red}} = T^*_{g.\text{phys}}\dfrac{288}{T^*_{\text{in}}}$

The reduction formulae convert current physical parameters to the sea-level conditions. In subareas ***II*** and ***III*** the reduction relations are true only qualitatively. This means that if two DCs in reduced parameters are known at two points of the "*H* versus *M*" plane, then the DC parameters for any intermediate conditions *H* and *M* are situated between the appropriate parameters of those two extreme dynamic characteristics. In other words, the DC at any point of the plane $\{H, M\}$ can be obtained by means of interpolation. The error of the obtained characteristic depends on the positions and the number of operating points chosen in the subareas ***I***, ***II*** and ***III***.

3.9 Interpolation Between Operating Points

The DC of the engine at the operating points will be referred to as the operating point dynamic characteristics. In the qualitative level, the reduction formulae can be used within ***Re*** independent areas; as a result, the difference between reduced parameters of the operating point is less than between physical parameters at the same operating points. Therefore, the error of interpolation of reduced parameters is also less than that of physical parameters. In practice, operating point DCs are more suitable, because of their more precise results.

In each practical case, the choice of the positions and the number of the operating points depends on the operation range of the engine; however, there are some major points, common for all turbine engines; see Figure 3.2.

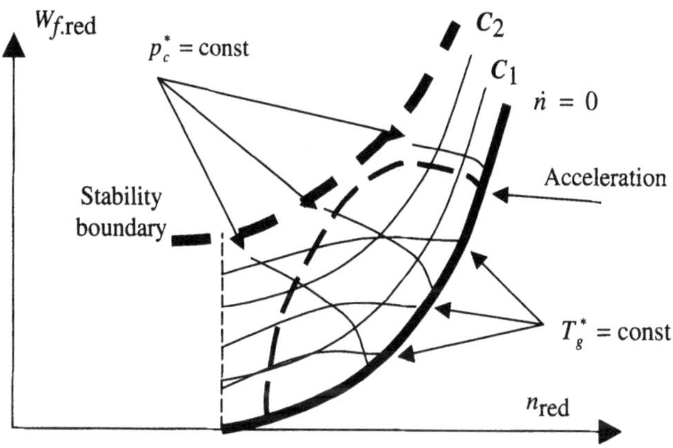

Figure 3.2. Dynamic characteristic at point 2

Operating point 1 corresponds to the ground conditions ($H = 0$, $M = 0$). Operating point 2 represents the Mach number M, where the critical mode of the gas efflux in the jet nozzle exists, starting from windmill. The dynamic characteristic at this point is characterised by minimum values of reduced fuel flow at the static line and the lines of constant accelerations:

$$\frac{d\bar{n}_{\text{red}}}{dt} = const \tag{3.20}$$

Operating point 2 corresponds to the boundary of the stratification of DC lines by the flight speed. Operating point 5 corresponds to the number $M = M_{\text{cr}}$ and the altitude H, where the number Re starts to influence the characteristics of the engine units at start-up ($Re \approx Re_{\text{cr}}$). Operating points 3 and 4 correspond to the high bound of the flight envelope; usually this area is independent of the Re number. Operating point 6 corresponds to the beginning of the influence of the Re number at engine start-up and acceleration with maximum Mach number. DCs at operating points 5 and 6 correspond to subarea I. The indicated points at the plane $\{H\text{-}M\}$ are used to define the dynamic characteristics. The line between operating points 5 and 6 can join the boundary of DC application, which corresponds to operating point 2. DC in operating points 1 and 2 describe subarea *II*; DC of operating point 2 describes subarea *I*; DC of operating points 1, 2, 4 and 5 describe subarea *III*.

At each point of the area of the engine operation, the parameters of the dynamic characteristic can be defined approximately within given conditions by means of interpolation between operating point DCs. For instance, see Figure 3.2, which shows the dynamic characteristic at point 2. In this case the limits for important engine parameters are determined by the following method. The grids of the constant reduced temperature $T^*_{t.\mathrm{red}}$ and the boundary of gas dynamic stability are drawn at the operating point DCs. The reduced gas temperature limits are determined based on known physical values at given flight conditions.

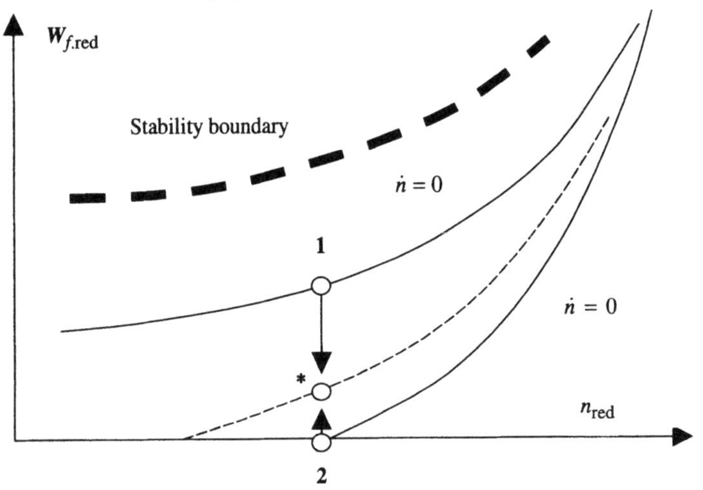

Figure 3.3. Interpolation of DC with reduction to sea-level conditions

The reduced limits are put on the DC in the form of grids of constant values. Figure 3.3 shows an example of DC of a single-shaft turbine engine at operating points 1 and 2. The static and dynamic properties of the engine are analysed using a number of DCs. This includes the analysis of the lines of start-up and acceleration with various factors (starter power, airflow speed, *etc.*) and different schedules of fuel supply taking into account the limits of maximum gas temperature and stability of the plant. This allows the search for optimum solutions when designing the control system for engine start-up and acceleration in a broad range of the flight conditions.

3.9 Linear Dynamic Models

The dynamic characteristic (DC) graphically presents the major nonlinear dynamic and static properties of a turbine engine. However, in order to allow conventional control techniques to be utilised and critical criteria to be assessed, *e.g.*, stability and robustness of control, it is necessary to perform analysis based on linear models. This analysis is essential to prove to the certification authorities that the control strategy meets the requirements for performance and safety.

Nonlinear static and dynamic characteristics of aero engines are continuous and differentiable functions during its normal operation. This property of the engine characteristics enables its dynamics to be described by a linear dynamic model (LDM) in a neighbourhood of the steady-state conditions with sufficient accuracy. Typically, this area is characterised by 5 to 10% deviation of shaft speeds from the considered operating point. The following forms of linear dynamic models are usually used:

- a system of linear differential and algebraic equations (usually in the state space);
- a matrix of transfer functions;
- a matrix of frequency responses.

In this chapter an example state-space model is given for a twin-shaft turbo jet.

3.10 Obtaining Linear Dynamic Models

Methods for obtaining the LDM are based on two principles:

- linearisation of the system of nonlinear differential and algebraic equations making up the detailed mathematical model of the engine [5];
- identification of the model parameters from observations of inputs and outputs using the model structure known *a priori* [6].

In solving the problem of identification, the realisations of the engine input and output coordinates can take various forms:

- a set of dynamic conditions of the engine (where any derivative of any state coordinate is not zero);
- a set of step responses of the engine coordinates;
- Bode diagrams, *etc.*

A method for linearisation of the performance-based nonlinear system of differential and algebraic equations using a Taylor series expansion is described below.

3.11 Estimation of LDM Parameters Using Detailed Static Model

For engine control purposes, the inertia of the rotating masses of the turbocompressor rotors is the most important factor of the energy accumulation, which determines the engine dynamic behaviour [7]. When accounting for this factor only, the LDM structure in the state space is given by the following:

$$
\begin{aligned}
\Delta\dot{\mathbf{X}} &= \mathbf{A}\Delta\mathbf{X} + \mathbf{B}\Delta\mathbf{U} \\
\Delta\mathbf{Y} &= \mathbf{C}\Delta\mathbf{X} + \mathbf{D}\Delta\mathbf{U}
\end{aligned}
\tag{3.21}
$$

where $\Delta\mathbf{X} = [\Delta n_1, \Delta n_2, ..., \Delta n_r]^T$ is the state vector;

$\Delta\mathbf{U} = [\Delta u_1, \Delta u_2, ... , \Delta u_m]^T$ is the control vector;

$\Delta\mathbf{Y} = [\Delta y_1, \Delta y_2, ... , \Delta y_s]^T$ is the observation vector;

$\mathbf{A} = [a_{ij}]_{r\times r}$, $\mathbf{B}=[b_{ij}]_{r\times m}$, $\mathbf{C}=[c_{ij}]_{s\times r}$, and $\mathbf{D}=[d_{ij}]_{s\times m}$ are the matrices whose elements are the parameters to find;

r is the number of the turbocompressor stages;

m is the number of control factors;

s is the number of observable coordinates;

$\Delta\mathbf{X} = \mathbf{X} - \mathbf{X}_0$, $\Delta\mathbf{U} = \mathbf{U} - \mathbf{U}_0$, and $\Delta\mathbf{Y} = \mathbf{Y} - \mathbf{Y}_0$ are the vectors of deviations from the steady-state conditions.

The LDM describes the engine dynamics in a neighbourhood of the static point characterised by the vectors $\mathbf{X}_0$, $\mathbf{U}_0$ and $\mathbf{Y}_0$.

The system in Equation (3.2) for an r-shaft engine includes r equations of power balance:

$$
\begin{aligned}
&\eta_{\text{mech.}1} NT_1 - N_{c1} + \Delta N_{\text{dyn.}1} = 0 \\
&\eta_{\text{mech.}2} NT_2 - N_{c2} + \Delta N_{\text{dyn.}2} = 0 \\
&\qquad \cdots \\
&\eta_{\text{mech.}r} NT_r - N_{c.r} + \Delta N_{\text{dyn.}r} = 0
\end{aligned}
\tag{3.22}
$$

All remaining conditional dynamic additions are supposed to be zero, *i.e.*, $\Delta W_{a.\text{dyn}} = \cdots = \Delta Q_{\text{dyn}} = \cdots = 0$. The control program is defined as $\mathbf{U} = \mathbf{S}$, where the elements of the vector $\mathbf{S}$ are constant. Calculation of the following static conditions on a model assigned in the form shown in Equations (3.8) and (3.9) is conducted in view of Equation (3.16):

$$
\begin{aligned}
&0)\ U = U^0 = S^0;\ \Delta N_{\text{dyn }1} = \Delta N_{\text{dyn }2} = \cdots = \Delta N_{\text{dyn }r} = 0 \ \text{(initial static conditions)}; \\
&1)\ U = U^0 = S^0;\ \Delta N_{\text{dyn }1} \neq 0;\ \Delta N_{\text{dyn }2} = \cdots = \Delta N_{\text{dyn }r} = 0; \\
&2)\ U = U^0 = S^0;\ \Delta N_{\text{dyn }2} \neq 0;\ \Delta N_{\text{dyn }1} = \Delta N_{\text{dyn }3} = \cdots = \Delta N_{\text{dyn }r} = 0;
\end{aligned}
\tag{3.23}
$$

$i)\ U = U^0 = S^0;\ \Delta N_{\text{dyn}\ i} \neq 0;\ \Delta N_{\text{dyn}\ 1} = \Delta N_{\text{dyn}\ i-1} = \Delta N_{\text{dyn}\ i+1} = \cdots = \Delta N_{\text{dyn}\ r} = 0;$

$r)\ U = U^0 = S^0;\ \Delta N_{\text{dyn}\ r} \neq 0;\ \Delta N_{\text{dyn}\ 1} = \cdots = \Delta N_{\text{dyn}\ r-1} = 0.$

At each ith calculated operating point ($i = \overline{1,r}$), a system of equations is built with the following structure:

$$
\begin{aligned}
0 &= a_{11}(n_{1i} - n_{10}) + a_{12}(n_{2i} - n_{20}) + \cdots + a_{1r}(n_{ri} - n_{r0});\\
&\ \ldots\\
0 &= a_{i-1,1}(n_{1i} - n_{10}) + a_{i-1,2}(n_{2i} - n_{20}) + \cdots + a_{i-1,r}(n_{ri} - n_{r0});\\
\dot{n}_i &= a_{i1}(n_{1i} - n_{10}) + a_{i2}(n_{2i} - n_{20}) + \cdots + a_{ir}(n_{ri} - n_{r0});\\
&\ \ldots\\
0 &= a_{r1}(n_{1i} - n_{10}) + a_{r2}(n_{2i} - n_{20}) + \cdots + a_{rr}(n_{ri} - n_{r0});
\end{aligned}
\tag{3.24}
$$

$$y_{il} - y_{i0} = c_{l1}(n_{1i} - n_{10}) + c_{l2}(n_{2i} - n_{20}) + \cdots + c_{lr}(n_{ri} - n_{r0}),\ l = \overline{1,s},$$

where

$$\dot{n}_i = \frac{\Delta N_{\text{dyn}\,l}}{n_i J_i (\pi/30)^2} \tag{3.25}$$

By changing $i = \overline{1,r}$, the system of $r(r+s)$ linear equations is obtained. The system consists of $r(r+s)$ independent subsystems with r equations each. Each subsystem determines the coefficients a_{ij} ($i, j = \overline{1,r}$) and c_{ij} ($l = \overline{1,s}; j = \overline{1,r}$).

In order to define the coefficients b_{ij} and c_{ij} ($l = \overline{1,s}; j = \overline{1,r}$), the operating points are calculated at the following condition:

$$\Delta N_{\text{dyn}.1} = \Delta N_{\text{dyn}.2} = \cdots = \Delta N_{\text{dyn}.r} = 0 \tag{3.26}$$

Elements of the control vector $\mathbf{U}_0$ are increased by steps $\Delta \mathbf{U}_i$, $i = \overline{1,m}$. At the ith static point:

$$\mathbf{U}^0 = \left[u_1^0, u_2^0, \ldots, u_i^0 + \Delta u_i, u_{i+1}^0, \ldots, u_m^0\right]^T, \tag{3.27}$$

the following system is obtained:

$$
\begin{aligned}
0 &= a_{11}(n_{1i} - n_{10}) + a_{12}(n_{2i} - n_{20}) + \cdots + a_{1r}(n_{ri} - n_{r0}) + b_{1i}\Delta u_i\\
&\ \ldots\\
0 &= a_{j1}(n_{1i} - n_{10}) + a_{j2}(n_{2i} - n_{20}) + \cdots + a_{jr}(n_{ri} - n_{r0}) + b_{ji}\Delta u_i
\end{aligned}
\tag{3.28}
$$

...

$$0 = a_{r1}(n_{1i} - n_{10}) + a_{r2}(n_{2i} - n_{20}) + \cdots + a_{rr}(n_{ri} - n_{r0}) + b_{ri}\Delta u_i$$

$$y_{il} - y_{i0} = c_{l1}(n_{1i} - n_{10}) + c_{l2}(n_{2i} - n_{20}) + \cdots + c_{lr}(n_{ri} - n_{r0}) + \Delta u_i, l = \overline{1, s}$$

where the coefficients b_{ji} ($j = \overline{1, r}$) and d_{li} ($l = \overline{1, s}$) are determined. Similarly, all operating points $i = \overline{1, m}$ are explored; the coefficients b_{ji} ($j = \overline{1, r}; i = \overline{1, m}$) and d_{li} ($l = \overline{1, s}; i = \overline{1, m}$) are obtained.

3.12 Accounting for Additional Gas Flow Dynamics

In addition to the inertial dynamics of the engine, it is also necessary to consider gas flow dynamics. The use of the conditional dynamic additions $\Delta W_{a.\text{dyn}}$, ΔW_g, ΔQ leads to building the LDM accounting for air/gas and heat accumulation in the engine volumes.

Conditional dynamic additions in the power balance equations describing the turbocompressor rotors can be determined by changing the nominal mechanical efficiency $\eta^0_{\text{mech}.i}$ ($i = \overline{1, r}$). The static conditions calculated at $\eta^0_{\text{mech}.i} + \Delta\eta_{\text{mech}.i}$ correspond to the dynamic condition of the engine, where the first derivative of the ith shaft speed is calculated by the formula:

$$\dot{n}_i = \frac{N_{ti}\Delta\eta_{\text{mech}.i}}{n_i J_i (\pi/30)^2} \tag{3.29}$$

By introducing the extended vector $\mathbf{W} = [\Delta\mathbf{N} \mid \Delta\mathbf{U}]^T$ with $\Delta\mathbf{N}_{\text{dyn}} = [\Delta N_{\text{dyn}.1}, \Delta N_{\text{dyn}.2}, \cdots, \Delta N_{\text{dyn}.r}]^T$, the description for the initial static operating conditions takes the form $\mathbf{W}^0 = [\Delta\mathbf{N}^0 \mid \Delta\mathbf{U}^0]^T$, $\Delta\mathbf{N}^0 = \mathbf{0}$. Note that the dimension of the vector $\mathbf{W}$ is $(r + m)$; and the number of unknown LDM coefficients is $(r + m)(r + s)$. Hence, the number of the necessary linearly independent equations should be $(r + m)(r + s)$ as well. Any calculated operating point gives $(r + s)$ equations. So, to obtain a complete system of linearly independent equations, $(r + s + 1)$ operating points should be calculated. In that system, one initial operating point is chosen, from which the deviations are considered.

In the method described above, the points for calculation were chosen by the orthogonal design of experiments. This procedure is set by the vector $\mathbf{W}_j = \mathbf{W}^0 + \Delta\mathbf{W}\delta_{ji}$, where $\Delta\mathbf{W} = [\Delta N_{\text{dyn}.1}, \Delta N_{\text{dyn}.2}, \ldots, \Delta N_{\text{dyn}.r}, \Delta u_{r+1}, \Delta u_{r+2}, \ldots, \Delta u_{r+m}]^T$; j is the operating point number; i is the element's number within the vector $\mathbf{W} = \{w_i\}, i = \overline{1, r+m}$; δ_{ji} is the Kronecker delta. The vector $\Delta\mathbf{W}$ is set so that the deviation of coordinates do not exceed the linearity area. The coefficients b_{ij} and d_{lj} are easily determined from the calculated operating points. Additional equations can be added to reflect desired control laws. Generally, the choice of a set of the calculated points should result in the maximum conditionality of the set of equations. Moreover, a system can be generated with the number of equations being more than the number of variables. Such a system is resolved by the least squares method or by optimisation methods.

3.13 Twin-shaft Turbo Jet Example

A linear dynamic model is constructed for a twin-shaft turbo jet. The engine coordinates are Δn_{LP}, Δn_{HP}, $\Delta T_c{}^*$, $\Delta T_t{}^*$, $\Delta p_c{}^*$, ΔW_a, ΔF. The control factor is the fuel flow ΔW_f. The operating point is the maximum thrust at sea-level static (SLS). The model structure is the following:

$$\begin{aligned} \Delta \dot{\mathbf{n}} &= \mathbf{A}\Delta\mathbf{n} + \mathbf{B}\Delta\mathbf{U} \\ \Delta\mathbf{Y} &= \mathbf{C}\Delta\mathbf{n} + \mathbf{D}\Delta\mathbf{U} \end{aligned} \tag{3.30}$$

where $\Delta\mathbf{n} = [\Delta n_{LP}, \Delta n_{HP}]^T$; $\Delta\mathbf{U} = \Delta W_f$; $\Delta\mathbf{Y} = [\Delta T_{comb}{}^*, \Delta T_t{}^*, \Delta p_c{}^*, \Delta W_a, \Delta F]^T$.

The extended vector $\mathbf{W}$ is $\mathbf{W} = [\Delta N_{dyn.LP}, \Delta N_{dyn.HP}, W_f]^T$. The initial vector of the maximum-thrust conditions is $\mathbf{W} = [0; 0; 0.957]^T$.

Each equation of the model has three unknown variables; hence the minimum number of the calculated operating points to define these factors is to be four.

The engine operation is calculated corresponding to the following extended vector of the input coordinates:

$$\mathbf{W}_1 = [20864; 0; 1.001]^T;\ \mathbf{W}_2 = [49731; 32775; 1.074]^T;\ \mathbf{W}_3 = [0; 0; 0.864]^T \tag{3.31}$$

The results of the calculations are presented in Table 3.3. Also, shown are the coordinates' deviations from the initial conditions and the derivatives corresponding to conditional dynamic additions.

After substitution of the deviations and derivatives calculated for the points 1, 2 and 3, a set of equations is obtained to define the LDM coefficients. For example, in order to find the coefficients of the first equation:

$$\dot{n}_{LP} = a_{11}\Delta n_{LP} + a_{12}\Delta n_{HP} + b_{11}\Delta W_f \tag{3.32}$$

the following system is built:

$$\begin{aligned} 473.7 &= 0\,a_{11} + 73.0\,a_{12} + 0.044\,b_{11} \\ 880.2 &= 0\,a_{11} + 48.4\,a_{12} + 0.117\,b_{11} \\ 0 &= -336.4\,a_{11} - 213.4\,a_{12} - 0.093\,b_{11} \end{aligned} \tag{3.33}$$

Resulting from its solution, the coefficients are defined:

$$a_{11} = -3.24;\ a_{12} = 1.56;\ b_{11} = 8.144 \tag{3.34}$$

Table 3.3. Calculation results

Parameter	Point 1	Point 2	Point 3
$\overline{n}_{LP}$	1.000	1.000	0.970
$\overline{n}_{HP}$	1.007	0.996	0.980
$\overline{W}_{f}$	1.046	1.122	0.903
$\overline{p}_{t.LP}$	1.021	1.054	0.933
$\overline{p}_{t.HP}$	1.022	1.048	0.936
$\overline{T}^{*}_{comb}$	1.022	1.056	0.953
$\overline{T}^{*}_{t}$	1.024	1.062	0.951
$\overline{p}^{*}_{t}$	1.013	1.0333	0.955
$\overline{p}^{*}_{c.HP}$	9.893	9.774	9.153
$\overline{T}^{*}_{c}$	1.004	1.006	0.98
$\overline{W}_{a}$	1.0	1.0	0.982
$\overline{p}$	1.02	1.051	0.932
ΔN^{dyn}_{LP}, kW	–20864	–49731.12	0
ΔN^{dyn}_{HP}, kW	0	–32775.03	0
Δn_{LP}, rpm	0	0	–336.4
Δn_{HP}, rpm	73	–48.4	–213.4
ΔW_{f}, kg/hr	0.044	0.117	–0.093
$\Delta \dot{n}_{LP}$, rpm/s	473.7	880.19	0
$\Delta \dot{n}_{HP}$, rpm/s	0	552.7	0
ΔT^{*}_{comb}, K	25.2	64.44	–53.74
ΔT^{*}_{t}, K	20.95	53.18	–41.6
Δp^{*}_{t}, kPa	0.031	0.078	–0.104
$\Delta p^{*}_{c.HP}$, kPa	0.333	–0.221	–0.361
ΔT^{*}_{c}, K	2.45	3.38	–14.16
ΔW_{a}, kg/s	0.006	–0.009	–1.188
$\Delta \overline{F}$, kg	82	207.4	–2.787

Similarly, the other coefficients are determined from the corresponding sets of equations:

$$\mathbf{A} = \begin{pmatrix} -0.324 & 1.56 \\ 0.39 & -2.23 \end{pmatrix}; \quad \mathbf{B} = \begin{pmatrix} 8144.0 \\ 3706.0 \end{pmatrix} \tag{3.35}$$

$$\mathbf{C} = \begin{pmatrix} -0.00044 & 0.01 \\ -0.0098 & 0.01 \\ 0.00011 & 0.000018 \\ -0.0015 & 0.0027 \\ 0.025 & 0.012 \\ 0.0034 & 0.0001 \\ 0.3 & 0.044 \end{pmatrix}; \quad \mathbf{D} = \begin{pmatrix} 555.0 \\ 459.0 \\ 0.67 \\ 3.0 \\ 34.0 \\ -0.034 \\ 1791.0 \end{pmatrix} \tag{3.36}$$

The advantage of the considered method is that LDM can be obtained at the stage of calculating the engine static characteristics (altitude-to-speed and fuel-scheduling characteristics) using standard programs for computer-aided design.

3.14 Concluding Remarks

In this chapter, a number of models have been introduced which are typically used within gas turbine engine control system development. Firstly, detailed static and dynamic performance models were derived and the relationships between the two models were highlighted. The importance of optimisation in achieving accuracy was also discussed. These thermodynamic models give accuracy for performance analysis at the expense of computational complexity.

In recent years, with improvements in processing capability, it is now possible to use such thermodynamic models directly in the control law design phase; however, from a control engineer's viewpoint much of the detailed information is unnecessary. Therefore, simplified dynamic characteristic models are often extracted which give a graphical view of engine performance. This is both useful in the design of gas turbine engines and in the investigation of control system transient behaviour. The chapter thus shows how the dynamic characteristic model can be derived and introduces the concepts of parameter reduction and interpolation over the engine's operating regime.

Finally, the chapter shows how linear dynamic models can be derived around steady-state conditions considering both inertial dynamics and gas dynamics. The advantage of these linear models is that they allow conventional linear control theory to be used to analyse the stability and robustness of the control system. This is particularly important for certification. Although, this type of model can be derived directly from performance-based models, it is common practice to derive these from the dynamic characteristic model. An example is given of a twin-shaft turbo jet engine showing how inertial dynamics can be derived in state-space form.

It should be noted that linear dynamic models only permit performance around specific operating points to be considered. In order to consider large-scale transient deviations from these operating points and between operating points, it is necessary to have a piecewise linear dynamic model. A key advantage of this type of model is that it enables transient performance to be assessed in real time. This is important when considering on-line use of a model in applications such as Hardware-In-the-Loop testing. In the next chapter, derivation of piecewise linear dynamic models is described.

Chapter 4
On-line Models

4.1 Introduction

In the previous chapter complex thermodynamic modelling was introduced. It was highlighted that the complexity of these models results in a high computational burden. It was then shown how simpler models can be derived from this complex model that represent the key dynamics of the engine suitable for control system development. The chapter ended with the development of linear dynamic models. These allow performance at steady-state operating points to be considered and controller stability to be assessed for certification. However, the engine dynamics are known to be nonlinear. A limitation of linear models is that they only consider a small neighbourhood around steady-state conditions. In order to assess transient performance over the operating range of the engine, another form of model is required. Full thermodynamic nonlinear models tend to be too complicated for "real-time" calculations, although they are the most exact. Linear models are computationally faster, but can only be used over a narrow range of operating conditions.

Hence another form of model is generally derived, which is able to operate in real time, that adequately reflects the main characteristics of the engine. Such models can be used in testing to exercise controller hardware during systems integration. It is usual to construct a real-time piecewise linear dynamic model (RPLDM) combining the accuracy of the nonlinear thermodynamic model and the speed of the linear dynamic model [1]. This approach extends the range of application of linear models to the entire area of the engine operation. The model is based on a set of linear models connected with the nonlinear static line. During simulation, linear interpolation of the linear model parameters is performed for the nearest point of the static line.

4.2 Conflicting Requirements of Simplicity, Speed and Accuracy

In control systems design, dynamic models at the steady-state conditions are *simplified* to transfer functions and frequency responses. Such models reflect the main properties of the plant [2, 3].

An aero engine is a nonlinear nonstationary plant. Therefore, it is difficult to describe the engine dynamics by a set of linear models in all areas of operating conditions. Moreover, when covering the whole area of engine dynamic properties by linear models, the space between these points is left not specified. Hence, it is necessary to combine linear models into a nonlinear model in order to describe the whole area of the engine operation. In this case, the knowledge of characteristics at isolated operating points is distributed over the whole area.

The more the model is simplified, the faster its computational *speed* is. The time factor is important in analysis, synthesis, and simulation of control systems. High-speed models are necessary in resolving the problem of optimisation of the engine operation, and also in monitoring and diagnosis of the GTE and its control system.

Control systems are designed based on an idealised model of the GTE. The real engine, however, may deviate from the idealised engine, and hence the controller may not be optimum for a specific engine. There are great potential advantages to implementing optimising controllers that work on-line. Currently, this is not common practice due to the certification difficulties introduced by using an adaptive control strategy. The book concentrates on the derivation and use of models throughout the control system design life cycle. Certification is a topic for further research.

Other potential uses for on-line models are for in-flight monitoring and diagnosis. There is much activity in this area with a variety of different approaches to modelling [4-10]. A key requirement is that models reflecting engine properties for good and/or faulty conditions need to work in real time. One possible solution is again to use linear dynamic models since they have high computational speed, mainly because they rely on fast addition and multiplication operations. It should be noted, however, that simplification of the model inevitably results in some decrease in accuracy of calculated output parameters. This restricts the usefulness of the model for fault diagnosis and needs to be taken into consideration.

Linear models reflect properties of the engine in the neighbourhood of a certain linear zone, which usually embraces 3-5% of controlled coordinates. The range of linearity is determined, as a rule, experimentally. In piecewise linear models, the indicated range of controlled parameters covers the whole area of the engine operation, and the dynamics of the plant are described with the accuracy 15-20%.

GTE dynamic models need to satisfy a variety of contradicting requirements. Note that the problems of *accuracy* and adequacy should be considered in the context of the subsequent application of the models. For example, the design models of the plant correspond to some general engine. Therefore, the control system designed using these models should possess enough reserves of stability. Moreover, even the most "precise" nonlinear thermodynamic model always differs from individual performances of a particular engine. Therefore, the final choice of the type of dynamic model is determined by the optimum combination of the factors mentioned above, analysed from the viewpoint of the model application. The fast computation and transient performance of RPLDM lends itself to controller design and for GTE simulation in test bed facilities.

4.3 Stages of Building RPLDM

The source data for building an RPLDM are the nonlinear static characteristics derived from the detailed performance-based static model, and a set of LDM coefficients describing the engine dynamics around the steady-state conditions. The source data is usually represented in a tabular or graphical form.

The method for constructing a RPLDM includes the following assumptions and statements. The engine dynamic model is simplified into a combination of a static line and a set of LDM in order to keep the main properties of the plant and to achieve high computational speed. When simulating the engine dynamics at any point of the state space, a linear model is used which corresponds to the nearest steady-state conditions. In order to find such static points, the nonlinear static line is substituted by a straight line. The perpendicular to that line is built through the current point in the state space. The position of this perpendicular in the state space is uniquely determined by the operating parameter. The static line is approximated by a polyline. At each time instant, the parameters of the static line and the LDM coefficients are determined by means of interpolation, then a system of differential equations is numerically integrated, and all the stages are repeated for a new current state of the plant.

The detailed description of each stage in constructing the model and an example of the engine simulation are described below.

4.4 Combination of Static Line and Set of Linear Models

The method for constructing the dynamic model with variable coefficients is based on the idea of combination of LDM parameters with nonlinear static characteristics. The accuracy of static and dynamic modelling is determined by the accuracy of the source static characteristics, the accuracy of LDM parameters and their quantity.

The static line, or the line of steady-state conditions, is set in the space:

$$\{\mathbf{X}, \mathbf{U}, \mathbf{Y}\} \tag{4.1}$$

where $\mathbf{X}$ is the state vector, $\mathbf{U}$ is the control vector and $\mathbf{Y}$ is the observation vector. For example, the static line of a single-shaft engine can be represented with the following coordinates:

$$\{n, W_f, A_n, T^*, p^*\} \tag{4.2}$$

Here the speed of the turbo-compressor shaft rotation n is selected as a state coordinate. The control inputs are fuel feed W_f and nozzle area A_n. The temperatures T^* and pressures p^* in the engine are referred to as observable coordinates.

The static model of a twin-shaft engine includes a slip line of shaft speeds. It describes the steady-state operation of the engine in the coordinates "high pressure shaft speed *versus* low pressure shaft speed."

The static line can be obtained from experiments or can be calculated using nonlinear models: fuel-scheduling and altitude-speed characteristics or the dynamic characteristic. As a result, the nonlinear static model of the engine is represented in a graphical or tabular form. Assume that the static characteristic coordinates and LDM parameters are known at N points of the operation line:

$$\{\mathbf{X}^{(k)},\mathbf{U}^{(k)},\mathbf{Y}^{(k)}\} \tag{4.3}$$

where $k = 1,..., N$ is the operating point number.

The linear dynamic model of the engine can be obtained by identification methods or by linearisation of the nonlinear thermodynamic model. In our case LDM at each operating point is represented in the Cauchy form:

$$\begin{aligned} \Delta\dot{\mathbf{X}}(t) &= \mathbf{A}^{(k)}\Delta\mathbf{X} + \mathbf{B}^{(k)}\Delta\mathbf{U} \\ \Delta\mathbf{Y}(t) &= \mathbf{C}^{(k)}\Delta\mathbf{X} + \mathbf{D}^{(k)}\Delta\mathbf{U} \end{aligned} \tag{4.4}$$

where $\mathbf{A}^{(k)}$, $\mathbf{B}^{(k)}$, $\mathbf{C}^{(k)}$, and $\mathbf{D}^{(k)}$ are the matrices of LDM coefficients of the dimensions $r \times r$, $r \times m$, $l \times r$ and $l \times m$; $\Delta\mathbf{X} = \mathbf{X} - \mathbf{X}^{(k)}$, $\Delta\mathbf{U} = \mathbf{U} - \mathbf{U}^{(k)}$, and $\Delta\mathbf{Y} = \mathbf{Y} - \mathbf{Y}^{(k)}$ are deviations of the state vector $\mathbf{X}(r \times 1)$, control vector $\mathbf{U}(m \times 1)$ and observation vector $\mathbf{Y}(l \times 1)$ from the kth steady-state point. Such a static point is further referred to as an operating point.

4.5 LDM at the Nearest Static Point

A linear model can describe the engine dynamics in a small neighbourhood of an operating point. This model represents a linearisation of the nonlinear plant characteristics around the selected point. Usually such linearisation is performed with respect to the nearest static point, although the dynamic transient operation point can also be selected for this purpose.

Examples of dynamic processes in space $\{x, u\}$ and the concept of a *neighbourhood* of an operating point are demonstrated in Figure 4.1.

In order to simplify further discussion, the LDM is used in describing the engine dynamics in the form of deviations Δ from the steady-state conditions (the static point). Suppose that the dynamic processes during transients are best described by the LDM at the nearest point of the static line as opposed to the linear models at all remaining static conditions. Hence, the distance from a required point on the static line to the current point should be minimum:

$$\mathbf{X}^{st} = \underset{\mathbf{X}^{st}}{\arg\min} \| \mathbf{X}(t) - \mathbf{X}^{st} \| \quad (4.5)$$

The geometrical interpretation of the nearest static point represents constructing a perpendicular to the static line.

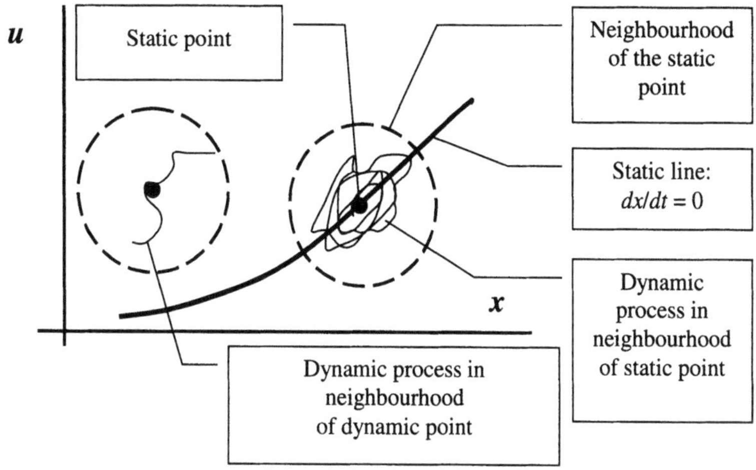

Figure 4.1. Dynamic process and operating point neighbourhood

4.6 Piecewise Linear Approximation of the Static Line

In constructing RPLDM, piecewise linear approximation of nonlinear functions is used several times. A complex relation is approximately substituted by sections of straight lines, and the function values between operating points are determined by means of interpolation.

The idea of linear interpolation consists of the following. Consider some monotonic function $u = f(x)$ receiving values in the range of $u \in [u_1, u_2]$ at the interval $x \in [x_1, x_2]$. Then, the function u in all intermediate points of the interval $[x_1, x_2]$ takes intermediate values within the interval $[u_1, u_2]$. Therefore, the qualitative character of the relation between u and x is kept when the nonlinear function $f(x)$ is approximated by a straight line $g(x)$ (see Figure 4.2). Linear interpolation produces an acceptable level of accuracy when enough operating points are selected. Such approximated calculations are widely applied in engineering practice. The computational relations have the following form:

$$
\begin{aligned}
u = f(x) \approx g(x) &= u_1 + \frac{x - x_1}{x_2 - x_1}(u_2 - u_1) \\
&= \left(1 - \frac{x - x_1}{x_2 - x_1}\right)u_1 + \frac{x - x_1}{x_2 - x_1}u_2 = (1 - p)u_1 + pu_2
\end{aligned}
\tag{4.6}
$$

where $p = (x - x_1)/(x_2 - x_1)$ is the interpolation parameter. The parameter value p determines the position of the current point x on the straight line. The points $\{x_1, u_1\}$ and $\{x_2, u_2\}$, between which the interpolation is made, are referred to as the interpolation operating points.

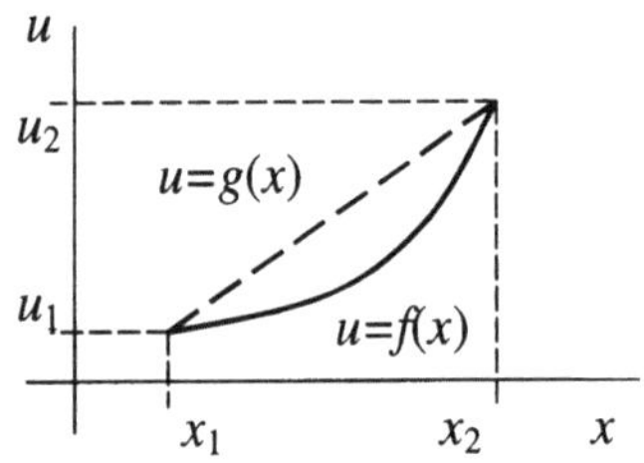

Figure 4.2. Linear interpolation

Such an approach is used for fast, automated calculations on nonlinear algebraic or transcendental equations. Interpolation is convenient when the source function is known only approximately and is presented in a tabular or graphical form. Interpolation enables a nonparametric model at a set of points to be transformed into the parametric form. Linear interpolation is simpler and quicker when compared with nonlinear interpolation using high-order polynomials.

Piecewise linear approximation of the static engine characteristics between operating points makes up a parametric form:

$$
\begin{aligned}
\mathbf{X}_{\mathrm{st}} &= (1 - p)\mathbf{X}^{(k)} + p\mathbf{X}^{(k+1)} \\
\mathbf{U}_{\mathrm{st}} &= (1 - p)\mathbf{U}^{(k)} + p\mathbf{U}^{(k+1)} \\
\mathbf{Y}_{\mathrm{st}} &= (1 - p)\mathbf{Y}^{(k)} + p\mathbf{Y}^{(k+1)}
\end{aligned}
\tag{4.7}
$$

where k is the operating point number: k = 1, 2,..., N–1; p is the interpolation parameter: $p \in [0,1]$.

Piecewise linear relations based on several operating points approximate the static characteristics. The parameters of the static line between the operating points are determined by means of linear interpolation. The coefficients of linear dynamic equations will also depend on the interpolation parameter p. In a simple case, interpolation of the static line is made concerning fuel feed or shaft speed of the engine.

4.7 Operating Parameter and Perpendicular to Static Line

Linear models for several operating points are used for describing the engine dynamics at each point of the space {**X**, **U**}. Deviations of the coordinates from steady-state values are used in the LDM equations. Therefore, the steady-state values are to be determined. Moreover, the LDM coefficients correspond to the model in the nearest static point. The distance to the nearest point on the static line is minimum. This distance is defined as:

$$\rho = \sqrt{\sum_{i=1}^{r} \alpha_i (x_i - x_i^{\text{st}})^2 + \sum_{j=1}^{m} \beta_j (u_j - u_j^{\text{st}})^2} \tag{4.8}$$

where α_i ($i = 1,...,r$) and β_j ($j = 1,...,m$) are the weighting factors depending on the engine characteristics. The static characteristics are approximated by polylines, and the coordinates of the nearest static point $\{\mathbf{X}_{\text{st}}, \mathbf{U}_{\text{st}}\}$ can be found for the current point {**X**, **U**} by a number of different methods. For example, if the static line is concave, then at the beginning the nearest operating point is sought:

$$\{\mathbf{X}^{(k)}, \mathbf{U}^{(k)}\},\ k = 1, 2, \ldots, N \tag{4.9}$$

A perpendicular is built to the straight line passing through $\{\mathbf{X}^{(k)}, \mathbf{U}^{(k)}\}$ and $\{\mathbf{X}^{(k+1)}, \mathbf{U}^{(k+1)}\}$. The coordinates of the point of their intersection $\{\mathbf{X}^*, \mathbf{U}^*\}$ are calculated using the formulae:

$$\rho^{(k)} = \sqrt{\sum_{i=1}^{r} \alpha_i (x_i^{(k+1)} - x_i^{(k)})^2 + \sum_{j=1}^{m} \beta_j (u_j^{(k+1)} - u_j^{(k)})^2} \tag{4.10}$$

$$\mathbf{X}^* = \mathbf{X}^{(k)} + [(\mathbf{X} - \mathbf{X}^{(k)}, \mathbf{X}^{(k+1)} - \mathbf{X}^{(k)})\alpha / (\rho^{(k)})^2](\mathbf{X}^{(k+1)} - \mathbf{X}^{(k)})$$

$$\mathbf{U}^* = \mathbf{U}^{(k)} + [(\mathbf{U} - \mathbf{U}^{(k)}, \mathbf{U}^{(k+1)} - \mathbf{U}^{(k)})\beta / (\rho^{(k)})^2](\mathbf{U}^{(k+1)} - \mathbf{U}^{(k)})$$

where

$$(\mathbf{X} - \mathbf{X}^{(k)}, \mathbf{X}^{(k+1)} - \mathbf{X}^{(k)})\alpha = \sum_{i=1}^{r} (x_i - x_i^{(k)})(x_i^{(k+1)} - x_i^{(k)})$$

$$(\mathbf{U} - \mathbf{U}^{(k)}, \mathbf{U}^{(k+1)} - \mathbf{U}^{(k)})\beta = \sum_{j=1}^{m} (u_j - u_j^{(k)})(u_j^{(k+1)} - u_j^{(k)})$$

are scalar products. The graphic interpretation of this method is shown in Figure 4.3.

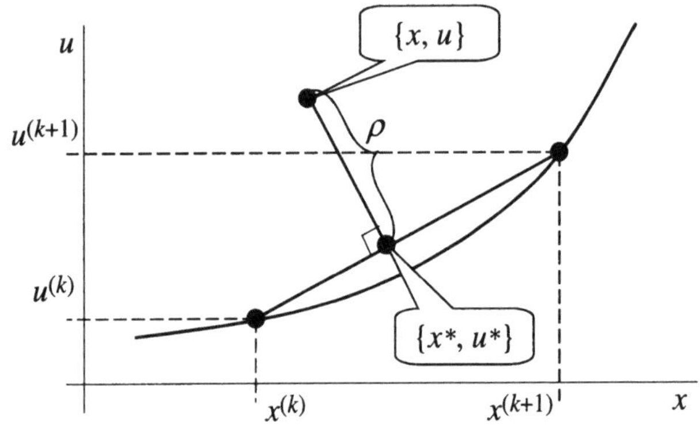

Figure 4.3. Distance to static line

Calculations of the distance are complex and substantially reduce computational speed. In order to simplify the search process for the nearest static point {**X***, **U***}, the following assumptions are made:

- the dynamic model parameters at the point {**X**, **U**} depend only on the state coordinates **X**;
- the state space can be divided into subspaces, where the LDM parameters are constant.

An auxiliary vector **Z** is introduced:

$$\mathbf{Z} = \mathbf{X}^{(N)} - \mathbf{X}^{(1)} \tag{4.11}$$

connecting the first and the last points of the static line.

The perpendicular through the current point to the static polyline is replaced by the perpendicular to the vector **Z**. The intersection of this perpendicular with a piecewise linear static line gives the required "nearest" static point, where current LDM parameters are determined. This approach is illustrated in Figure 4.4.

Perpendiculars to the vector **Z** from any point of the state space make up a set of parallel lines. The LDM coefficients along each line are considered constant and equal to those at the intersection of the "perpendicular" and the piecewise linear static line. They have identical inclination and are uniquely connected to the known coordinates of the vector **Z**. Thus, the position of each of these parallel lines can be uniquely set by one scalar parameter. The operating parameter η is introduced for this purpose; it represents a linear combination of the state coordinates:

$$\eta(t) = (\mathbf{Z}, \mathbf{X}(t)) = \sum_{i=1}^{r} z_i x_i(t) \tag{4.12}$$

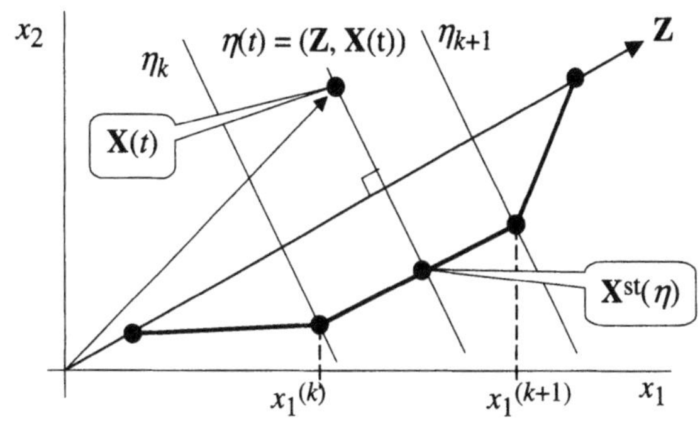

Figure 4.4. Operating parameter η and "perpendicular" to a static line

The equation η = const represents a parametric expression for the "perpendicular" to the static line. All dynamic states belonging to this "perpendicular" will be described by the same LDM.

4.8 Piecewise Linear Approximation of LDM Parameters

LDM coefficients change with the engine operating point as a nonlinear function. It is replaced by a piecewise linear approximation, like the static line. The dynamic parameters for any point on the static operating line $\{\mathbf{X}_{st}, \mathbf{U}_{st}\}$ are determined by means of piecewise linear approximation of the known matrices of LDM coefficients. The formulae for interpolation between operating points are the following:

$$
\begin{aligned}
\mathbf{A}_{st} &= (1-p)\,\mathbf{A}^{(k)} + p\,\mathbf{A}^{(k+1)} \\
\mathbf{B}_{st} &= (1-p)\,\mathbf{B}^{(k)} + p\,\mathbf{B}^{(k+1)} \\
\mathbf{C}_{st} &= (1-p)\,\mathbf{C}^{(k)} + p\,\mathbf{C}^{(k+1)} \\
\mathbf{D}_{st} &= (1-p)\,\mathbf{D}^{(k)} + p\,\mathbf{D}^{(k+1)}
\end{aligned}
\tag{4.13}
$$

The introduced parametric relationships analytically describe the static characteristics of GTE and LDM parameters. The engine dynamic model can be represented as follows:

$$
\begin{aligned}
\Delta\dot{\mathbf{X}} &= \mathbf{A}_{st}\Delta\mathbf{X} + \mathbf{B}_{st}\Delta\mathbf{U} \\
\Delta\mathbf{Y} &= \mathbf{C}_{st}\Delta\mathbf{X} + \mathbf{D}_{st}\Delta\mathbf{U}
\end{aligned}
\tag{4.14}
$$

where $\Delta\mathbf{X} = \mathbf{X} - \mathbf{X}_{st}$; $\Delta\mathbf{U} = \mathbf{U} - \mathbf{U}_{st}$; $\Delta\mathbf{Y} = \mathbf{Y} - \mathbf{Y}_{st}$.

The interpolation parameter p and the operating point number k determine the static line approximation $\mathbf{X}_{st}$, $\mathbf{U}_{st}$ and $\mathbf{Y}_{st}$.

The piecewise linear static characteristics together with piecewise linear relations of LDM factors make up a unified RPLDM. The computational formulae link the matrices of the linear model coefficients **A**, **B**, **C** and **D** and the static line coordinates $\mathbf{X}_{st}$, $\mathbf{U}_{st}$ and $\mathbf{Y}_{st}$ with the operating parameter η. Finally, the piecewise linear model has the following form:

$$\begin{aligned}\dot{\mathbf{X}}(t) &= \mathbf{A}(\eta)(\mathbf{X}(t)-\mathbf{X}_{st}(\eta))+\mathbf{B}(\eta)(\mathbf{U}(t)-\mathbf{U}_{st}(\eta))\\ \mathbf{Y}(t) &= \mathbf{C}(\eta)(\mathbf{X}(t)-\mathbf{X}_{st}(\eta))+\mathbf{D}(\eta)(\mathbf{U}(t)-\mathbf{U}_{st}(\eta))+\mathbf{Y}_{st}(\eta)\end{aligned} \tag{4.15}$$

where

$$\eta = (\mathbf{Z}, \mathbf{X}(t)) = \sum_{i=1}^{r} z_i x_i(t)$$

The current value of the operating parameter η determines the interpolation parameter p, which is used in calculating current vectors $\mathbf{X}_{st}(\eta)$, $\mathbf{U}_{st}(\eta)$, and $\mathbf{Y}_{st}(\eta)$ and matrices $\mathbf{A}(\eta)$, $\mathbf{B}(\eta)$, $\mathbf{C}(\eta)$ and $\mathbf{D}(\eta)$:

$$p = \frac{\eta - \eta^{(k)}}{\eta^{(k+1)} - \eta^{(k)}} \tag{4.16}$$

where $\eta^{(k)}$ is the operating parameter in the kth operating point calculated as a scalar product of two vectors:

$$\eta^{(k)} = (\mathbf{Z}, \mathbf{X}^{(k)}),\ k = 1,\dots, N \tag{4.17}$$

Figure 4.5 demonstrates linear interpolation applied to static and dynamic parameters of the model. The block diagram of the algorithm for simulating the piecewise linear dynamic model is shown in Figure 4.6. The simulation consists of calculations of the operating parameter η, static characteristics and LDM with the subsequent integration of a system of differential equations.

The main steps of simulation are the following:

1. find the current operating parameter value η ;
2. establish two nearest operating points $\eta^{(k)}$ and $\eta^{(k+1)}$, where the current η is situated;
3. calculate the interpolation parameter p ;
4. compute the static point $\mathbf{X}_{st}$, $\mathbf{U}_{st}$ and $\mathbf{Y}_{st}$ with the parameters η and p;
5. find the **A**, **B**, **C** and **D** matrices using η and p values;
6. calculate the derivatives of the state coordinates;
7. numerically integrate the system of differential equations (one time step).

During simulation, at every new step all the operations from 1 to 7 are repeated.

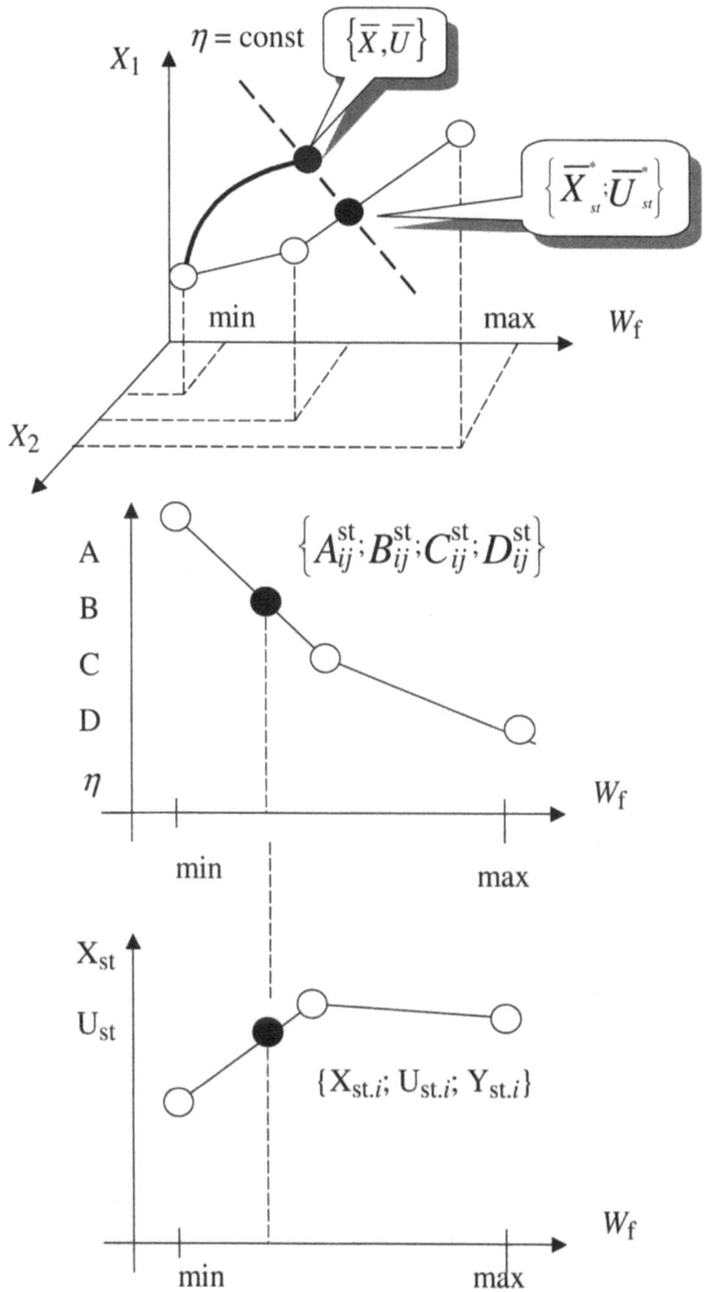

Figure 4.5. Linear interpolation of RPLDM parameters

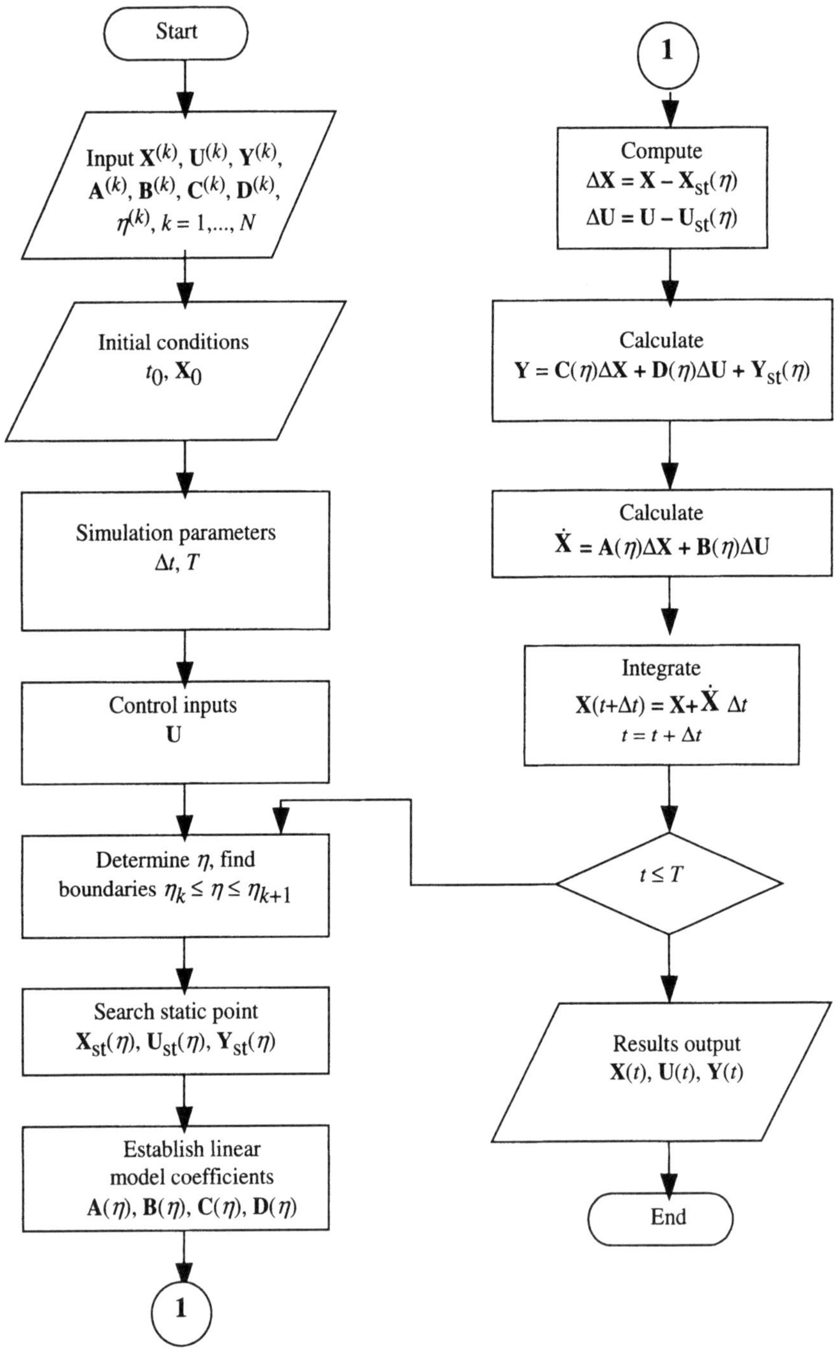

Figure 4.6. Block diagram of RPLDM simulation algorithm

4.9 Example of Turbo Jet Modelling

As an example, the problem of building a dynamic model of a twin-shaft turbo jet engine is considered. Modelling is performed in the operating range of the normalised fan speed from 0.7 to 1.2 at the sea-level conditions ($H = 0,\ M = 0$). The following structure of the model is used:

$$\begin{aligned}
\dot{n}_f &= a_{11}\Delta n_f + a_{12}\Delta n_c + b_{11}\Delta W_f + b_{12}\Delta A_n \\
\dot{n}_c &= a_{21}\Delta n_f + a_{22}\Delta n_c + b_{21}\Delta W_f + b_{22}\Delta A_n \\
p_c^* &= c_{11}\Delta n_f + c_{12}\Delta n_c + d_{11}\Delta W_f + d_{12}\Delta A_n \\
p_t^* &= c_{21}\Delta n_f + c_{22}\Delta n_c + d_{21}\Delta W_f + d_{22}\Delta A_n \\
T_t^* &= c_{31}\Delta n_f + c_{32}\Delta n_c + d_{31}\Delta W_f + d_{32}\Delta A_n
\end{aligned} \tag{4.18}$$

The model should reflect static and dynamic characteristics of the engine in a ±20% neighbourhood of the operating line with the error less than 3-4%. The source static model represents fuel-scheduling characteristics. The analysis of the static characteristics revealed that it is necessary to set not less than three operating points. This will ensure the required accuracy of the piecewise linear approximation in the given operating range. Four operating points are selected:

$$n_f^{(k)}=\{0.7;\ 0.88;\ 1.1;\ 1.2\} \tag{4.19}$$

The static characteristics and LDM parameters are set as follows:

$$\mathbf{X} = \begin{pmatrix} n_f \\ n_c \end{pmatrix};\ \mathbf{U} = \begin{pmatrix} W_f \\ A_n \end{pmatrix};\ \mathbf{Y} = \begin{pmatrix} p_c^* \\ p_t^* \\ T_t^* \end{pmatrix} \tag{4.20}$$

The state coordinates are the fan and compressor speeds; the control coordinates are the fuel feed and the nozzle area; the observed coordinates are the compressor pressure, turbine pressure and temperature.

Figure 4.7 shows the results for the transient from the "nominal" operation point $\bar{n}_{f.\text{red}} = 0.7$ to "maximum thrust" operation $\bar{n}_{f.\text{red}} = 1.2$.

A linear ramp in the control coordinate W_f was applied to the piecewise linear model. For the comparison, the dashed line shows transition processes obtained with a nonlinear detailed dynamic model. The difference in transition processes does not exceed 4%.

The piecewise linear model can be used for modelling nonlinear multivariable systems in computer-aided design (CAD) of control systems. The block diagram of the model in the CAD system SIMULINK is shown in Figure 4.8. Similar structures are used in digital and analogue test bed facilities for investigating control systems. In a multiprocessor on-board computer, the model can be realised using parallel processors if required.

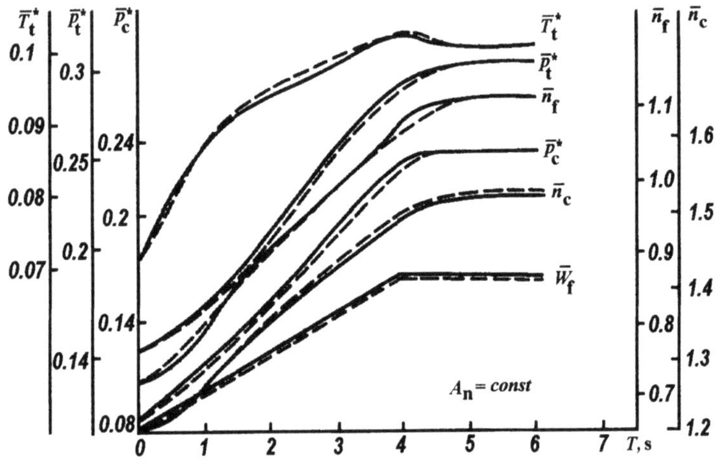

Figure 4.7. Example of transient simulation

4.10. Accounting for Atmospheric Conditions of Engine Operation

Known formulae for reducing the engine parameters to the sea-level conditions are often used for calculating GTE dynamic model parameters in a broad range of the external operating conditions. The formulae operate with the inlet temperature T_{in}^* and pressure p_{in}^*. Hence, RPLDM can be defined in the reduced coordinates in the following way:

$$
\begin{aligned}
\dot{\mathbf{X}}^{\text{red}} &= \mathbf{A}^{\text{red}}(\eta^{\text{red}})(\mathbf{X}^{\text{red}} - \mathbf{X}_{\text{st}}^{\text{red}}(\eta^{\text{red}})) + \mathbf{B}^{\text{red}}(\eta^{\text{red}})(\mathbf{U}^{\text{red}} - \mathbf{U}_{\text{st}}^{\text{red}}(\eta^{\text{red}})) \\
\mathbf{Y}^{\text{red}} &= \mathbf{C}^{\text{red}}(\eta^{\text{red}})(\mathbf{X}^{\text{red}} - \mathbf{X}_{\text{st}}^{\text{red}}(\eta^{\text{red}})) + \mathbf{D}^{\text{red}}(\eta^{\text{red}})(\mathbf{U}^{\text{red}} - \mathbf{U}_{\text{st}}^{\text{red}}(\eta^{\text{red}})) + \mathbf{Y}_{\text{st}}^{\text{red}}(\eta^{\text{red}})
\end{aligned}
\tag{4.21}
$$

where $\eta^{\text{red}} = (\mathbf{Z}\mathbf{X}^{\text{red}})$; $\mathbf{A}^{\text{red}} = \mathbf{P}_{\dot{\mathbf{X}}}^{-1}\mathbf{A}\mathbf{P}_{\mathbf{X}}$; $\mathbf{B}^{\text{red}} = \mathbf{P}_{\dot{\mathbf{X}}}^{-1}\mathbf{B}\mathbf{P}_{\mathbf{U}}$; $\mathbf{C}^{\text{red}} = \mathbf{P}_{\mathbf{Y}}^{-1}\mathbf{C}\mathbf{P}_{\mathbf{X}}$; $\mathbf{D}^{\text{red}} = \mathbf{P}_{\mathbf{Y}}^{-1}\mathbf{D}\mathbf{P}_{\mathbf{U}}$.

The diagonal matrices $\mathbf{P}_{\dot{\mathbf{X}}}$, $\mathbf{P}_{\mathbf{X}}$, $\mathbf{P}_{\mathbf{Y}}$ and $\mathbf{P}_{\mathbf{U}}$ connect physical and reduced parameters:

$$
\dot{\mathbf{X}} = \mathbf{P}_{\dot{\mathbf{X}}}\dot{\mathbf{X}}^{\text{red}};\ \mathbf{X} = \mathbf{P}_{\mathbf{X}}\mathbf{X}^{\text{red}};\ \mathbf{Y} = \mathbf{P}_{\mathbf{Y}}\mathbf{Y}^{\text{red}};\ \mathbf{U} = \mathbf{P}_{\mathbf{U}}\mathbf{U}^{\text{red}}
\tag{4.22}
$$

The reduction matrices for the considered example are the following:

$$\mathbf{P}_{\dot{\mathbf{X}}} = \begin{pmatrix} \dfrac{p_{\text{in}}^*}{10133} & 0 \\ 0 & \dfrac{p_{\text{in}}^*}{10133} \end{pmatrix}; \quad \mathbf{P}_{\mathbf{X}} = \begin{pmatrix} \sqrt{\dfrac{T_{\text{in}}^*}{288.17}} & 0 \\ 0 & \sqrt{\dfrac{T_{\text{in}}^*}{288.17}} \end{pmatrix} \tag{4.23}$$

$$\mathbf{P}_{\mathrm{U}} = \left(\begin{array}{c:c} \dfrac{p_{\text{in}}^*}{10{,}133}\sqrt{\dfrac{T_{\text{in}}^*}{288.17}} & 0 \\ \hdashline 1 & 1 \end{array}\right); \quad \mathbf{P}_{\mathrm{Y}} = \begin{pmatrix} \dfrac{p_{\text{in}}^*}{10{,}133} & 0 & 0 \\ 0 & \dfrac{p_{\text{in}}^*}{10{,}133} & 0 \\ 0 & 0 & \dfrac{T_{\text{in}}^*}{288.17} \end{pmatrix}$$

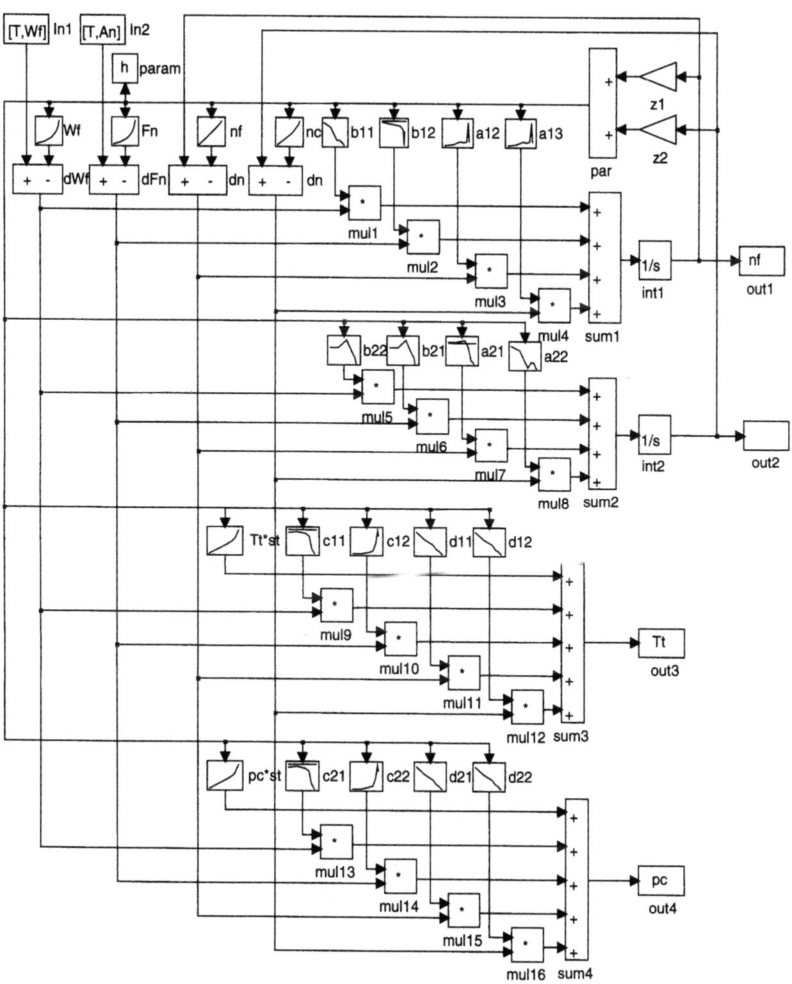

Figure 4.8. SIMULINK diagram of RPLDM realisation

4.11 Concluding Remarks

A method for building real-time piecewise linear dynamic models (RPLDM) has been considered. RPLDM combines the accuracy of nonlinear models with the speed of linear models. When building a RPLDM, several assumptions and simplifications are introduced. This leads to a substantial increase in the computational speed at the cost of some decrease in modelling accuracy, as compared with nonlinear thermodynamic models. An example of a jet engine demonstrates practical details of RPLDM realisation.

RPLDMs are widely used for control engineering purposes. Control algorithms are often initially tested with such models and testing of electronic control system units on test beds may well be against a RPLDM. The ability to operate these models in real time also offers the opportunity for their use in on-board engine monitoring systems to give "normal" operation models. Here the loss of accuracy of the RPLDM becomes a critical issue in the performance of the diagnostic system.

The structure of the model also allows construction of analytical models. For this purpose, the static characteristics and the matrices of the linear model coefficients **A**, **B**, **C** and **D** are approximated by analytical functions, for example, by polynomials. This would allow keeping the RPLDM model speed while improving the model accuracy. Moreover, approximation with smooth functions eliminates sharp changes in derivatives around the operating points. Based on the discussed approach, a variety of simplified models can be constructed with different degrees of approximation to the nonlinear thermodynamic model in various application areas.

PART II. GAS TURBINE SYSTEM IDENTIFICATION

Chapter 5
Linear System Identification

5.1 Introduction

In general there are two ways of arriving at models of physical processes:

Physical principles modelling. Physical knowledge of the process, in the form of first principles, is employed to arrive at a model that will generally consist of a multitude of differential / partial differential / algebraic relations between physical quantities. The construction of a model is based on presumed knowledge about the physics that governs the process. The first principles relations concern, *e.g.*, the laws of conservation of energy and mass and Newton's law of movement.

Experimental modelling, or system identification. Measurements of several variables of the process are taken and a model is constructed by identifying a model that matches the measured data as well as possible.

In this chapter an introduction is provided to the fundamental concepts relating to the modelling and estimation of linear time invariant systems in the time and frequency domain. Estimators are described for parametric time-domain modelling and for both nonparametric and parametric frequency-domain modelling. A comparison is made between discrete-time and continuous-time models and their applications are discussed. A discussion is presented of the key issues in experiment design. Particular attention is paid to the selection of the input excitation and the use of periodic signals is advocated. Finally, the properties of multisine and maximum length binary sequences are discussed.

5.2 Linear Models

A causal, *linear time-invariant* (LTI) system can be described by the following equation, which expresses the output of such a system as a convolution of its impulse response and the input:

$$y(t) = \int_{\tau=0}^{\infty} h(\tau)\,u(t-\tau)\,d\tau = h(\tau) * u(\tau),\ \tau \geq 0 \qquad (5.1)$$

where $u(t)$ is the input to the system, $y(t)$ the output and $h(t)$ the system unit impulse response. The impulse response provides a complete nonparametric characterisation of a LTI system in the time-domain since it directly relates the output $y(t)$ to the input $u(t)$ of the system. Since most physical relationships are expressed in terms of differential equations it is natural to work in the continuous-time domain. Under steady-state conditions the Laplace transform of the above equation gives:

$$Y(j\omega) = H(j\omega) \cdot U(j\omega) \tag{5.2}$$

which if solved for $H(s)$ by substituting $j\omega \Rightarrow s$:

$$H(s) = \frac{Y(s)}{U(s)} \tag{5.3}$$

This is termed the continuous transfer function, which describes all the dynamic properties of a LTI system in the frequency domain, independently of the input. The continuous transfer function can be expressed as a rational polynomial function in s, with the number of zeros less than or equal to the number of poles ($nb \le na$). A further restriction that the transfer function must be strictly proper can be extended to ($nb < na$) implies that there is no direct term relating the system input and output. Note that the exponential term in the equation models the pure time delay T_d of the system.

$$H(s) = \frac{b_0 + b_1 s + \cdots + b_{nb} s^{nb}}{a_0 + a_1 s + \cdots + a_{na} s^{na}} e^{-j\omega T_d} = \frac{N(s)}{D(s)} \tag{5.4}$$

In practice, the system output will be observed at discrete sampling instants:

$$y(kT_s) = \int_{\tau=0}^{t} h(\tau) u(kT_s - \tau) d\tau, \; k = 1, 2, \ldots \tag{5.5}$$

where T_s is the sampling interval. If the input signal is kept constant between sampling instants, termed *piecewise constant*, such that

$$u(t) = u(kT_s), \; kT_s \le t < (k+1)T_s \tag{5.6}$$

then the convolution integral of Equation (5.1) can be replaced, without approximation, by a discrete summation:

$$y(kT_s) = \sum_{l=1}^{t} h_{T_s}(l)u(kT_s - l) \quad \text{where} \quad h_{T_s}(l) = \int_{\tau=(l-1)T_s}^{lT_s} h(\tau)d\tau \tag{5.7}$$

which defines the system output at the discrete time points kT_s [1]. The term h_{Ts} is the discrete impulse response of the system. There is no approximation involved in the discretisation if it is assumed that a zero-order-hold (ZOH) device is used to apply the signal to the continuous system. This is termed the ZOH assumption. If a backward shift operator z^{-1} is defined where:

$$z^{-1}u(kT_s) = u((k-1)T_s) \tag{5.8}$$

then Equation (5.7) can be rewritten as:

$$y(kT_s) = \left[\sum_{l=1}^{t} h_{T_s}(l)z^{-1}\right]u(kT_s) = H_{T_s}(z)u(kT_s) \tag{5.9}$$

where $H_{Ts}(z)$ can be expressed in closed form as a rational polynomial function z, termed the discrete transfer function:

$$H_{T_s}(z) = \frac{b_1 z^{-1} + b_2 z^{-2} + \cdots + b_{nb} z^{-nb}}{1 + a_1 z^{-1} + a_2 z^{-2} + \cdots + a_{na} z^{-na}} = \frac{B(z)}{A(z)} \tag{5.10}$$

It is apparent from the above equation that the parameters of the discrete transfer function are dependent on the sampling interval T_s. A continuous transfer function $H(s)$ can be transformed to a discrete transfer function $H_{T_s}(z)$ without approximation, under the ZOH assumption. The *impulse invariant transform* is defined such that the impulse response of the discrete system is equal to that of the underlying continuous system at the sampling instants. The following relation can map the continuous poles to the discrete time domain:

$$z = e^{sT_s} \tag{5.11}$$

It is clear from the above that the selection of the discrete sampling frequency is crucial in system identification. If the sampling frequency is too low, information about the continuous system will be lost. If, for example, the sampling frequency is less than twice the break-point frequencies of the continuous poles, then aliased poles will appear in the z-domain, at frequencies lower than their true values. On the other hand, sampling too quickly, can lead to numerical problems, since as T_s approaches zero in Equation (5.11) the z-domain poles will converge to a single point at $z = 1$, irrespective of the values of the s-domain poles. Very fast sampling can also give rise to nonminimum phase discrete models (with zeros outside the unit circle), of minimum phase continuous systems (with all the zeros in the left-hand s-plane).

5.3 Model Estimation

The process of model estimation involves three key elements: data gathering, selection of the model set from which to estimate, and selection of a rule by which the most appropriate model can be chosen. Once a model has been chosen a further decision must be made, with regard to the quality of the model for its intended application, termed validation. A deficiency in the model may be due to a variety of reasons, which are summarised by Ljung [1] as:

- the numerical procedure failed to find the best model;
- the selection criterion was not well chosen;
- the model set was not appropriate;
- the data set was not informative enough.

5.3.1 Time-Domain

A complete characterisation of an LTI system in the time-domain can be obtained by the impulse response defined in Equation (5.1). This can be carried out by subjecting the system to an impulsive input and measuring the output, or by using binary sequences and correlation methods [2].

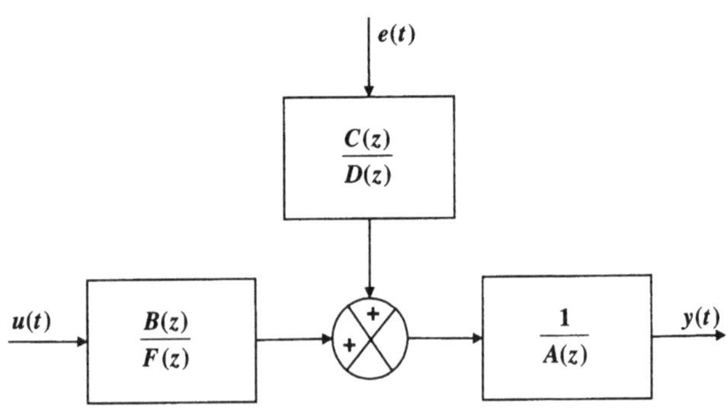

Figure 5.1. A general discrete transfer function model

The general structure for the class of discrete LTI transfer function models is shown in Figure 5.1, which can be expressed as:

$$A(z)y(t) = \frac{B(z)}{F(z)}u(t) + \frac{C(z)}{D(z)}e(t) \quad \text{where} \tag{5.12}$$

$$A(z) = 1 + a_1 z^{-1} + \cdots + a_{na} z^{-na} \tag{5.13}$$

$$B(z) = b_1 z^{-1} + \cdots + a_{nb} z^{-nb} \tag{5.14}$$

and $F(z)$ has the same form as $A(z)$, while $C(z)$ and $D(z)$ have the same form as $B(z)$. The system input is $u(t)$, its output $y(t)$ and the noise sequence $e(t)$ is assumed to be normally distributed and independent of the input signal, with zero mean and variance λ_0. Simpler model structures can be achieved by setting the orders of certain polynomials to zero. Two basic model classes exist, the first being *equation error* models, where the noise term enters as a direct error in the difference equations. Models of this class include:

- *Auto Regressive with eXogenous input* (ARX): $nc = nd = nf = 0$, so called because the predictor can be defined as a linear regression.
- *Auto Regressive Moving Average with eXogenous input* (ARMAX): $nd = nf = 0$, where $C(z)$ acts as a moving average filter of the noise.

The second class of models is called *output error* structures where, as the name suggests, the noise term is added to the system output. Such models include:

- *Output Error* (OE): $\underline{na} = nc = nd = 0$.
- *Box–Jenkins* (BJ): $na = 0$; which allows greater flexibility in the choice of the noise model at the cost of increased complexity.

Further details of the different structures which can be employed are given in Ljung [1, Chapter 4]. The general model can be compactly expressed as

$$y(t) = G(z,\mathbf{p})u(t) + H(z,\mathbf{p})e(t) \tag{5.15}$$

where $G(z,\mathbf{p})$ and $H(z,\mathbf{p})$ are filters of finite order and functions of a parameter vector $\mathbf{p}$:

$$\mathbf{p} = [a_1 \cdots a_{na},\ b_1 \cdots b_{nb},\ c_1 \cdots c_{nc},\ d_1 \cdots d_{nd},\ f_1 \cdots f_{nf}]^T \tag{5.16}$$

The one-step-ahead prediction for Equation (5.15) is denoted as $\hat{y}(t)$:

$$\hat{y}(t) = H^{-1}(z,\mathbf{p})G(z,\mathbf{p})u(t) + [1 - H^{-1}(z,\mathbf{p})]y(t) \tag{5.17}$$

The errors between the actual process output and the predicted model output can be calculated by subtracting Equation (5.17) from Equation (5.15) to yield:

$$\varepsilon(t,\mathbf{p}) = y(t) - \hat{y}(t) = H^{-1}(z,\mathbf{p})[y(t) - G(z,\mathbf{p})u(t)] \tag{5.18}$$

These are termed the *residuals*, *prediction errors* or *innovations*, and the family of estimators which use this approach are called *prediction error methods* (PEM).

If the filter $H(z)$ is set to unity then the predicted error calculated according to Equation (5.18) becomes

$$\varepsilon(t,\mathbf{p}) = y(t) - G(z,\mathbf{p})\,u(t) \tag{5.19}$$

which is the difference between the measured output $y(t)$ and the noise-free model output, corresponding to the OE structure described above. The most common approach is to determine the estimates of $G(z)$ and, if appropriate, $H(z)$ by minimising a quadratic cost function:

$$V_N(\mathbf{p}) = \sum_{t=1}^{N} \varepsilon^2(t,\mathbf{p}) \tag{5.20}$$

$$\hat{\mathbf{p}}_N = \arg\min V_N(\mathbf{p}) \tag{5.21}$$

where arg min means the argument which minimises the function and $\hat{\mathbf{p}}_N$ is the estimated parameter vector. An analytical solution can be found to this problem for the ARX model, but this is not the case for the other model structures and a nonlinear least-squares technique may be used, such as the Gauss-Newton approach [1]. A pure time delay can be incorporated in the model by delaying the input by a certain number of sampling intervals. An empirical approach to select this delay is suggested by Ljung [3], which involves estimating a second-order ARX model and finding the delay which minimises the cost function. This delay is then used for further modelling, using alternative structures and model orders, and once a satisfactory model has been found the delay is once again varied until the best model fit is obtained.

The properties of PEM estimators have been extensively studied and it has been shown that they are maximum likelihood estimators (MLE) in the presence of Gaussian innovations [1]. Provided that the true model is part of the model set and that the data are informative enough, then the estimates will converge in a mean square sense to the true parameter values $\mathbf{p}_0$ (as $N \to \infty$), so that the estimator is *asymptotically unbiased*. If the true system is not part of the model set then the estimates will converge to the best approximation available in the set. If the measurements are independent and identically distributed (iid) then the MLE is strongly consistent and the covariance matrix of the estimates reaches the Cramer–Rao bound asymptotically, so that the estimator is asymptotically efficient [4]. The identical distribution of the measurements implies that the disturbing noise has stationary statistical properties over the measurement interval. The covariance matrix can be approximated by the Cramer–Rao bound, under the assumption that the estimator has reached its asymptotic properties (N is sufficiently large). This can be estimated from the data using

$$\mathbf{C}_{CR} = \hat{\lambda}_N \left[\frac{1}{N} \sum_{t=1}^{N} \psi(t,\mathbf{p})\psi^T(t,\mathbf{p}) \right]^{-1} \tag{5.22}$$

which is the inverse of the covariance matrix of the predictor gradients $\psi(t,\mathbf{p})$, normalised by the innovations variance $\hat{\lambda}_N$ [1]. The variances of the parameter estimates are found on the diagonal of this matrix. The predictor gradients are defined as:

$$\psi(t,\mathbf{p}) = \frac{d\ \hat{y}(t\,|\,\mathbf{p})}{d\,\mathbf{p}} = -\frac{d\ \varepsilon(\mathrm{t},\mathbf{p})}{d\,\mathbf{p}}\Big|_{\mathbf{p}=\mathbf{p}_0} \tag{5.23}$$

which is a column vector of the partial derivatives of $\hat{y}(t\,|\,\mathbf{p})$ with respect to each of the parameters. This should be evaluated at the true parameter values $\mathbf{p}_0$, but since these are unknown the estimated parameters must be used in their place. Since $\psi(t,\mathbf{p})$ is the gradient of $\hat{y}(t\,|\,\mathbf{p})$, then Equation (5.22) clearly implies that the asymptotic accuracy of a given parameter is related to the sensitivity of the prediction with respect to that parameter. The innovations variance can be estimated as

$$\hat{\lambda}_N \frac{1}{N}\sum_{t=1}^{N} \varepsilon^2(t,\hat{\mathbf{p}}_N) \tag{5.24}$$

Strictly speaking, the expression in Equation (5.22) is only valid if the true model lies within the set of possible models which can be estimated. This is rarely the case; however, the expression can be considered as a useful approximation of the parameter variances, as long as the modelling errors are not too severe.

The maximum likelihood property also implies that the estimates are asymptotically normally distributed, which allows parameter confidence bounds to be calculated in a straightforward manner. The parameter variances can also be translated into confidence ellipsoids of the poles and zeros, using the linear approximation technique of Gauss. This is particularly useful, since the uncertainties of the model poles and zeros are far easier to interpret than the uncertainty of individual model parameters. The transformation must be treated with some caution; however, since the behaviour of estimated poles and zeros across several experiments with noisy data cannot always be modelled in a linear fashion. The estimators and algorithms discussed have been implemented as part of the MATLAB System Identification Toolbox, written by Ljung [3].

5.3.2 Frequency-Domain

A general *error-in variables* (EV) model for frequency-domain identification of LTI systems is shown in Figure 5.2, where the measured Fourier coefficients of the input $U_m(j\omega)$ and output $Y_m(j\omega)$, are corrupted by noises $M(j\omega)$ and $N(j\omega)$. It is assumed that the noise sequences are zero mean, have a complex normal distribution, and that the noise is uncorrelated between frequency points, such that $E\{M(j\omega_i)\cdot M(j\omega_k)\} = 0$ for $i \neq k$, and similarly for $N(j\omega)$. Brillinger [5] has shown that, under mild conditions (which include stationarity), the frequency-domain noise observed via the discrete Fourier transform will approach these properties asymptotically, as the number of time samples increases. It has also been shown that the asymptotic properties are well approximated for values of N as small as 512 samples [6].

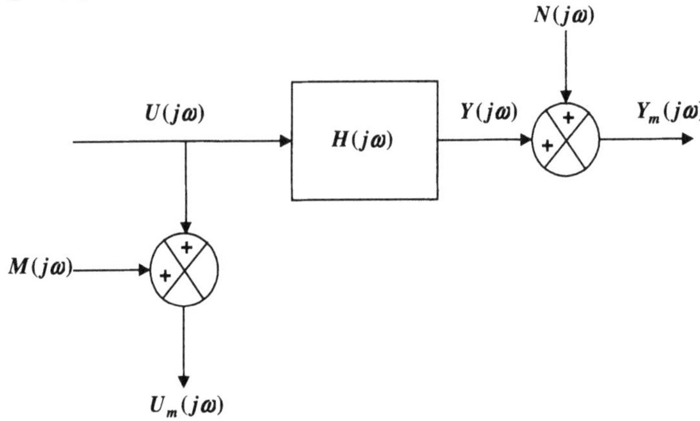

Figure 5.2. A general error-in variables model

Most commercially available frequency response analysers calculate the nonparametric FRF by exciting the system with a random signal and estimating the frequency response from ratios of the cross- and auto-spectra of the input and output signals. The two approaches most commonly implemented are the classical H_1 and H_2 estimators:

$$H_1(j\omega) = \frac{\frac{1}{M}\sum_{m=1}^{M} U_m^*(j\omega)Y_m(j\omega)}{\frac{1}{M}\sum_{m=1}^{M} U_m^*(j\omega)U_m(j\omega)} = \frac{G_{UY}(j\omega)}{G_{UU}(j\omega)} \tag{5.25}$$

$$H_2(j\omega) = \frac{\frac{1}{M}\sum_{m=1}^{M} Y_m^*(j\omega)Y_m(j\omega)}{\frac{1}{M}\sum_{m=1}^{M} Y_m^*(j\omega)U_m(j\omega)} = \frac{G_{YY}(j\omega)}{G_{YU}(j\omega)} \tag{5.26}$$

where $U_m(j\omega)$ and $Y_m(j\omega)$ are measured across M independent time blocks of equal length. Time-domain windowing is normally employed if a random excitation signal is used, in order to minimise the leakage errors introduced by the *fast Fourier transform* (FFT) [7]. Assuming that the input and output noises are not correlated with each other and with the input and output signals, the following results can be obtained:

$$H_1(j\omega) = H(j\omega)\frac{1}{1+[G_{MM}(\omega)/G_{UU}(\omega)]} \tag{5.27}$$

$$H_2(j\omega) = H(j\omega)\left(1 + \frac{G_{NN}(\omega)}{G_{YY}(\omega)}\right) \tag{5.28}$$

where $G_{MM}(\omega)$ and $G_{NN}(\omega)$ are the auto-spectra of the input and output noise. The equations show that the H_1 estimator is biased, in magnitude only, in the presence of input noise and H_2 is similarly biased for noise on the output. The bias depends on the *signal-to-noise ratio* (SNR) at the input and output, respectively, and hence on the second-order moments (*i.e.*, power) of the noise [8]. The quality of the FRF estimates is usually assessed by calculating the squared coherence function, defined as:

$$\gamma^2(\omega) = \frac{\left|G_{UY}(j\omega)\right|^2}{G_{UU}(\omega)G_{YY}(\omega)} \tag{5.29}$$

which is the H_1 estimator divided by the H_2 estimator. It can be shown that:

$$\left|G_{UY}(j\omega)\right|^2 \leq G_{UU}(\omega)G_{YY}(\omega) \tag{5.30}$$

and hence the coherence can vary between zero and one, where a value of one indicates an identical linear relationship between $U_m(j\omega)$ and $Y_m(j\omega)$ for each of the measurements [8]. Since the cross- and auto-spectra are averaged quantities; the coherence can be thought of as a measure of the deterministic nature of the system frequency response. If the estimated frequency response is constant for each measurement the coherence values of less than one will be caused by:

- uncorrelated noise in the measurements;
- nonlinearities in the system;
- leakage errors in the FFT

Each of these effects will cause a random (stochastic) influence on the estimated response, the first due to its inherent nature and the second and third due to the nature of the input signal. Thus, when random signals are used, the coherence can be interpreted as a measure of the fraction of the output spectrum which is linearly related to the input spectrum at a given frequency.

However, care must be taken in the interpretation of the coherence function if periodic input signals are used, with averaging carried out over complete signal periods. The nonlinear contributions at the output will now be deterministic, since the response of the nonlinearity to the signal period will be the same, and there will be no drop in the coherence. Thus, if periodic signals are used, the coherence will only allow the detection of stochastic effects and will not allow the linear and nonlinear behaviour of the system to be distinguished.

The use of periodic input signals allows an alternative estimator to be defined, which is simply the ratio of the mean values of the output and input Fourier coefficients, at the discrete test frequencies ω_k:

$$\hat{H}_{EV}(j\omega_k) = \frac{\frac{1}{M}\sum_{m=1}^{M} Y_m(j\omega_k)}{\frac{1}{M}\sum_{m=1}^{M} U_m(j\omega_k)} = \frac{\overline{Y}(j\omega_k)}{\overline{U}(j\omega_k)} \tag{5.31}$$

This has been termed the EV estimator and it has been shown that it is an MLE if the input and output noises have a complex normal distribution, even if they are mutually correlated [9]. The bias of the estimator depends on the fourth-order moments of the noise and, with synchronised measurements of the input and output signals, the estimator is both asymptotically unbiased and efficient. It is also important to consider the uncertainty of the FRF estimates. Schoukens *et al.* [10] have shown that using linear approximations, each of the estimators described above has the same uncertainty. Defining the variance of the estimated FRF:

$$\sigma_H^2(\omega) = E\{\,(H(j\omega) - E\{H(j\omega)\}\,)\,(H(j\omega) - E\{H(j\omega)\})^*\,\} \tag{5.32}$$

which can be written as:

$$\sigma_H^2(\omega) = \sigma_{|H|}^2(\omega) + |H(j\omega)|^2 \sigma_{\angle H}^2(j\omega) \tag{5.33}$$

the approximate variance of the FRF estimates is given by:

$$\begin{aligned} \sigma_H^2(\omega_k) &= 2\frac{|\hat{H}(j\omega_k)|^2}{M}\left(\frac{\sigma_M^2(\omega_k)}{|U(j\omega_k)|^2} + \frac{\sigma_N^2(\omega_k)}{|Y(j\omega_k)|^2}\right) \\ &= \frac{2}{M\,|U(j\omega_k)|^2}(|\hat{H}(j\omega_k)|^2\,\sigma_M^2(\omega_k) + \sigma_N^2(\omega_k)) \end{aligned} \tag{5.34}$$

where the terms $\sigma_M^2(\omega_k)$ and $\sigma_N^2(\omega_k)$ are the variances of the real or imaginary parts of $M(j\omega_k)$ and $N(j\omega_k)$. If the input and output noises are cross-correlated, then an additional factor must be introduced, such that:

$$\begin{aligned} \sigma_H^2(\omega_k) &= \frac{2}{M\,|U(j\omega_k)|^2}(\,|\hat{H}(j\omega_k)|^2\,\sigma_M^2(\omega_k) + \cdots + \\ &\sigma_N^2(\omega_k) - 2\Re\{\sigma_{MN}(j\omega_k)H^*(j\omega_k)\,) \end{aligned} \tag{5.35}$$

where $\sigma_{MN}(j\omega_k)$ is the cross-covariance. Correlated noise may be the result of input noise passing through the system or of the system operating within a feedback loop. It can be seen from Equation (5.35) that the variance of the estimated FRF is inversely proportional to the number of measurements and the power of the input harmonics and proportional to the noise variances *referred to the system output*.

Parametric identification in the frequency domain involves selecting the parameters of an *s*- or *z*-domain model, with pure time delay T_d:

$$H(s) = \frac{b_0 + b_1\Omega + \cdots + b_{nb}\Omega^{nb}}{a_0 + a_1\Omega + \cdots + a_{na}\Omega^{na}} e^{-j\omega T_d} \tag{5.36}$$

where Ω is interchangeably *s* or *z*. A range of approaches have been proposed, which are discussed in detail by Pintelon *et al.* [11]. Employing the EV model in Figure 5.2, Schoukens, *et al.* [12] derived the maximum likelihood estimator of $H(\Omega)$ which was subsequently developed by Pintelon, *et al.* [11] to arrive at a cost function

$$\begin{aligned} K(\mathbf{p}) = \frac{1}{2}\sum_{k=1}^{F} \Big\{ & \left| e^{-j\omega_k T_d} N(j\omega_k, \mathbf{p})\bar{U}_M(j\omega_k) - D(j\omega_k, \mathbf{p})\bar{Y}_M(j\omega_k) \right|^2 \Big/ \\ & \sigma_N^2 \left| D(j\omega_k, \mathbf{p}) \right|^2 + \sigma_M^2 \left| N(j\omega_k, \mathbf{p}) \right|^2 - \\ & 2\Re\left[\sigma_{MNy} e^{-j\omega_k T_d} N^*(j\omega_k, \mathbf{p}) D(j\omega_k, \mathbf{p})\right] \Big\} \end{aligned} \tag{5.37}$$

where **p** is the parameter vector:

$$\mathbf{p} = [a_0, \cdots, a_{na}, b_0, \cdots, b_{nb}]^T \tag{5.38}$$

and *F* is the number of frequencies used in estimation. It can be seen that a nonparametric noise model is employed, in contrast to the time-domain approach, and the noise variances and covariance are required as *a priori* information. These can be obtained from the data before estimation, provided several independent measurements are available. This cost function can be interpreted as a least squares formulation, with each frequency point weighted by a noise variance term. The larger the noise variance at a given frequency the less that frequency will influence the cost function. The estimator is an MLE under the assumption of complex normal noise. Under this assumption, double the cost function is the sum of 2*F* squared Gaussian variables and is hence χ^2 distributed with $2F-n_p$ degrees of freedom, where n_p is the number of free parameters. Consequently, its expected value will be:

$$K_{\min} = F - \frac{n_p}{2} \tag{5.39}$$

This cost function is insensitive to the multiplication of the parameters by a scalar and hence a constraint must be placed on the scaling of the parameter vector. This can be achieved by fixing one nonzero parameter value or by fixing the overall norm of the parameter vector itself. The pure time delay T_d can also be included as a free parameter for estimation, which is an attractive feature of the frequency-domain approach, since its value is not fixed to multiples of the sampling interval. The delay is estimated in seconds for s-domain models and fractional multiples of the sampling period for z-domain models. The parameters which minimise the cost function can be estimated using nonlinear optimisation techniques such as Gauss–Newton, Levenberg–Marquardt or singular value decomposition. Since the estimator is an MLE the covariance matrix can once again be approximated by the Cramer–Rao bound, under the assumption that the true system lies within the range of possible models. The uncertainty of the estimated parameters can then be translated to uncertainties on the estimated poles and zeros, using the analytical technique described by Guillaume, *et al.* [9].

The estimator is now available as part of a MATLAB Frequency-domain System Identification Toolbox, written by Kollár [13] and its derivation and properties are reviewed in a paper by Kollár [14]. Further information about system identification in the frequency-domain, can be found in an excellent book by Pintelon and Schoukens [15].

5.3.3 Model Order Selection and Validation

The selection of an appropriate model order and the validation of that model involves techniques which are common to both time- and frequency-domain approaches. The most fundamental is to monitor the behaviour of the cost function as the model order is increased. If good convergence is achieved then a graph of the cost function against model order will usually show a definite knee. This is illustrated in Figure 5.3 for a typical example using the frequency-domain estimator discussed in the previous section.

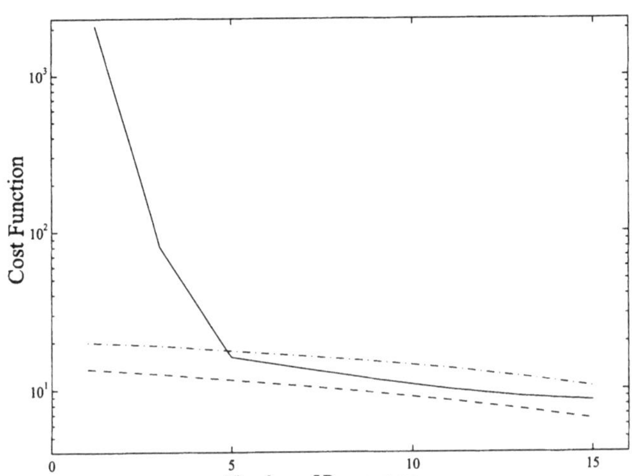

Figure 5.3. Cost function versus number of model parameters (solid), with its expected value (dashed) and the 95% bound on that value (dash-dot) (reproduced from Evans [16] with permission)

The cost function is plotted against the number of model parameters, along with its expected value and the 95% bound on that value.

Once a model structure has been selected in this way it can be further checked by looking at the statistical properties of the residuals. The time-domain residuals defined in Equation (5.18) should be approximately independent and normally distributed, with a zero mean and a given variance. A visual inspection of the residuals themselves is also useful, since large or untypical values may indicate bad data or *outliers*.

Evaluation of the residuals is not straightforward in the frequency-domain case, since their calculation requires the estimation of the true input and output Fourier coefficients, which are termed the *nuisance parameters*. It is thus preferable to study the residuals of the cost function itself, also termed the residuals of the noise referred to the output, as discussed for Equation (5.35). If it is assumed that the true input and output Fourier coefficients can be obtained from the measured data then the frequency-domain residuals can be also obtained as the difference between the estimated FRF and the model frequency response. Any modelling errors which are present stem from one of three sources: stochastic effects, under-modelling of the linear dynamics or unmodelled nonlinear dynamics. Schoukens, *et al.* [17] have proposed a technique to distinguish between these errors, based on exciting the system with a random-phased broadband multisine and examining the normalised autocorrelation of the residuals. The normalised autocorrelation of the frequency-domain residuals is defined as:

$$R_{ee}(j\omega_m) = \frac{1}{F-m}\sum_{k=1}^{F-m}\frac{(\hat{H}(j\omega_k) - H(j\omega_k,\mathbf{p}))(\hat{H}(j\omega_{k+m}) - H(j\omega_{k+m},\mathbf{p}))^*}{\sigma(j\omega_k)\sigma(j\omega_{k+m})} \qquad (5.40)$$

$$-\delta(m-0)$$

where $\hat{H}(j\omega_k)$ is the estimated FRF, $\sigma(j\omega_k)$ is its standard deviation and $H(j\omega_k,\mathbf{p})$ is the frequency response of the model. The autocorrelation is a complex quantity, only the magnitude of which is of interest. Appropriate statistical bounds for the magnitude can be derived if R_{ee} is assumed to have a complex normal distribution. If there are no modelling errors then R_{ee} should be zero for all values of the lag m. Taking into consideration the estimator cost function K, the residuals R_{ee}, there are four possible scenarios:

Correct model. The cost function is small, approaching $K_{\min}$ the residuals are small and R_{ee} is close to zero for all lags.

Stochastic effects. The cost function is small, the residuals are large and R_{ee} is small, with a sharp form concentrated at $R_{ee}(0)$. The cost function is reduced by the large noise variances in the denominator of Equation (5.37). The residuals are large due to the scatter of the FRF estimates, but they are uncorrelated, resulting in small values of R_{ee} at all but the zero lag. The values of R_{ee} are further reduced by the large value of standard deviation in the numerator of Equation (5.40).

Under-modelling. The cost function is large, the residuals are large and R_{ee} is also large, with significant values at nonzero lags. The broad shape of R_{ee} indicates that the residuals are correlated. This suggests the difference between the model and data are smoothly varying, which points to unmodelled linear dynamics.

Nonlinear distortion. The cost function is large, the residuals are large and R_{ee} is also large with a sharp form concentrated $R_{ee}(0)$.The shape of R_{ee} indicates that the residuals are uncorrelated. This suggests that the model errors are due to nonlinear distortions, so that increasing the order of the linear model will not significantly improve the fit.

The final and decisive test for any estimated model is *cross-validation*, which involves repeating the estimation using a completely new data set and comparing the model parameters. The parameters should be identical, to within the prescribed uncertainty bounds.

5.3.4 Comparison of Models and Approaches

A comparison is made in Table 5.1 of the different properties of *s*-domain and *z*-domain models. It can be seen that *z*-domain models have important advantages for applications such as control system design and system simulation. On the other hand, if a precise description of the system is required, then *s*-domain models are preferred. The choice of model has a direct impact on the type of identification employed, since the estimation of *s*-domain models is not straightforward in the time domain. On the other hand, frequency-domain identification allows *s*-domain models to be directly identified. In addition, the effects of nonlinearities are more easily detected in the frequency domain, by the presence of output harmonics in addition to those present in the input signal. It is thus important to distinguish the advantages and disadvantages of the two approaches.

Table 5.1. A comparison of *s*-domain and *z*-domain models (reproduced from Evans [16], with permission)

Property	*s*-domain	*z*-domain
Exact representation of a continuous system	Yes	Only if input is ZOH
Exact representation of the pure time delay	Yes	Only if a multiple of T_s
Simulation	Approximate	Yes
Predictive models	No	Yes
Recursive estimation	No	Yes
Controller design	Limited	Extensive

Systematic errors. A restriction of frequency-domain identification is that if the data are not periodic, leakage will be introduced in the frequency data by the use of the FFT. Time-domain windowing, however, reduces the error but does not eliminate it completely, so periodic excitations should be given preference when frequency-domain methods are used.

Experiment design. Contrary to the time-domain identification, identification in the frequency domain requires the sampling of both the input and output of the system.

Transient errors. Transients in the data will cause error in the frequency domain and have to be removed before conversion from the time domain. These transients do not cause any problem in time-domain identification.

Data reduction. Frequency-domain identification allows reduction in the amount of data used for estimation since this is dependent of the excitation frequencies, regardless of the number of samples.

Time delay. An accurate estimate of the pure time delay is not possible with time domain identification since this delay is estimated as a multiple of the sampling interval. Identification in the frequency domain allows direct estimation of the pure time delay as a fraction of the sampling interval.

Input signal. Time-domain identification is based on the assumption that the input signal is piecewise constant, in other words, it is passed through a sample-and-hold device (ZOH assumption). This assumption can be violated in cases where the system under consideration is driven by an actuator, which acts like a low-pass filter, and the input data has to be measured at the output of the actuator to exclude its dynamics from the model. This can cause an error in the estimated model.

As a result of the above, the choice of the identification method depends on the measurement conditions for the particular system, and the purpose of the model. If a model is to be used for control purposes, z-domain techniques are preferred since most of the controllers are time discrete. Time domain identification provides the easiest solution to this identification problem, since only the output of the system needs to be sampled and there is no restriction in the choice of excitation signals. On the other hand if the model is to be given a physical interpretation, frequency techniques are more advantageous since a differential equation can be directly derived from the continuous s-domain model.

5.4 Experiment Design

The basic assumption for the identification of a continuous s-domain model is that the data has been sampled without aliasing. In order to avoid aliasing the *band-limited* (BL) assumption must hold. This demands that the input and output signals have been band-limited before sampling and sampled at a frequency f_s, which is more than twice the highest frequency in the signal f_{max}.

Another requirement for the identification of s-domain models is that the data should be gathered in dynamic steady state, which implies that time is allowed for all the transients to die away. Figure 5.4 shows a suitable test-setup for continuous s-domain model identification. It can be seen that an analogue reconstruction filter is placed at the output of the ZOH device to remove the higher-frequency components generated by the ZOH. These components usually referred to as frequency butterflies because of their shape, are a product of convolution in the frequency domain and are centred at multiples of the sampling frequency.

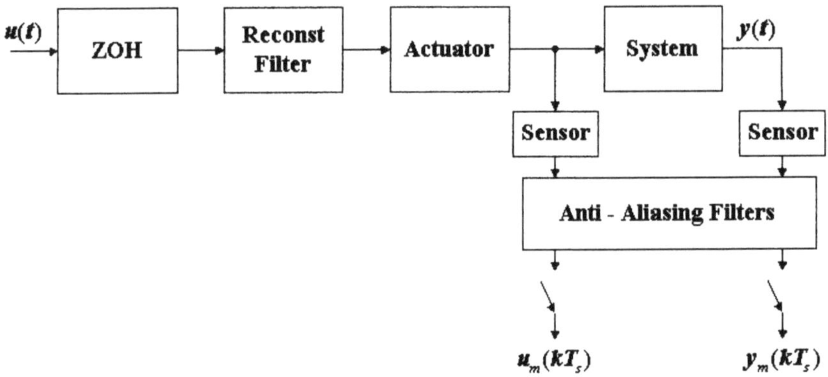

Figure 5.4. Experimental set-up under BL assumption

The input and output signals are band-limited by analogue *anti-aliasing* (AA) filters before being sampled. The selection of the cut-off frequencies of these filters is essential since it must be ensured that no vital system information is lost and at the same time sufficient attenuation at the sampling Nyquist frequency is provided. A useful guideline given in Evans [16] is to sample the signal at a frequency at least ten times the maximum frequency of the signal f_{max}, and then set the AA cut-off frequency at $2.5f_{max}$.

A discrete model can only be an exact representation of a continuous system if the input signal is piecewise constant between sampling intervals (Equation (5.6)). This implies that the signal applied to the system should be generated by a ZOH, which means that the input signal must not be band-limited. In addition, no AA filters should be used otherwise important information will be lost. These can lead to a more simplified set-up as shown in Figure 5.5, especially if the ZOH is assumed to be perfect. It is clear from Figure 5.5 that the model fitted between $u(t)$ and $y(t)$ will include the ZOH, as well the system under test. This presents a problem if an accurate model for the system is required, it is thus necessary to exclude the influence of the ZOH by measuring the input signal at the output of the actuator. If the actuator itself band-limits the input signal, then the ZOH assumption is violated, which will introduce errors into the estimated discrete model [18]. This phenomenon has been observed by Hill [19] and Evans [16] when estimating discrete models for the Rolls Royce gas turbine, and it will be discussed further in the next sections.

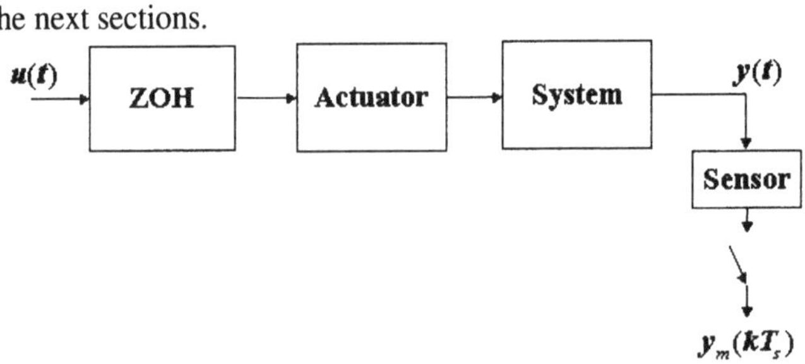

Figure 5.5. Experimental set-up under ZOH assumption

5.4.1 Input Signals

The choice of excitation signal usually depends on the type of identification technique employed and the available *a priori* knowledge for the system under test. Classical techniques make use of transient signals such as impulse and step inputs. These signals are susceptible to noise and in certain cases their adequate use is limited by the constraints placed on the input signal amplitude that can be applied to the system. A constraint on the amplitude in linear system identification may be placed in order not to drive the system into its nonlinear region or if the operating range of the system is inherently limited.

The use of random signals provided an alternative to transient signals especially because of the attractive statistical properties of Gaussian signals. The use of such signals suffers from poor SNR and the need for power spectrum averaging to obtain accurate FRF estimates. In addition, the application of random signals to real processes might be difficult due to the use of actuators that cannot cope with the sudden fluctuations that can occur in these signals. The most attractive alternative to random signals is provided by pseudorandom binary signals (PRBS) whose properties are summarised by Godfrey [20]. These are periodic binary signals which change level at predetermined event points. PRBS have similar autocorrelation functions as Gaussian signals; they are periodic and can be easily generated if they are based on maximum-length sequences.

In general, in linear system identification the input signals should be able to excite all the modes of the system in the range of amplitude for which the experiment is designed. This implies that the frequency spectrum of the signal should cover the bandwidth of the system under test. More specifically, a signal is said to be *persistently exciting* (PE) of order n if its spectral density is nonzero for at least n distinct points in the interval $(-\pi, \pi)$, which can be generated by the sum of $(n + 1)/2$ sinusoids. Van den Bos [21] showed that a signal of at least $(n + 1)/2$ sinusoids is required to identify an s-domain model with n parameters and the same result was derived for z-domain models by Ljung [1]. It is recommended to design signals with more than $(n + 1)/2$ harmonics in order to allow room for an increase in the model order, or the detection of model errors through the comparison of the model frequency response with the estimated FRF.

Periodic Signals

Periodic signals are essential for frequency-domain identification if systematic errors are to be avoided, and they also offer a number of advantages for time-domain identification, the most important summarised as follows.

Improvement in the signal-to-noise ratio. The crest factors of periodic signals are generally much better than those of random or transient signals, which allows more power to be injected for a given time-domain amplitude. Since the signals have discrete spectra it is also possible to select only those frequency lines at which power is injected and discard the noise lines.

Model validation. Periodic signals can be transformed to the frequency domain without leakage errors and if the SNR is good a high-quality estimate of the FRF can be obtained. This allows the validation of both continuous and discrete parametric models by comparing the frequency response of the model with the estimated FRF.

Combination of data sets. Data sets covering different frequency ranges, gathered at different sampling rates, can be transformed in the frequency domain and combined into one single data set. Such combination is not possible with nonperiodic time-domain data.

Improvement of finite sample behaviour. Only those frequencies at which power is injected need to be used in estimation. For time-domain estimation this can be achieved by transforming the signals to the frequency domain, setting the noise contributions at all other frequencies to zero and then calculating the inverse transform. This has been shown to have a large impact on the convergence region of both time- and frequency-domain estimators.

Data resampling. The minimum sampling interval for time-domain modelling is usually set by the sampling frequency used in data gathering. The use of periodic signals allows the sampling interval to be decreased in a straightforward manner, by calculating the FFT of the signal and then taking the inverse Fourier transform at a faster sampling rate.

In what follows, the two most widely used periodic signals in system identification are discussed. These are the multisine signals and binary signals.

Multisine Signals

A multisine signal is the sum of an arbitrary collection of cosines, which are harmonically related:

$$u(t) = \sum_{k=1}^{F} A(k)\cos(\mathbf{i}(k)\omega_0 t + \phi(k)) \tag{5.41}$$

The harmonic vector $\mathbf{i}$ can be any set of integer numbers, which makes the signal design very flexible in terms of the frequency content. This is an important feature of multisine signals since by appropriate selection of input harmonics certain nonlinear features of the system can be detected at the output. If the fundamental and all the harmonics over a given bandwidth are included in the signal it is termed a *concecutive multisine*. A signal composed of only those harmonics which are odd multiples of the fundamental is termed an *odd multisine*.

In order to maximise the amount of power to be injected into the system the signal should be compressed in the time domain by optimising the phases. One measure of the compression of the signal is the crest factor given by:

$$CF = \frac{\max(|u(t)|)}{u_{rms}(t)} \tag{5.42}$$

Several methods have been developed to minimise the crest factor. The most empirical method was proposed by Schroeder [22], who observed that phase modulated signals had low peak factors and derived the formula:

$$\phi(1) = 0 \qquad \phi(k) = 2\pi \sum_{l=1}^{k-1} (k-l) p_l \qquad k = 2, 3, \ldots, F \tag{5.43}$$

where p_l is the power of the lth harmonic and $\Sigma p_l = 1$.The above formula can be simplified for signals with flat amplitude spectrum to give:

$$\phi(1) = 0 \qquad \phi(k) = \pi \frac{k^2}{F} \qquad k = 2, 3, \ldots, F \tag{5.44}$$

The method that produced signals with the lowest CF achieved to date was derived by Guillaume *et al.* [23] based on the observation that the CF can be expressed as:

$$CF = \frac{l_\infty(u)}{l_2(u)} \tag{5.45}$$

where $l_\infty(u)$ is the Chebyshev norm of $u(t)$ and $l_2(u)$ its RMS value. Since l_2 is independent of ω the problem reduces to the minimisation of the l_∞ norm with respect to the phases. Since this is nondifferentiable, the l_{2p} norm its minimised in its place, using a Levenberg–Marquardt algorithm. The norm is initially minimised with a small value of p, which is then repeatedly incremented, with the phases of the previous minimisation being used as starting values for each new step. This method, referred to as the $L\infty$ method, is fast enough for practical application, and it can be used to minimise CFs of signals at the input and the output of a linear system.

In order to measure the FRF of a system a frequency range must be selected which adequately covers the system breakpoints. Figure 5.6 shows an 30 odd harmonic multisine signal with a fundamental of 0.005 Hz and a bandwidth of 0.005-0.0295 Hz. This signal is only suitable to measure the dynamics of a system that has breakpoints within the 0.005-0.03 Hz band. The multisine shown in Figure 5.6 has been compressed using the $L\infty$ method, resulting in a CF of 1.35.

In Evans [16] the quality of single sine and multisine signals was examined in relation to the estimation of the FRF. It was concluded that by using multisine signals a reduction in the order of 80% in test times can be achieved when compared with single sine tests. The comparison was made under the constraint that the relative accuracy of both single sine and multisine tests should be equal.

Binary Signals

Binary signals are widely used in linear system identification because of the fact that they are easily generated, they are suitable for a wide range of actuators that work using only two levels and they allow the injection of the maximum possible power for a given time domain amplitude. Out of a range of binary signals, maximum length binary sequences (MLBS) are the most popular. These are easily generated by shift registers of different length with appropriate feedback [20]. The length of the sequence N is determined by the number of registers n:

$$N = 2^n - 1 \tag{5.46}$$

The amplitude spectrum of an MLBS with a peak amplitude of a is given by:

$$A(f_k) = a\frac{\sqrt{N+1}}{N}\operatorname{sinc}\left(\frac{k\pi}{N}\right) \qquad k = \pm 1, \pm 2, \pm 3, \ldots \tag{5.47}$$

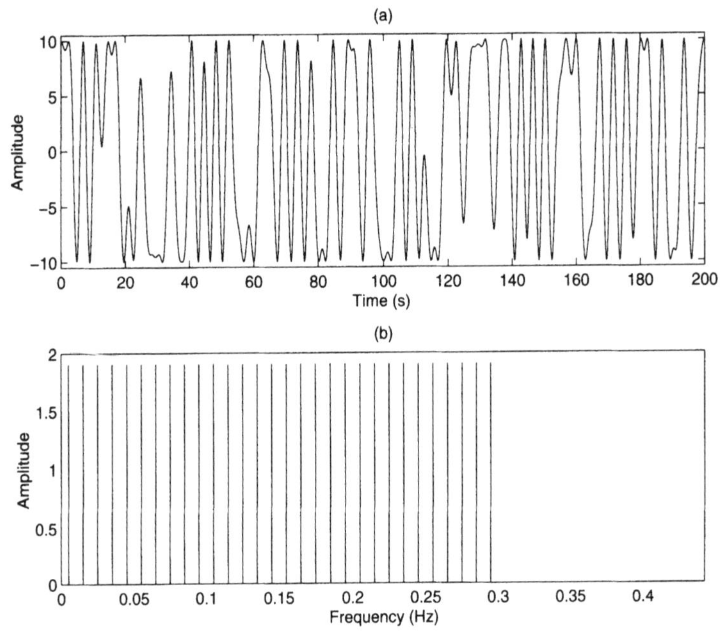

Figure 5.6. Multisine signal in (a) time domain and (b) frequency domain

The frequency of the harmonics is determined by the bit period δt:

$$f_k = \frac{k}{N \cdot \delta t}, \quad k = 1, 2, 3, \ldots \tag{5.48}$$

and the half-power bandwidth of the signal occurs at approximately $0.443/\delta t$. It is thus possible to change N and dt to alter the frequency range of the signal.

The time-domain design criteria employed with MLBS signals are based on their use with correlation techniques to measure the impulse response of a system. The autocorrelation of an MLBS sequence is approximately impulsive and this can be used to deconvolve the Wiener–Hopf equation and estimate the system impulse response [2]. A binary signal composed of only odd harmonics can be generated by the modulo-2 addition of the bit sequence of an MLBS signal with a sequence of alternating ones and zeros [20]. If the addition is carried out across two periods of the original MLBS signal an *inverse repeat* (IR) sequence will be generated, with the required odd harmonic properties. An IRMLBS signal is shown in Figure 5.7. It can be seen that it is not possible to have a perfectly flat spectrum in the frequency range of interest, and that a significant amount of power falls outside the $0.443/\delta t$ half-power bandwidth.

The test times associated with FRF estimates when using a binary signal are similar to the test times achieved when using multisines [16]. The disadvantage of binary signals compared to multisines is that they are far less flexible in their frequency content and a lot of the power will fall outside the frequency band of interest.

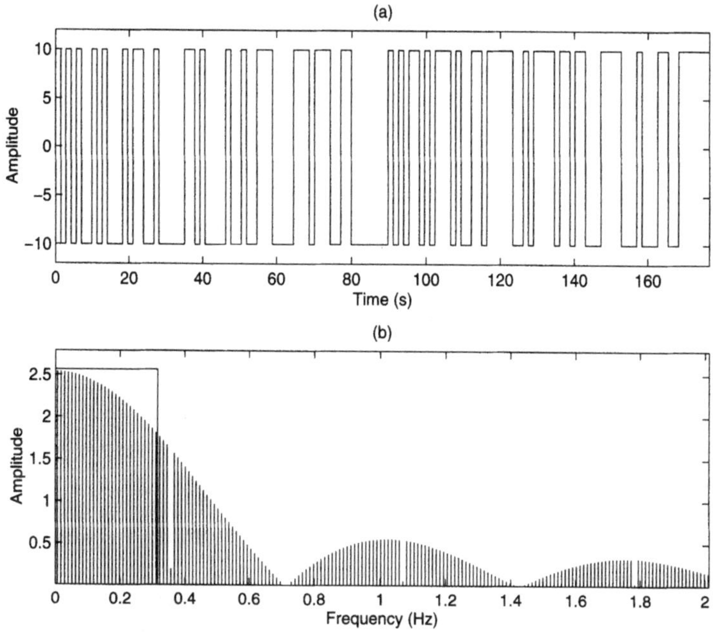

Figure 5.7. IRMLBS signal in (a) time domain and (b) frequency domain

5.4.2 Analysing Periodic Data

The use of periodic signals greatly simplifies the analysis of the measured data. The presence of errors such as poor sampling synchronisation, process drift and measurement outliers can be detected from the time records. Provided the data are sampled across complete periods, at a sampling frequency which is an integer multiple of the fundamental frequency, then no leakage errors will be introduced by the FFT. Statistical information about the frequency data can then be obtained by applying a period by period analysis. This allows a straightforward assessment of the levels of noise and nonlinearities from the frequency-domain records. The following outline owes to the excellent paper by Kollár *et al.* [24].

Synchronisation. It is important that the signal reconstruction and sampling clocks are synchronised when conducting measurements for frequency-domain identification, since a deviation in either clock will result in incomplete signal periods being sampled. The quality of synchronisation can be checked by examining the circular covariance of the measured signals:

$$c(nT_s) = T_s \sum_{m=0}^{N-1} u(m)\, u(n \ominus m) \qquad n = 0,1, \dots, N-1 \tag{5.49}$$

where T_s is the sampling period, N the number of samples and $u(n \ominus m)$ corresponds to a circular shift of m samples to the right. This can be calculated via the inverse FFT of the signal power spectra, with the dc component supressed [24]. The circular covariance of periodic data will have peaks at intervals corresponding to the signal period. Any problems with the sampling synchronisation will cause these peaks to appear at irregular intervals.

Drift and repeatability. The presence of a drift in the operating point during the experiment can be checked by calculating the mean values of each period of the measured input and output. The presence of significant drift would affect the frequency-domain estimation, since the low-frequency transients would be introduced which would corrupt the Fourier coefficients of the measured data. Time-domain estimation of discrete models will also be adversely affected.

The application of periodic signals also allows the repeatability of the system response to be checked, by plotting the outputs period by period, with their mean values removed. This also allows the visual detection of outliers which can be then eliminated by substituting values calculated by linear interpolation between previous and next data points.

Noise analysis. A first indication about the noise levels present in the data can be obtained by transforming the complete data lengths into the frequency domain using the FFT. Since the frequency resolution of the FFT is inversely proportional to the total time of the data record ($\omega_f = 1/T_{\text{test}}$), this will give the greatest resolution achievable with the data. It is then possible to calculate the total power in the signal P_{tot}, and the power at the excited harmonics P_{test}. An initial estimate of the SNR can then be obtained by treating all the power at the nonexcited frequencies as noise:

$$\text{SNR}_{td} = 10\log_{10}[P_{\text{test}}/(P_{\text{tot}} - P_{\text{test}})] \tag{5.50}$$

This is a measure of the total excitation power over the total remaining power in the signal and is hence a calculation of the raw time-domain SNR. The next step in the noise analysis is to estimate the means, variances and covariance of the input and output Fourier coefficients. These can be used to estimate the uncertainty on the estimated FRF and as *a priori* information for the frequency-domain estimator. For this purpose, the data must be transformed to the frequency domain on a period-by-period basis. The following estimates can then be made

$$\overline{V}(j\omega_k) = \frac{1}{M} \sum_{m=1}^{M} V_m(j\omega_k) \tag{5.51}$$

$$\sigma_v^2(j\omega_k) = \frac{1}{M-1}\sum_{m=1}^{M}\left(V_m(j\omega_k) - \bar{V}(j\omega_k)\right)\left(V_m(j\omega_k) - \bar{V}(j\omega_k)\right) \tag{5.52}$$

where $V(j\omega_k)$ is interchangeably $U(j\omega_k)$ or $Y(j\omega_k)$. Since the Fourier coefficients are complex the variance calculated according to Equation (5.65) will be the sum of the variances of the real and imaginary parts. The covariance can be estimated as:

$$\sigma^2{}_{UY}(j\omega_k) = \frac{1}{M-1}\sum_{m=1}^{M}\left(U_m(j\omega_k) - \bar{U}(j\omega_k)\right)\left(Y_m(j\omega_k) - \bar{Y}(j\omega_k)\right) \tag{5.53}$$

A significant covariance indicates that the noise is not simply present in the input measurement channel but also passes through the system and hence correlated with the output noise. This can be caused by noise on the measure input variable, such as turbulence in a measured flow, or by the presence of a feedback loop. The variances and covariance calculated in this way are double the quantities required for Equations (5.34) and (5.37), where the variances and covariance of the real or imaginary parts are used. It has been shown that the estimated models are strongly consistent, for an increasing number of data points in each experiment, if the number of independent measurements M is greater than or equal to four [25].

Since the estimation is conducted only at the excited frequencies it is now possible to exclude all the other frequency points, which are termed *noise lines*, from the data. Assuming the presence of stationary white noise, the improvement to the SNR will be in the order of:

$$\text{SNR}_{ex} = \text{SNR}_{td} + 10\log_{10}\left(\frac{N}{2F}\right) \tag{5.54}$$

which depends on the degree of oversampling employed. Since the means and variances estimated from Equations (5.64) and (5.65) are a measure of the periodic and stochastic power, respectively, they can be used to calculate the actual SNR following the exclusion of the noise lines:

$$\text{SNR}_{Vex} = 10\log_{10}\left(\sum_{k=1}^{F}\left|\bar{V}(j\omega_k)\right|^2 \Big/ \sum_{k=1}^{F}\sigma_v^2(\omega_k)\right) \tag{5.55}$$

A further improvement in the SNRs is provided by averaging the data across M periods. The variance of the sample means is smaller than the sample variances estimated using Equation (5.53), by a factor of $1/M$. This will provide further improvement in the SNR of:

$$\text{SNR}_{Vav} = \text{SNR}_{Vex} + 10\log_{10}(M) \tag{5.56}$$

Detection of nonlinearities. It is possible to detect the system nonlinearity in the frequency domain. This can be achieved by detecting the presence of additional harmonics generated by the system nonlinearity in the output. In addition, the use of specially designed multisines allows the separation of the frequency contributions generated at the system output by the system nonlinearity [16].

If a signal contains only harmonics that are odd multiples of the fundamental (such as an IRMLBS or an odd multisine) then all the frequency contributions at the output resulting from any even-order nonlinearities will fall at even harmonics [26]. Thus the even nonlinearities can be detected just by inspection of the frequency content of the system input and output signals. Similarly if an odd-odd multisine is used (a signal where every other odd harmonic is also excluded) both even-order and odd-order nonlinearities can be detected, since the odd-order nonlinear contributions will fall at the omitted odd harmonics. A useful tool with which to assess the periodicity of the generated harmonics, and distinguish them from noise harmonics, is the squared coherence function:

$$\gamma_{nl}(\omega) = \left| \frac{1}{M} \sum_{m=1}^{M} Y_m(j\omega) \right|^2 \Bigg/ \frac{1}{M} \sum_{m=1}^{M} Y_m(j\omega) Y_m^*(j\omega) = \frac{|\bar{Y}(j\omega)|^2}{G_{YY}(\omega)} \tag{5.57}$$

where $Y_m(j\omega)$ is the output spectrum at the excited and nonexcited frequencies, $Y^*_m(j\omega)$ its complex conjugate and $G_{YY}(\omega)$ the autospectrum of the output [27]. The coherence represents the ratio of the periodic power to the total power at the output, and if there is no periodic power at the output then the coherence will assume a value of $1/M$. The periodic power due to nonlinearities can be detected since the coherence function at those frequencies will rise well above the $1/M$ bound.

5.5 Concluding Remarks

This chapter has served to introduce the modelling concepts and estimation techniques which will be used in the next chapter to estimate linear models for the gas turbine. A general discussion of linear modelling and time- and frequency-domain identification techniques has been presented. Each of the estimators discussed is an MLE if the disturbing noise is Gaussian. This means that the estimates are asymptotically unbiased, efficient and normal and that the Cramer–Rao lower bound can be used to approximate the covariance matrix of the estimates.

An overview has also been presented of the main issues relating to the design of the identification experiment. The implications for the experimental set-up of the basic assumptions underlying continuous and discrete modelling have been emphasised. The advantages of periodic signals for continuous and discrete modelling have been presented and the properties of multisine and MLBS signals have been discussed.

Chapter 6
Linear Gas Turbine Modelling

6.1 Introduction

In this chapter the engine test facility will now be described, and the signals used during the engine tests will be presented. Single-input, single-output s-domain models of the HP and LP dynamics will be estimated at different shaft speeds using frequency-domain techniques. The estimated models will be cross-validated using other models estimated from different tests at the same amplitude. The influence of the engine nonlinearity on the estimated models will be examined by analysing the frequency-domain residuals and the estimated models will be compared with the thermodynamic models at the different operating points. Discrete models will also be estimated in the time domain and their poles and zeros will be compared with those of the frequency-domain models by means of the impulse invariant transform.

6.2 Gas Turbine Testing

The engine tests were conducted at the Glen sea-level test facility of the Defence and Evaluation and Research Agency at Pyestock. A Rolls-Royce Spey Mk 202 engine was tested, which is a typical military two-shaft turbofan, with a low by-pass ratio and a variable reheat nozzle. The engine is controlled by varying the rate of fuel flow, the angle of the inlet guide vanes, the reheat nozzle area, the reheat fuel flow and the LP compressor bleed valve position. The reheat system was inoperative during the tests and the compressor bleed valve was closed. The angle of the inlet guide vanes and the reheat nozzle area were fixed to appropriate values for the duration of the tests.

The engine speed control was operated in open loop and a perturbed fuel demand signal was fed to the fuel feed system, which regulates the fuel flow to the engine by means of a stepper valve. A number of problems encountered during the measurement of the fuel flow and shaft speeds will be discussed in what follows. These problems are either inherent to gas turbines, others are due to the characteristics of the test facility, which is given diagrammatically in Figure 6.1.

The fuel feed system is composed of a valve driven by a stepper motor. These exhibit both linear and nonlinear characteristics and affect the actual fuel flow applied to the gas turbine.

The speed of the low-pressure shaft is measured by counting the rotations of the turbine blades and the speed of the high-pressure shaft by measuring the rotation of the gear attached to the shaft itself. The pulses are counted for a specific interval and the resulting speed is provided as the transducer output.

The dominant noise sources in the gas turbine are the turbulence in the fuel flow on the input and the vibration of the turbine blades and jitter in the gears on the output shaft speeds. The noise input propagates through the system and influences the output.

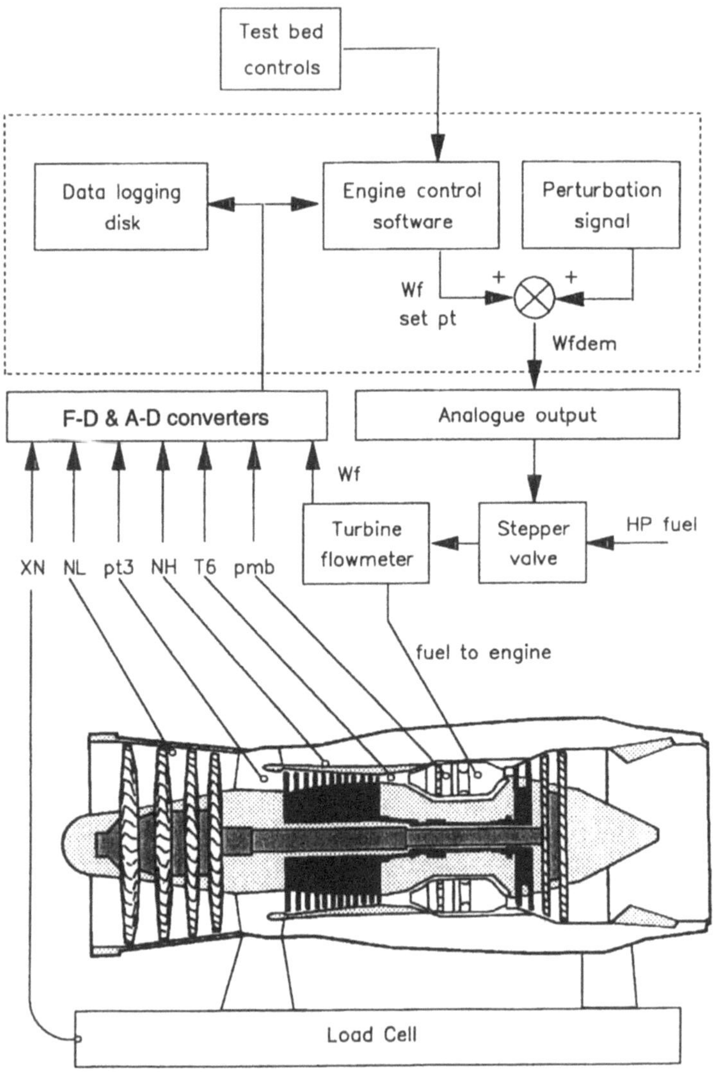

Figure 6.1. Engine test facility (reproduced from Hill [1], with permission)

6.2.1 Test Signal Designs

The tests conducted at Pyestock were focused on frequency-domain estimation of s-domain models. It was thus necessary to adhere to the BL assumption, which implies that the measured signals are band-limited before sampling, and then sampled at more than twice their highest frequency. The lack of AA filters in the gas turbine set-up might thus seem to preclude the use of frequency-domain techniques and it is important to address this point prior to discussing the signal designs.

Prior information gathered by previous tests was essential for the signal designs. It was clear from the results obtained by Evans [2] that a signal bandwidth of 0.09 – 0.35 Hz is sufficient to cover the breakpoints of the gas turbine poles, while at the same time being significantly lower than the breakpoints of the fuel feed poles. A wide range of signals was designed with the above consideration taken into account, along with the need for sufficient harmonics to allow CF minimisation and the detection of modelling errors. These signals were then applied to the engine at 55%, 65%, 75%, 80%, and 90% N_H. Table 6.1 summarises the multisine signals used in the tests. An IRMLBS, also was used in the tests. The signal was generated by a 7-bit shift register and had a fundamental of 0.0049 Hz and a sampling frequency of 20 Hz.

Table 6.1. Multisine signals used in the tests

Signal	Number of harmonics	Fundamental frequency (Hz)	Crest factor	Sampling frequency (Hz)
Multisine (1)	20	0.005	1.44	20
Multisine (2)	56	0.005	1.39	20
General-purpose (GP) multisine	27	0.01	1.10	25

6.3 Nonparametric Analysis

A detailed analysis was made of the gathered data before proceeding with parametric estimation. The data analysis was performed in both time- and frequency-domains. The nonlinearities present in the engine are assessed and the noise is quantified. The data were transformed to the frequency domain and the frequency response functions were estimated at different operating points.

6.3.1 Synchronisation

The quality of synchronisation can be checked by examining the circular covariance of the measured signals, defined in Equation (5.49) which is plotted for the multisine (1) signal in Figure 6.2. The periodicity of the input is clear, and the detailed plots of the first and third lags indicate that the synchronisation is very good. The maximum deviation between clocks must be less than 0.00315%, this being half a sampling interval (0.05/2) in four signal periods (4×4000×0.05).

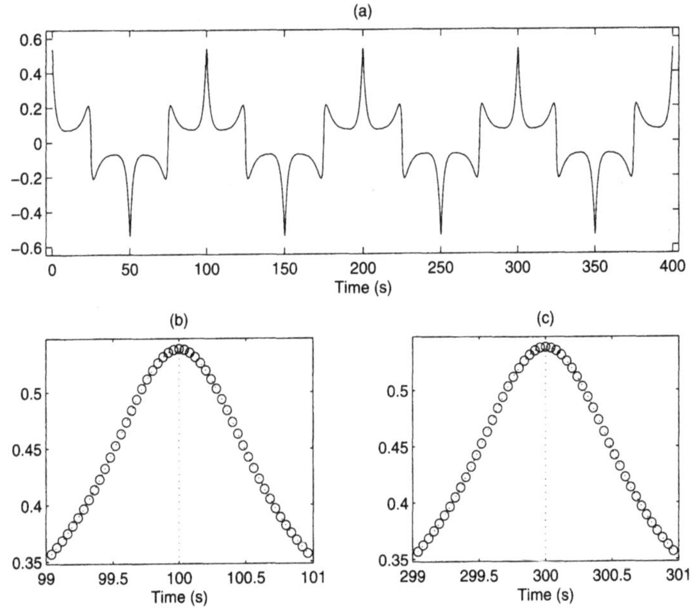

Figure 6.2. Circular covariance of input signal, (a) across complete data length, (b) first lag and (c) third lag

6.3.2 Drift and Repeatability

The possible presence of drift in the operating point during the experiment was checked by calculating the mean values of the measured fuel flow and the shaft speeds across each test. The results are presented in Table 6.2 for a series of tests in order to illustrate the effect on tests at different operating points. It can be seen that drift effects are not significant in this case since the deviation between tests is very small.

Table 6.2. Mean values of measured fuel flow and shaft speeds

Signal	$\overline{W_f}$ (cc/s)	$\overline{N_H}$ (% max N_H)	$\overline{N_L}$ (% max N_H)
Multisine (1)	139.7	53.8	26.3
IRMLBS	138.0	53.3	26
Multisine (1)	259.3	75.3	43.0
IRMLBS	258.2	75.1	42.9
Multisine (2)	259.1	75.2	43.0
Multisine (1)	389.9	88.7	53.0
IRMLBS	390.6	88.8	53.1

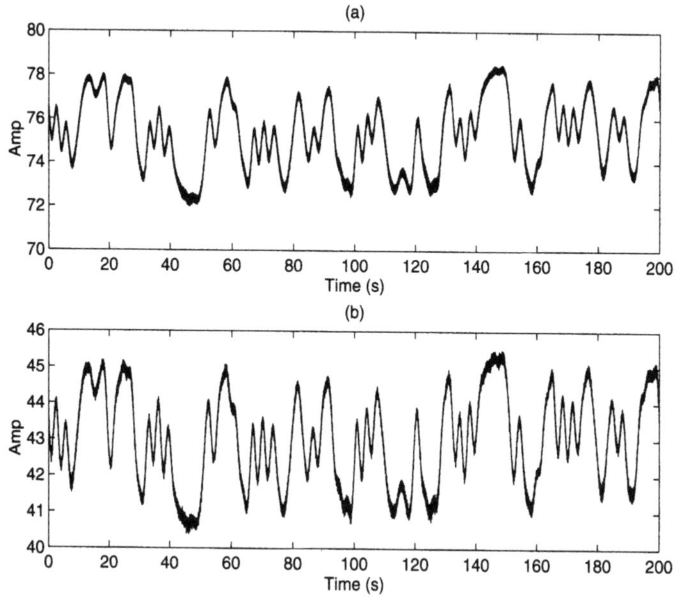

Figure 6.3. Outputs of (a) HP shaft and (b) LP shaft, plotted period by period

The application of periodic signals also allows the repeatability of the system response to be checked, by plotting the outputs period by period, with their mean values removed. This is shown in Figure 6.3 for multisine (1) at 75% N_H, where the repeatability of the response is clear. Plotting the data records in this way facilitated the easy detection of outliers, which were eliminated by substituting a value calculated through linear interpolation between the previous and subsequent data points.

6.3.3 Noise and Nonlinearities

The variances and covariance of multisine (1) HP shaft data at 75% N_H, are plotted in Figure 6.4. The dominant noise source at the input is process noise, which passes through the system and hence correlated with the output noise. Figure 6.5 shows the spectra of the HP shaft multisine (1) test at 75% N_H. The highest output power is at the frequencies excited by the input signal, referred to as the input frequencies. Additional power can be found at other frequencies which are multiples of the fundamental but were not excited by the input signal. They are a result of nonlinearities in the fuel feed, which create contributions at the input at frequency multiples of the fundamental. The improved SNRs achieved by excluding the noise lines and by averaging over four periods were then calculated using Equations (5.55) and (5.56), and the results are shown in Table 6.3. The SNRs are very good, with values of at least 43 dB in each case. The exclusion of the noise lines should have improved the time-domain SNRs by around 23 dB

($10.\log_{10}(4000/20)$), and an improvement of this order was obtained for the input frequencies.

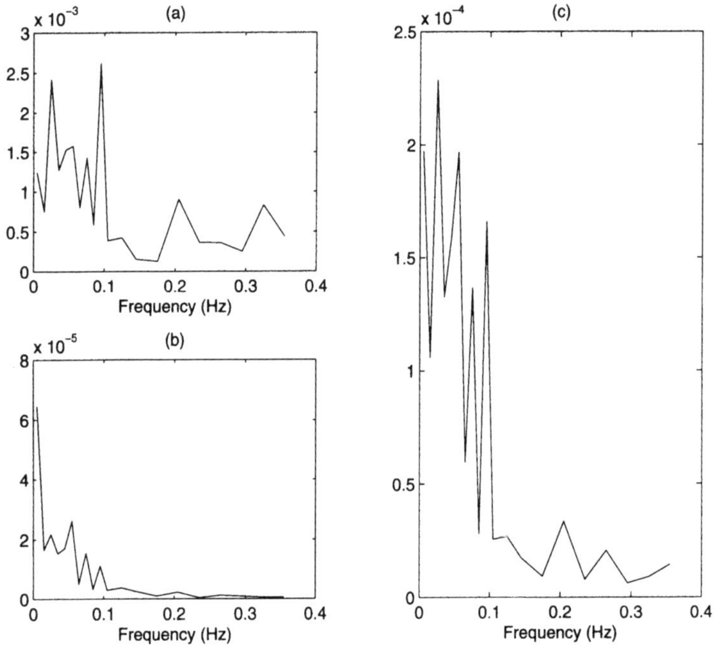

Figure 6.4. Noise variances on (a) input and (b) output, and (c) absolute value of covariance, for the GP multisine signal, HP shaft

Table 6.3. Signal-to-noise-ratios

SNR (dB)	Input	HP	LP
Time domain	16	22	23
Excluding noise lines	37	35	35
Averaging	43	41	41

It is possible to detect the presence of the engine nonlinearity by analysing the data at a single operating point. Evans *et al.* [3] used the coherence function to detect the existence of an even-order nonlinearity in the engine. At the same time there was no evidence for the existence of odd-order nonlinearities. The nonlinear coherence of an odd-odd multisine at an input amplitude of ±10% W_f is plotted in Figure 6.6, along with the $1/M$ bound, in order to assess the nonlinear contributions. It is seen that the coherence of the even harmonics in the input spectrum is close to the $1/M$ bound whereas the coherence of the even harmonics at the output is more significant. It can be also seen that the coherence of the omitted odd harmonics is consistently lower on both the input and output. This suggests the presence of a weak even-order nonlinearity in the engine, for small input amplitudes.

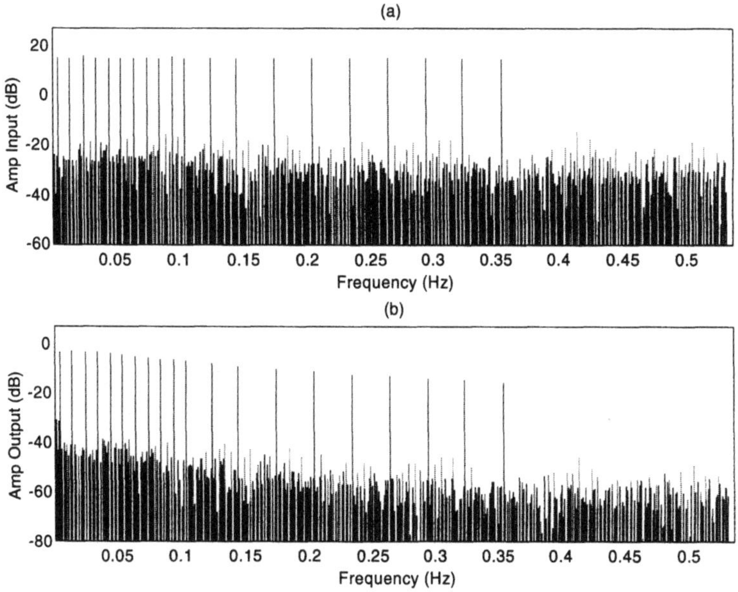

Figure 6.5. Spectra of the (a) fuel flow and (b) HP shaft speed. Excited harmonics (red), omitted harmonics: even (green), odd (blue), noise (black)

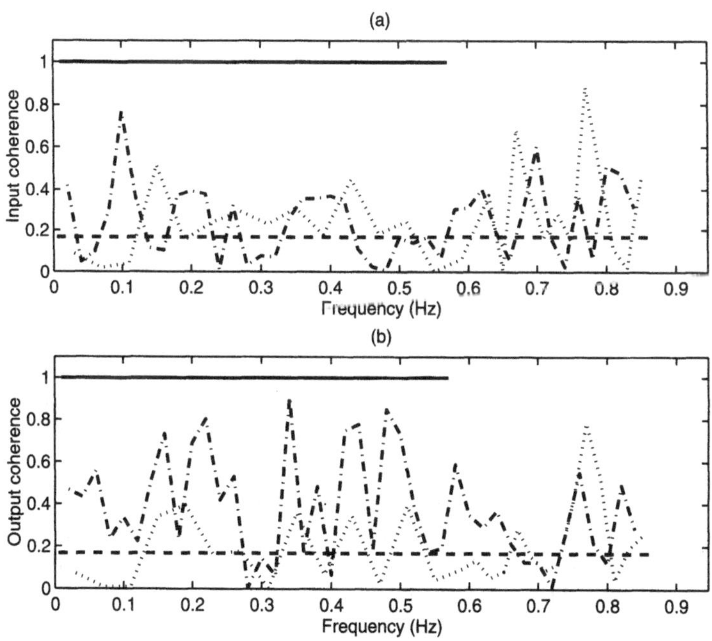

Figure 6.6. Nonlinear coherence of an odd-odd multisine at (a) input and (b) output. Input harmonics (solid), even harmonics (dashdot) and omitted odd harmonics (dotted). With 1/M bound (dashed).

6.3.4 Frequency Response Functions

The FRFs of the HP and LP shafts were estimated for each of the tests, using the HEV estimator defined in Equation (5.31). The FRFs for the HP and LP shaft for three tests at 80% N_H, and amplitudes of ±1%, ±2% and ±10% W_f are shown in Figures 6.7 and 6.8.

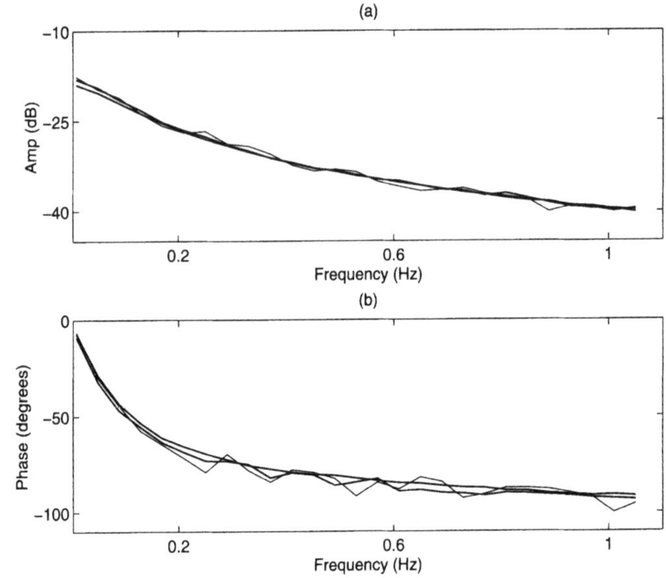

Figure 6.7. Estimated FRFs for HP shaft at an operating point of 80% N_H and amplitudes of ±1% W_f, ±2% W_f, and ±10% W_f

It can be seen that the bandwidth of the test signals was well chosen, covering a drop of 21 dB on the HP shaft and 17 dB on the LP shaft, with the phase shift varying between –10 and –90 degrees. It can be also seen that the FRFs estimated at different input amplitudes correspond very well, suggesting that the FRF does not depend on the amplitude of the input signal and also that the effect of nonlinearities for an amplitude up to ±10% W_f is minimal.

The accuracy of the FRFs was investigated by calculating the standard deviation of the estimates, according to Equation (5.35), for the general-purpose multisine at the three input amplitudes. The amplitudes for these FRF estimates along with the 2σ bounds are shown in Figures 6.9 and 6.10 for both HP and LP shafts. It can be seen for both shafts that the uncertainty at ±10% W_f is very small whereas it begins to become significant at ±2% W_f. At ±1% W_f the uncertainty is large and the FRF estimates are biased since the noise becomes significant at this amplitude. Figures 6.11 and 6.12 show the estimated FRFs of the HP and LP shafts at 5 different operating points estimated using the IRMLBS tests. It is interesting to see that the dynamics of both shafts become faster and the steady-state gains of both shafts decrease, as the operating point increases. This is evidence of the fact that the engine is nonlinear.

Nonparametric analysis of the engine data shows that the synchronisation between signal reconstruction and data sampling clocks is very good. The influence of drift on the engine data is negligible and the repeatability of the estimates is very good. The input and output noises are correlated, particularly at low frequency. The data gathered at input amplitude of $\pm 10\%$ W_f should be used for parametric linear modelling since a lower uncertainty on the estimated FRFs is achieved with data at this amplitude. Both shaft dynamics and steady-state gains change with operating point and an even-order nonlinearity has been detected for small amplitudes

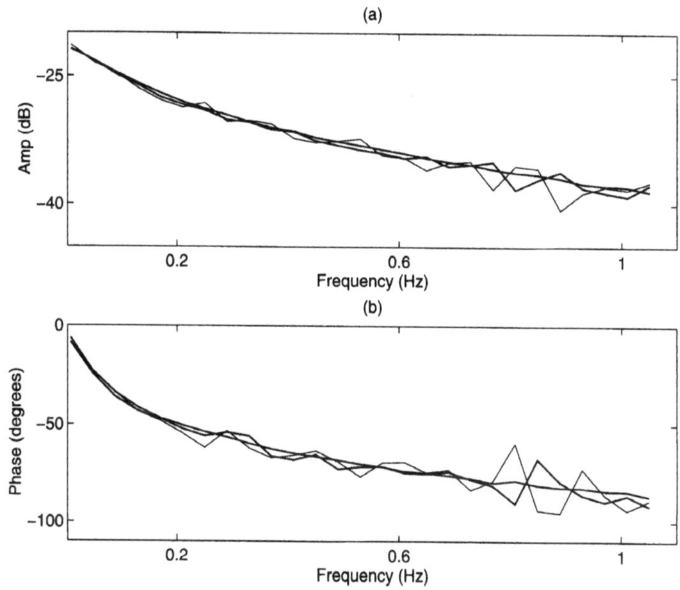

Figure 6.8. Estimated FRFs for LP shaft at an operating point of 80% N_H and amplitudes of $\pm 1\%$ W_f, $\pm 2\%$ W_f, and $\pm 10\%$ W_f

6.4 Frequency-domain Estimation

The frequency domain estimator discussed in Chapter 5 will be employed to estimate transfer function models of the HP and LP shaft dynamics at different operating points. The noise variances and covariance, which were calculated during the averaging of the frequency data, will be used as *a priori* information, and the pure time delay will be included as a free parameter in each estimation. The expected value of the estimator cost function varies as a function of the estimation frequencies and the number of free parameters, as defined in Equation (5.39). All results presented in the following sections refer to Test 6A, with the multisine (1) at input amplitude of $\pm$ 10% W_f and operating point of 75% N_H, unless otherwise stated. The estimated models will be validated using an IRMLBS test at input amplitude of $\pm$ 10% W_f and operating point of 75% N_H.

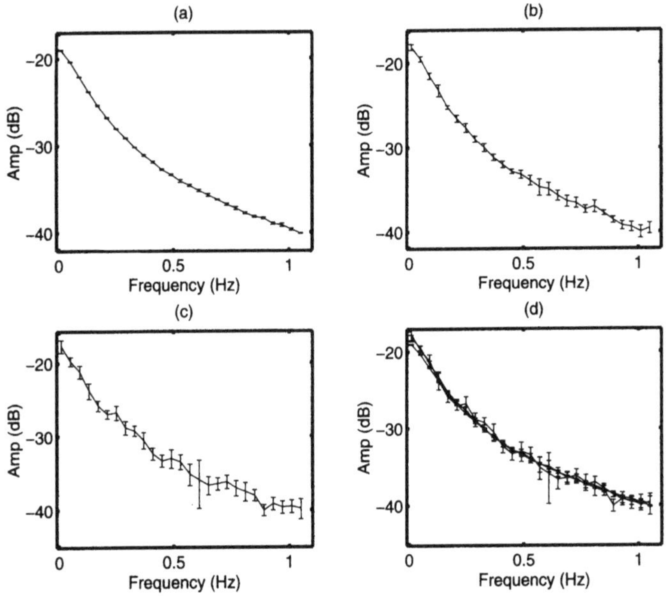

Figure 6.9. Amplitude of FRF estimates for HP shaft, using an odd-odd harmonic multisine, showing 2σ bounds. With input amplitude (a) $\pm 10\%\ W_f$, (b) $\pm 2\%\ W_f$, (c) $\pm 1\%\ W_f$ and (d) all plotted together.

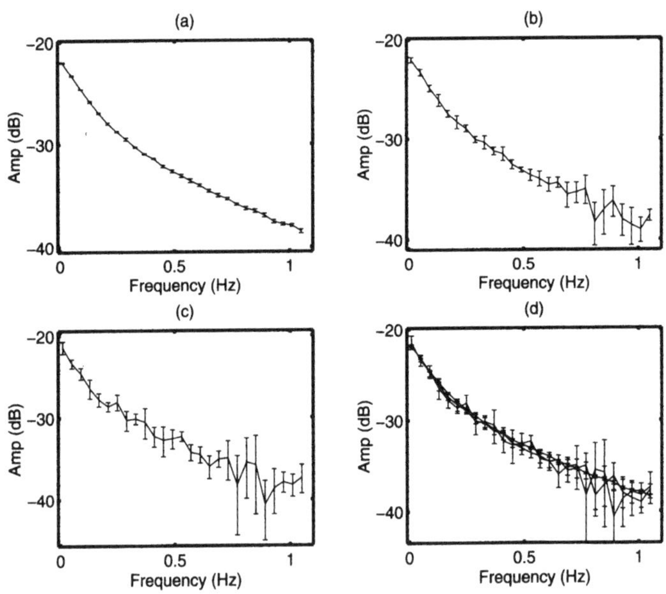

Figure 6.10. Amplitude of FRF estimates for an LP shaft, using an odd-odd harmonic multisine, showing 2σ bounds. With input amplitude (a) $\pm 10\%\ W_f$, (b) $\pm 2\%\ W_f$, (c) $\pm 1\%\ W_f$ and (d) all plotted together.

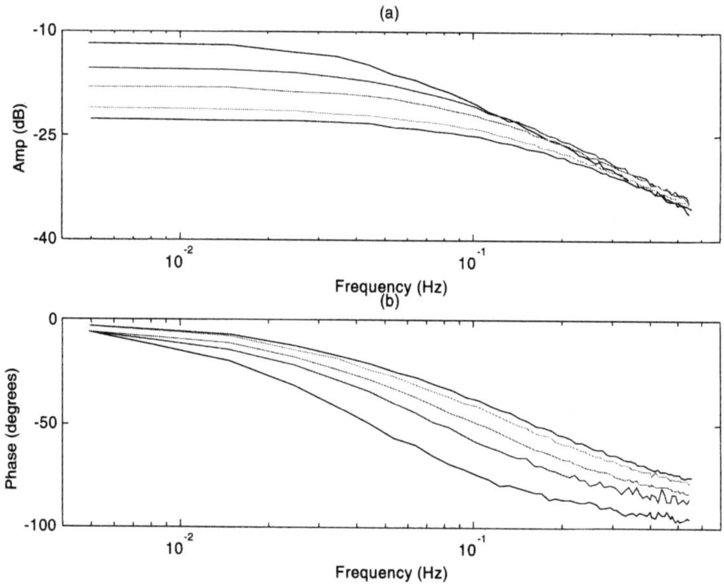

Figure 6.11. FRFs for the HP shaft at 55% (black), 65% (red), 75% (green), 85% (cyan), 89% (blue)

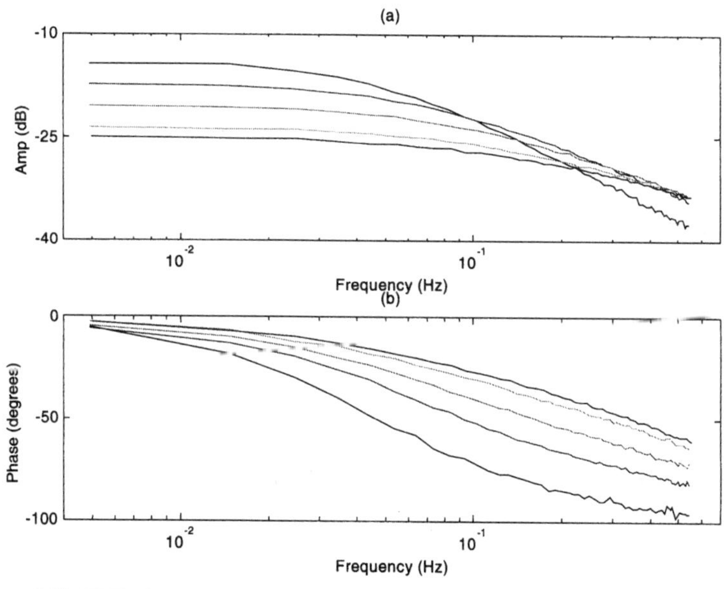

Figure 6.12. FRFs for the LP shaft at 55% (black), 65% (red), 75% (green), 85% (cyan), 89% (blue)

6.4.1 High-pressure Shaft

The variation of the cost function with model order for multisine (1) is shown in Table 6.4. A pole and zero are added for each increase in model order, which represents the addition of a freely specified mode and modal gain. The major drop in the cost function occurs with the one-zero, two-pole (1/2) model, suggesting that this is the best structure. Table 6.5 shows the pole and zero positions of the first three of these models with their standard deviations expressed as a percentage of their magnitudes.

Table 6.4. Cost function for HP shaft models, quasi-log multisine (1)

Order	Cost function (K)	K_{min}
0/1	740	18.5
1/2	30	17.5
2/3	26	16.5

Table 6.5. Estimation results for HP shaft, multisine (1)

Order	DC gain	Delay (ms)	Zeros	σ_z (%)	Poles	σ_p (%)
0/1	0.1227	2.95	-	-	–0.5512	0.12
1/2	0.1265	10.42	–0.2746	5.63	–0.2440 –0.6159	5.11 0.78
2/3	0.1265	–67	+46.3 –0.297	1306 7.02	–17.2 –0.6290 –0.2600	200 1.56 5.99

The 1/2 structure has a pole-zero pair close to the origin whose 2σ regions of uncertainty do, however, overlap. Adding another pole-zero pair gives a positive zero and a negative time delay, which is not a credible model. This is also clear from the uncertainties of the additional zero and pole, which are very high. The frequency responses of the 0/1 and 1/2 models are shown in Figures 6.13 and 6.14, where it can be seen that the additional pole and zero are seen to be modelling a localised low-frequency effect. The large drop in the cost function between the 0/1 and 1/2 models appears to contradict the very small improvement in the frequency fit shown in Figure 6.18. This is due to the fact that the noise variances appear in the denominator of the estimator cost function given in Equation (5.37). This means that small values of noise variance will tend to increase the value of the cost function, making it more sensitive to small modelling errors.

A powerful way to validate an estimated model is by cross-validation. This involves simulating the model on a different test data and comparing the model output with the measured output. A similar technique, is to estimate a model with a different data set and compare the two estimated models in terms of their zeros and poles. To this end, the estimated models were cross-validated using the IRMLBS signal data, the results for which are shown in Tables 6.6 and 6.7. It can be seen that the results obtained from the IRMLBS test show the same pattern as the results obtained from multisine (1) test. The major drop in the cost function occurs with the 1/2 model, having a dominant pole at –0.5927 and a pole-zero pair near the origin, the 2σ uncertainty regions of which once again overlap.

Table 6.6. Cost function for HP shaft models, IRMLBS

Order	Cost function (K)	K_{min}
0/1	265	54.5
1/2	100	53.5
2/3	110	52.5

Table 6.7. Estimation results for HP shaft, IRMLBS

Order	DC gain	Delay (ms)	Zeros	σ_z (%)	Poles	σ_p (%)
0/1	0.1226	7.16	-	-	–0.5601	0.19
1/2	0.1275	10.38	–0.1929	10.9	–0.1768 –0.5927	10.6 0.68
2/3	0.1269	3.42	–9.9500 –0.2520	41.3 10.8	–9.1100 –0.6000 –0.2270	35.8 1.17 10.1

It could be said that the close pole-zero pairs in the 1/2 models estimated with each data set are simply an indication of over-modelling and should be cancelled out. However, they may be highly correlated, such that they move together along the real axis, rather than cancel. They are also seen to model a real feature of the data in each case and the drop in the estimator cost function is significant. They are clearly modelling a localised low-frequency dynamic, though the difference between the models suggests that there is some uncertainty about the exact location of the pole and zero. It must be mentioned that Evans [2] obtained very similar models at 75% N_H using data obtained from different tests. This demonstrates once more the high quality of data obtained, and the repeatability of the measurements.

6.4.2 Low-pressure Shaft

The variation of the cost function for the LP shaft, estimated using multisine (1) is shown in Table 6.8. The large drop in the cost function between the 0/1 and 1/2 models shows that the dynamics are at least second-order. There is even a case for selecting a 2/3 model, though the drop in the cost function is not as significant as that for the 1/2 model. Table 6.9 shows that the 2/3 model has a very close pole-zero pair near the origin, the uncertainty regions of which do overlap. This close pole-zero pair is once again modelling a low-frequency effect in a similar way to the pole-zero pair in the 1/2 model of the HP shaft. Figure 6.15 shows the amplitudes of the model frequency responses. It can be seen that a 0/1 model is unable to model the LP shaft dynamics, whereas the 1/2 and 2/3 models perform very well with their difference being indistinguishable.

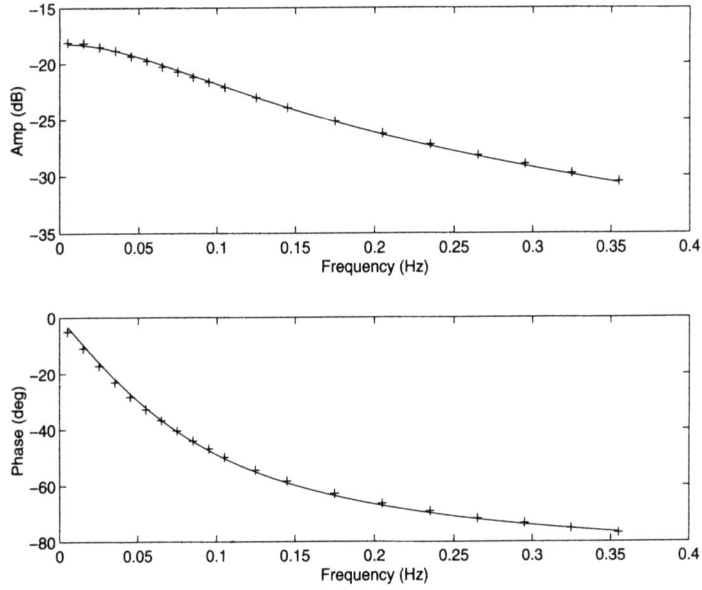

Figure 6.13. Frequency response of model (solid) and FRF (crosses), for 0/1 model, HP shaft

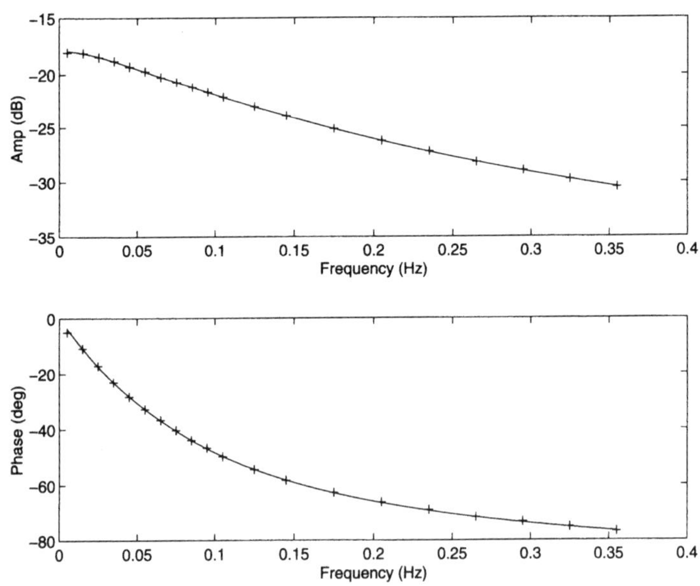

Figure 6.14. Frequency response of model (solid) and FRF (crosses), for 1/2 model, HP shaft

Table 6.8. Cost function for LP shaft models, multisine (1)

Order	Cost function (K)	K_{min}
0/1	3139	18.5
1/2	92.08	17.5
2/3	23.25	16.5
3/4	14.12	15.5

These models were again cross-validated using the IRMLBS signal data, the results of which are shown in Tables 6.10 and 6.11. There is again great similarity between the two sets of models, with the low-frequency pole-pairs being in the limit of modelling.

Table 6.9. Estimation results for LP shaft, multisine (1)

Order	DC gain	Delay (ms)	Zeros	σ_z (%)	Poles	σ_p (%)
0/1	0.0916	–84.1	-	-	–0.6949	0.19
1/2	0.0956	–4.64	–0.8080	2.69	–0.4461	1.22
					–1.4982	2.20
2/3	0.0978	4.42	–0.1061	17.99	–0.1001	17.66
			–1.0704	4.67	–0.5252	2.38
					–1.8185	3.95
3/4	0.1049	5.49	–0.0071	143	–0.0065	154
			–1.9200	23.9	–0.1800	22.8
			–1.1500	4.99	–0.5560	4.01
					–1.9100	4.99

Table 6.10. Cost function for LP shaft models, IRMLBS

Order	Cost function (K)	K_{min}
0/1	4365	54.5
1/2	272	53.5
2/3	93.3	52.5
3/4	72.3	51.5

Table 6.11. Estimation results for LP shaft, IRMLBS

Order	DC gain	Delay (ms)	Zeros	σ_z (%)	Poles	σ_p (%)
0/1	0.0890	–4.14	-	-	–0.7672	0.19
1/2	0.0959	3.30	–0.8318	2.17	–0.4511	1.03
					–1.5442	1.6
2/3	0.0993	9	–0.0372	13.76	–0.0351	2.39
			–1.0383	3	–0.5113	1.38
					–1.7837	2.39
3/4	0.111819	15	–0.0049	126	–0.0042	144
			–0.2360	14.8	–0.2130	13.7
			–1.3800	4.90	–0.6130	3.28
					–2.1600	3.94

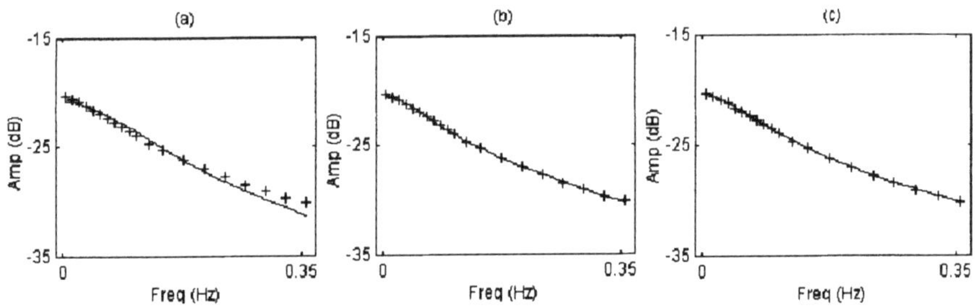

Figure 6.15. Amplitude response of LP shaft model (solid) and FRF (crosses), for model orders (a) 0/1, (b) 1/2 and (c) 2/3

6.4.3 Models at Different Operating Points

Frequency-domain parameter estimation was performed at data gathered for other operating points, the detailed results of which are shown in the Appendix. Figure 6.16 shows the locations of the poles and zeros of models estimated at different operating points using frequency-domain techniques. It is clear that the HP and LP shafts have different order dynamics. Cancelling pole-zero pairs suggest that the HP shaft is predominantly first-order, across most of the operating range, and that the LP shaft is second-order. It is also clear from Figure 6.16 that the position of the poles and zeros, in other words the dynamics of both shafts, change with the operating point. The dc gains of these models also decrease as the operating point is increased, as shown in Table 6.12. This shows once again that the gas turbine is nonlinear [4].

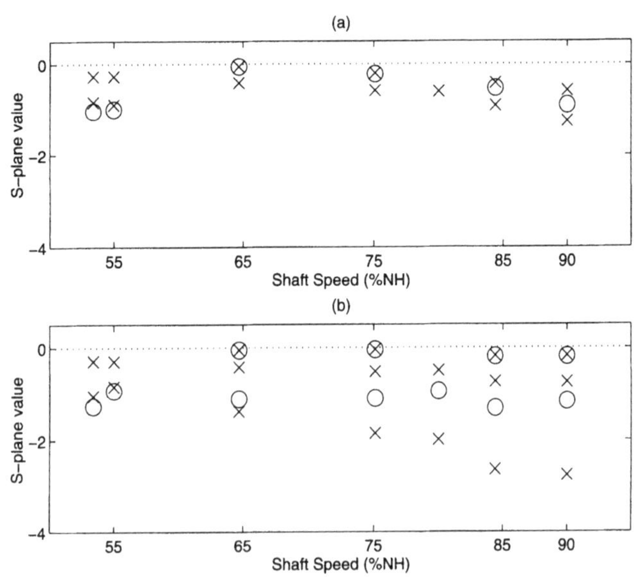

Figure 6.16. Poles and zeros of the models at different operating points (a) HP shaft and (b) LP shaft. Showing poles (crosses) and zeros (circles).

Table 6.12. Model dc gains at different operating points

Operating point	HP shaft dc gain	LP shaft dc gain
53	0.262	0.195
65	0.177	0.141
75	0.127	0.098
85	0.088	0.067
89	0.075	0.056

6.4.4 Low-frequency Mode

It is clear that the HP and LP shafts have different order dynamics, with the HP shaft being predominantly first-order and the LP shaft second-order. In most cases, the addition of a further pole-zero pair models a low-frequency effect and reduces the model cost function significantly. The influence of this low-frequency mode on the estimated models was assessed by Evans [2] who excluded a number of low-frequency harmonics from the estimation data set, re-estimated a range of models and monitored the cost function as the model order was increased. It was concluded in this case that omitting harmonics has little influence on the estimate of the dominant pole but has a significant influence on the location of the pole-zero pair. A similar pattern was observed for the transition between the 1/2 and 2/3 models for the LP shaft.

Pure Time Delay. The time delay was included as a freely varying parameter in the frequency-domain estimation. It is observed from the results that the value of this delay is dependent of the operating point. Typical values of the estimated delay were between 4 and 23 ms.

6.4.5 Influence of Engine Nonlinearity

The presence of unmodelled nonlinear dynamics can be detected using the normalised autocorrelation of the frequency-domain residuals defined in Equation (5.40) and denoted R_{ee}. The R_{ee} of the 0/1 and 1/2 HP shaft models estimated using multisine (1) are plotted in Figure 6.17 and for the IRMLBS in Figure 6.18. In both models it can be seen that R_{ee} is broad in shape and it is considerably minimised as the model order is increased. This indicates the presence of unmodelled linear dynamics in the 0/1 model, which are eliminated by increasing the model order. There is no indication from the frequency-domain residuals for any modelling errors coming from unmodelled nonlinear dynamics. Evans [2] detected the presence of unmodelled nonlinear dynamics on an engine model estimated using a consecutive multisine, whereas no such unmodelled dynamics were detected on a model estimated using an odd multisine. The results of this chapter confirm Evans' results since both models were estimated using odd harmonic signals, thus minimising the influence of the engine nonlinearity.

The variation of R_{ee} with model order for the LP shaft is shown in Figures 6.19 and 6.20. It can be seen that R_{ee} is considerably minimised when the model order is

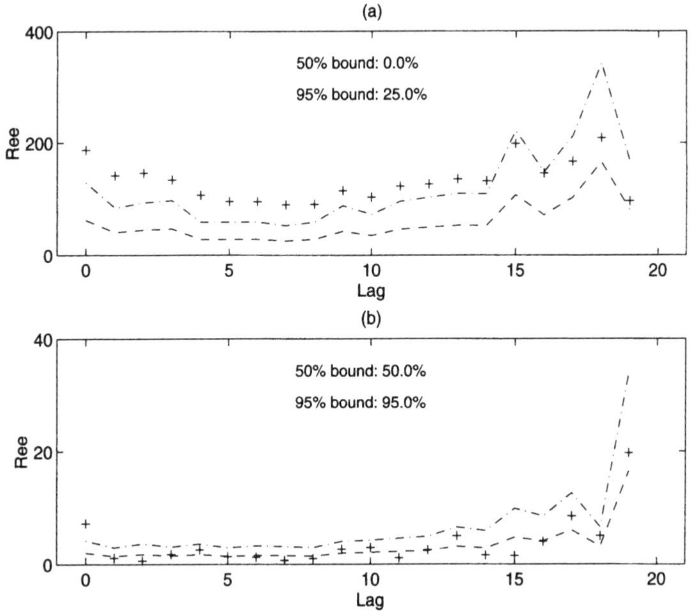

Figure 6.17. R_{ee} for quasi-log multisine, HP shaft data: (a) 0/1 model and (b) 1/2 model, with 50% (dashed) and 95% (dash-dot) confidence intervals

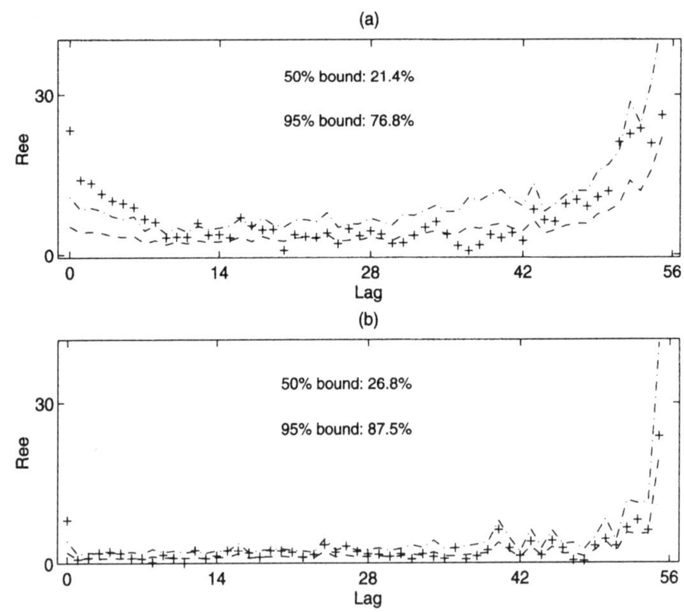

Figure 6.18. R_{ee} for IRMLBS, HP shaft data: (a) 0/1 model and (b) 1/2 model, with 50% (dashed) and 95% (dash-dot) confidence intervals

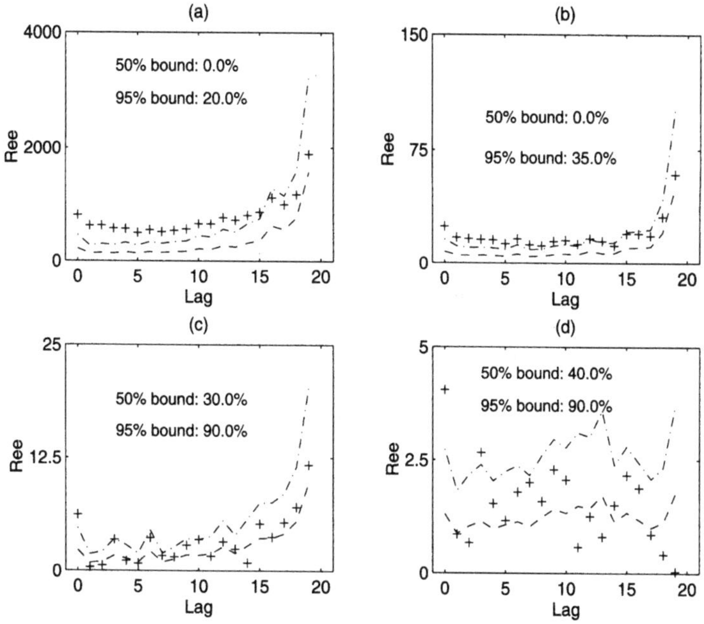

Figure 6.19. R_{ee} for quasi-log multisine, LP shaft data: (a) 0/1 model and (b) 1/2 model, (c) 2/3 model, (d) 3/4 model, with 50% (dashed) and 95% (dash-dot) confidence intervals

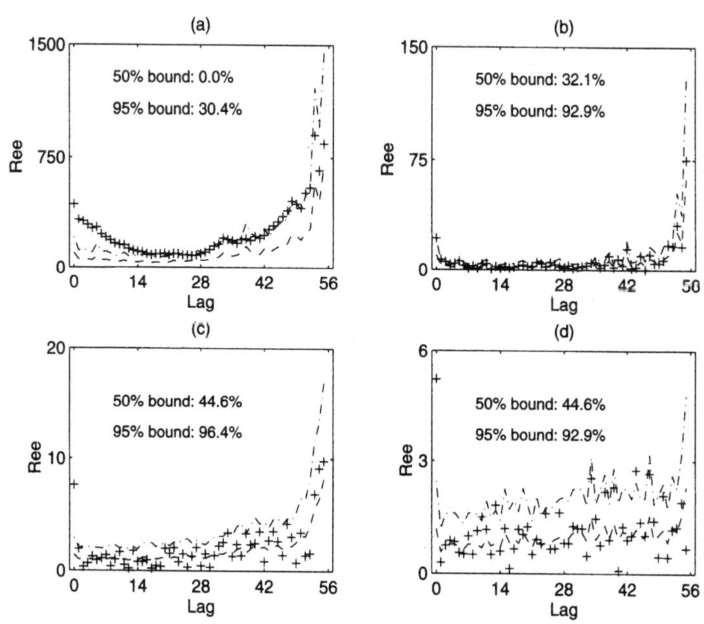

Figure 6.20. R_{ee} for IRMLBS, LP shaft data: (a) 0/1 model and (b) 1/2 model, (c) 2/3 model, (d) 3/4 model, with 50% (dashed) and 95% (dash-dot) confidence intervals

increased from 0/1 to 1/2. This once again suggests that the LP dynamics are at least second-order if not third-order.

The frequency-domain results indicate that high-quality models of the shaft speeds can be estimated using frequency-domain techniques. The HP shaft dynamics are first-order and the LP shaft dynamics are second-order. The shafts share a close, but not identical, dominant mode. The model fits can be further improved by the addition of low-frequency pole-pairs, which are thought to be modelling a fast heat soakage effect, such as a blade expansion. Both shaft dynamics and dc gains change with operating point indicating that the engine is nonlinear. The pure time delay changes with the operating point.

6.5 Time-Domain Estimation

Recent work by Hill [1, 5] examined the time-domain identification of discrete engine models. Each of the second-order transfer function models estimated in this way contained a negative real pole, which could not be transformed to the s-domain to allow comparison with the thermodynamic models. Evans [2] replicated this effect on a data obtained from a Spey engine at 75% N_H. A range of possible causes for this effect have been suggested and investigated:

- influence of noise;
- crude quantisation;
- model errors due to nonlinearity;
- pure time delay;
- over-modelling;
- violation of the ZOH assumption;
- model sensitivity.

Evans [2] studied the effects related to the correlated noise, the pure time delay, over-modelling and the violation of the ZOH assumption. It was shown through simulation that resampling at a frequency at least four times the original rate could reduce the errors due to the delay and the violation of the ZOH assumption.

In this section discrete *output error* (OE) models estimated in the time-domain for the available engine data will be presented. Results from the same tests used previously are going to be presented, these being multisine (1) and the IRMLBS tests at 75% N_H. The use of periodic signals allowed the SNRs to be improved by calculating the FFT of each data record, setting all the nonexcited frequency lines to zero, calculating the inverse FFT and using these new data sets in the estimation.

6.5.1 High-pressure Shaft

The HP shaft models estimated with the 20 harmonic multisine (1) are presented in Table 6.13. It can be seen that the cost function drops with the 1/2 model but increases with the 2/3 model which has a negative pole at the unit circle, which makes it a noncredible model. Figures 6.21 and 6.22 show the frequency responses of the 0/1 and 1/2 models, the amplitudes of which have a reasonable good fit to the estimated FRFs. The fact that the 0/1 and 1/2 model performances on the HP shaft data are very close is illustrated in Figures 6.23 and 6.24 where the time responses of the two models are shown to match closely the measured HP shaft output. It is interesting to note though, that the second mode of the 2/3 model at $z = 0.9696$ is very different from the second mode of the 1/2 model which is at $z = 0.8302$. This result will be analysed further in the next section where a comparison of the models estimated with frequency- and time-domain techniques is going to be presented.

The cross-validation results using the IRMLBS data are shown in Table 6.14, where it can been seen that they show the same pattern as the results obtained using multisine (1). A significant feature of the results obtained with the IRMLBS is the estimation of a single real negative pole in the 2/3 model which was discussed previously.

Table 6.13. Time-domain estimation results for HP shaft, multisine (1)

Order	Cost Fn.	Zeros	σ_z (%)	Poles	σ_p (%)
0/1	1.3617	-	-	0.9727	0.0025
1/2	0.6360	0.8809	0.0006	0.9746 0.8302	0.0026 0.1097
2/3	28.4138	0.9853 0.6726	0.0112 0.2596	0.9871 0.9696 −1.0000	0.0090 0.0077 0.1795

Table 6.14. Time-domain estimation results for HP shaft, IRMLBS

Order	Cost Fn.	Zeros	σ_z (%)	Poles	σ_p (%)
0/1	1.4990	-	-	0.9728	0.0025
1/2	0.8561	0.7220	0.1408	0.9739 0.4908	0.0023 0.4250
2/3	0.4232	0.9856 0.6505	0.0065 0.1457	0.9870 0.9708 −0.9924	0.0056 0.0050 0.2895

6.5.2 Low-pressure Shaft

The estimation results for the LP shaft are shown in Table 6.15 for multisine (1) and in Table 6.16 for the IRMLBS signal. For both tests the major drop in the cost function occurs with the 1/2 models. A further reduction of the cost function is observed with the 2/3 models, which suggests that these models should also be considered. The performance of the 1/2 and 2/3 models estimated using multisine (1) is illustrated in the time-domain in Figures 6.25 and 6.26. It is seen that the models have similar performance and they can both model the engine dynamics very well. It can also be again seen that a negative real pole close to the unit circle is estimated with the 2/3 models.

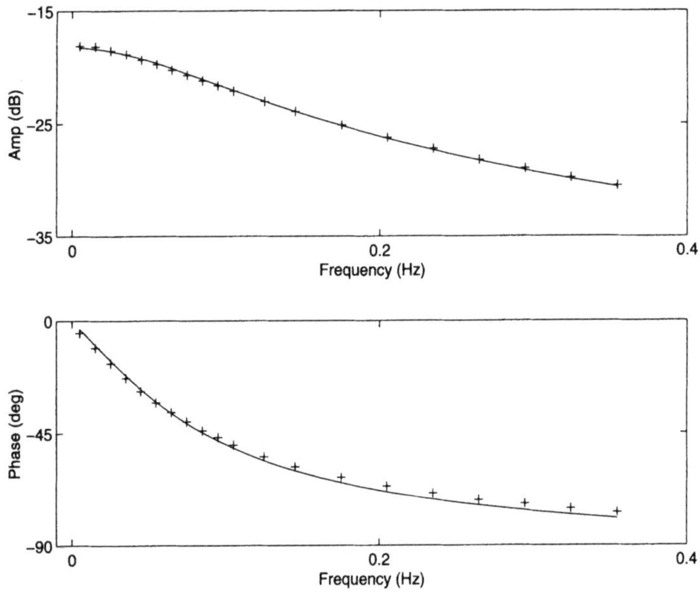

Figure 6.21. Frequency response of discrete model (solid) and FRF (crosses), for 0/1 model, HP shaft

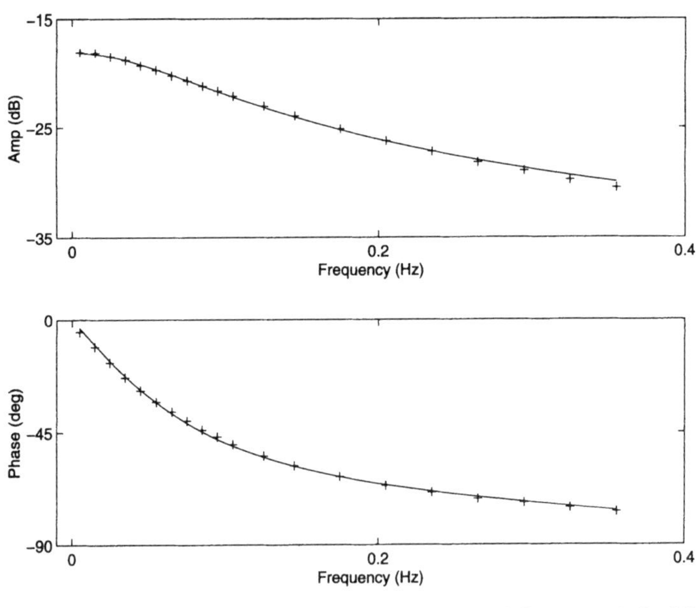

Figure 6.22. Frequency response of discrete model (solid) and FRF (crosses), for 1/2 model, HP shaft

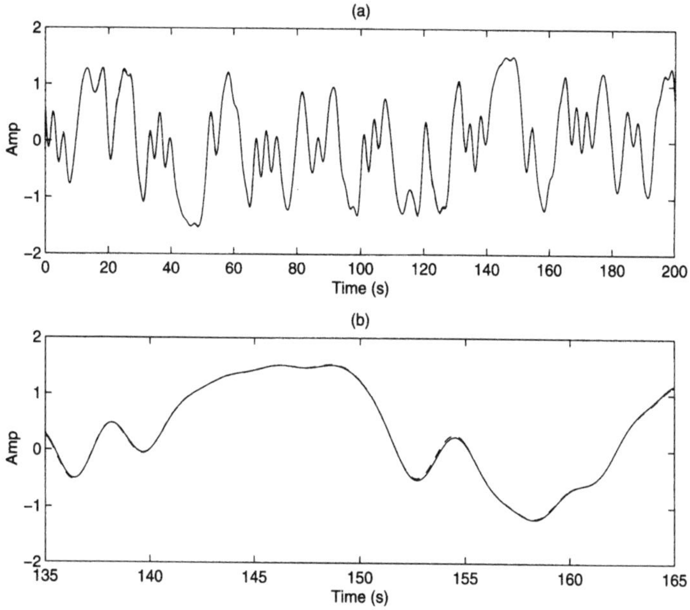

Figure 6.23. Discrete model output (solid) and measured engine output (dotted) for (a) one period and (b) a portion of that period, for 0/1 model, HP shaft

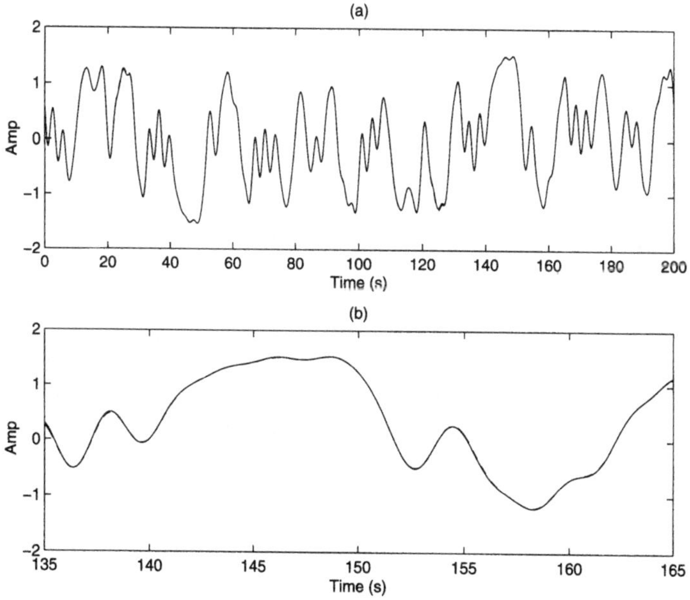

Figure 6.24. Discrete model output (solid) and measured engine output (dotted) for (a) one period and (b) a portion of that period, for 1/2 model, HP shaft

Table 6.15. Time-domain estimation results for LP shaft, multisine (1)

Order	Cost Fn.	Zeros	σ_z (%)	Poles	σ_p (%)
0/1	8.2058	-	-	0.9633	0.0108
1/2	0.5249	0.9337	0.0352	0.9746	0.0050
				0.8567	0.8960
2/3	0.1380	0.9764	0.0104	0.9828	0.0052
		0.7659	0.0766	0.9528	0.0117
				–0.9919	0.3781

The time-domain results indicate that discrete models with excellent approximation capabilities can be estimated using time-domain techniques. The HP shaft dynamics are first-order and the LP shaft dynamics are-second order. It was not possible to estimate the low-frequency pole-pair with time-domain techniques as was done with frequency-domain techniques. It is possible to estimate a single negative pole using time-domain techniques which, when transformed into the frequency domain, results to a single complex pole.

Table 6.16. Time-domain estimation results for LP shaft, IRMLBS

Order	Cost Fn.	Zeros	σ_z (%)	Poles	σ_p (%)
0/1	11.0130	-	-	0.9613	0.0122
1/2	0.9738	0.9140	0.0417	0.9726	0.0053
				0.8227	0.1020
2/3	0.1820	0.9709	0.0107	0.9807	0.0046
		0.7292	0.0784	0.9461	0.0132
				–0.9718	0.4628

6.6 Time- Versus Frequency-domain

In order to compare the models obtained with the two methods the discrete poles and zeros of the models estimated using time-domain techniques are transformed in the frequency domain using Equation (5.11) and compared with the continuous poles and zeros estimated using frequency-domain techniques, as shown in Table 6.17 for the HP shaft. It is clear from the table that the estimated dominant mode on the HP shaft is the same for the 0/1 models estimated with both techniques. This is not the case for the 1/2 models since the two poles estimated in the frequency-domain are different that the poles estimated in the time domain.

The most interesting feature of this comparison comes when looking at the 2/3 model estimated using time-domain techniques. It is easily seen that the estimated two poles of this model are very similar to the estimated poles of the 1/2 model using frequency-domain techniques. This suggests that the addition of an extra mode absorbs the error during time-domain estimation and allows the estimation of the "true" modes of the system.

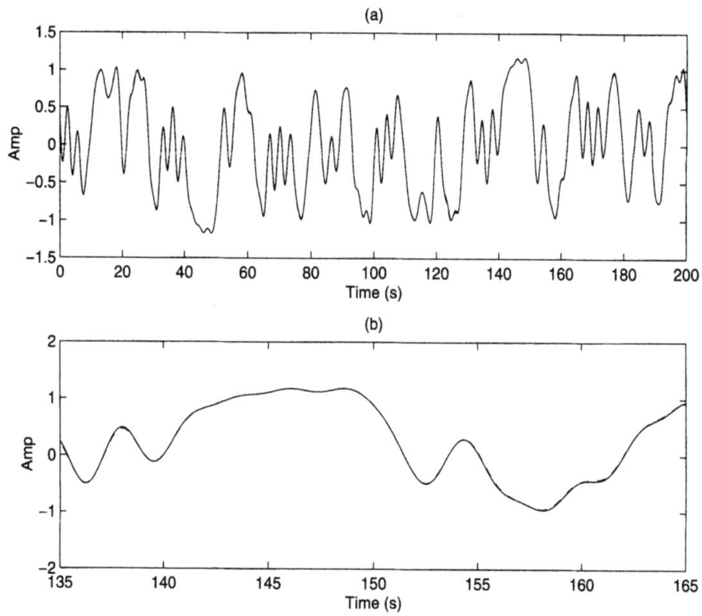

Figure 6.25. Discrete model output (solid) and measured engine output (dotted) for (a) one period and (b) a portion of that period, for 1/2 model, LP shaft

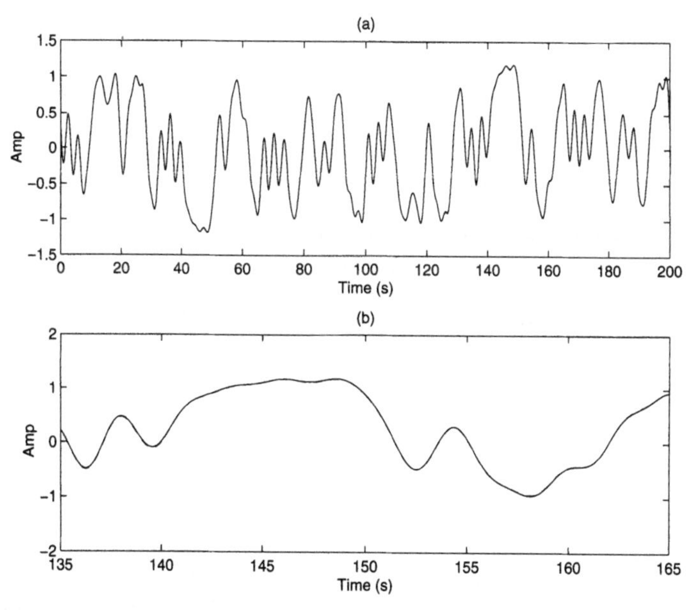

Figure 6.26. Discrete model output (solid) and measured engine output (dotted) for (a) one period and (b) a portion of that period, for 2/3 model, HP shaft

The same applies for the zeros of these models. The extra mode in this case leads to the estimation of a discrete real negative pole, which results in a single complex pole in the frequency-domain.

A comparison of the modes estimated using the two techniques for the LP shaft is presented in Table 6.18. In this case, the trend discussed above is not as clear as in the HP shaft case. More information can be gathered from Tables 6.19 and 6.20 where a comparison of the poles of the second-order models estimated using frequency-domain techniques, along with the two dominant poles of the second- and third-order models estimated using time-domain techniques for different operating points is presented. From Table 6.19 it can be seen that for the HP shaft the trend is clear for the models estimated at the operating points of 55%, 75% and 85% N_H whereas is not that clear for the models at the other two operating points. Similar trend is observed from Table 6.20 for the LP shaft with the two poles estimated for the 2/3 models approaching the two poles estimated with the 1/2 models using frequency-domain techniques.

Table 6.17. Comparison of time- and frequency-domain models at 75% N_H for HP shaft, multisine (1)

	Frequency-domain		**Time-domain**	
Model order	**Zeros**	**Poles**	**Zeros**	**Poles**
0/1	-	–0.5512	-	–0.5537
1/2	–0.2746	–0.2440 –0.6159	–2.5368	–0.5140 –3.7206
2/3	+46.3 –0.297	–17.200 –0.6290 –0.2600	–0.2962 –7.9309	–0.2606 –0.6164 –0.0010+62.8i

Table 6.18. Comparison of time- and frequency-domain models at 75% N_H for LP shaft, multisine (1)

	Frequency-domain		**Time-domain**	
Model order	**Zeros**	**Poles**	**Zeros**	**Poles**
0/1	-	–0.6949	-	–0.7488
1/2	–0.8080	–0.4461 –1.4982	–1.3724	–0.5145 –3.0928
2/3	–0.1061 –1.0704	–0.1001 –0.5252 –1.8185	–0.4781 –5.3346	–0.3476 –0.9672 –0.162+62.83i
¾	–0.0071 –1.9200 –1.1500	–0.0065 –0.1800 –0.5560 –1.9100	–0.0556 –1.3978+0.40i –1.3978–0.40i	–0.0706 –0.4546 –1.1661 –0.0281+62.83i

Table 6.19. Comparison of time- and frequency-domain models for HP shaft at different operating points

Model	Operating point (% N_H)				
	55	65	75	85	90
Frequency-domain	–0.8415	–0.8970	–0.2440	–0.9189	–0.5912
second order	–0.2728	–0.2717	–0.6159	–0.4241	-1.2632
Time-domain	–0.4330	–0.4011	–0.5140	–0.6820	–0.7797
second order	–0.3108	–16.29	-3.7206	–17.98	-18.7+78i
Time-domain	–0.2690	–0.0214+7.1i	–0.2606	–0.4074	–0.7339
third order	–1.2244	–0.4079	–0.6164	–0.8316	-2.2754

Table 6.20. Comparison of time– and frequency–domain models for LP shaft at different operating points

Model	Operating point (% N_H)				
	55	65	75	85	90
Frequency-domain	–0.2841	–0.8350	–0.4461	–0.5595	–0.5974
second order	–1.0430	–0.2907	–1.4982	–2.2399	–2.6306
Time-domain	–0.35+0.05i	–0.4345	–0.5145	–0.6879	–0.9202
second order	–0.35–0.05i	–6.5564	–3.0928	–6.6597	–5.7808
Time-domain	–0.2830	–0.2169	–0.3476	–0.4423	–0.4076
third order	–1.4020	–0.5331	–0.9672	–1.5500	–1.8875

6.7 Comparison with Thermodynamic Models

It is possible to compare the models estimated using frequency-domain techniques with the thermodynamic models derived from the engine physics. The poles and zeros estimated using frequency-domain techniques are compared with those of the thermodynamic models in Figure 6.28, for both shafts. It can be seen that the estimated dominant HP shaft pole matches the lower-frequency pole of the thermodynamic model quite well, up to an operating point of 80% N_H. It is also clear, that there is a small discrepancy between the estimated dominant pole and the lower-frequency pole of the thermodynamic model at higher operating points. An additional discrepancy is also observed between the estimated pole-zero pair and the pole-zero pair of the thermodynamic models. For the LP shaft, the estimated 1/2 models have a pole at low frequency, a second pole at a higher frequency and a zero which migrates between them as the shaft speed increases. This suggests that the low-frequency pole is dominant at lower operating points, while the dynamics of the higher-frequency pole are more significant at higher shaft speeds.

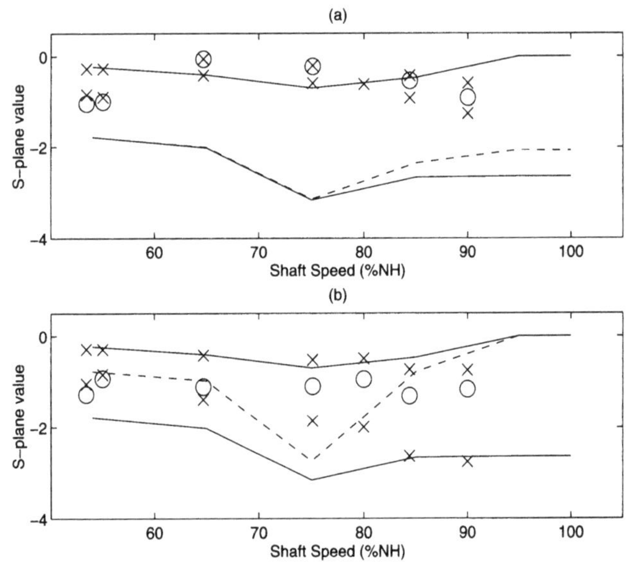

Figure 6.28. Variation of engine models for (a) HP shaft and (b) LP shaft. Estimated models shown as poles (x) and zeros (o) and thermodynamic models as poles (solid) and zeros (dashed).

Comparing the estimated LP models with the thermodynamic models, there is once again quite good agreement between the low-frequency poles, up to an operating point of 80% N_H. However, the second pole of the estimated models is at a lower frequency for most of the operating range and the variation of the zero with shaft speed is very different from that of the thermodynamic models. It is clear from these results that the thermodynamics models are not adequately representing the dynamics of either shaft at higher operating points and that the second-order dynamics of the LP shaft are also badly represented.

6.8 Concluding Remarks

High-quality models of both HP and LP shafts at different operating points were estimated using the broadband multisine data. The results were cross-validated on an IRMLBS data set and the estimated models were found to be very similar. Additional low-frequency effects were detected on both shafts which could be modelled by the addition of a close pole-zero pair. These weak modes suggest that they are modelling thermal effects, which are not incorporated in the thermodynamic models. The dynamics of both shafts were found to vary considerably at different speeds confirming that the engine is nonlinear.

This work has been extended in the multivariable case, where frequency techniques were used to estimate linear state-space SIMO models. The results obtained were consistent with the results obtained in this chapter and revealed similar discrepancies between estimated and thermodynamic engine models.

Chapter 7
Closed-Loop Control and System Identification

7.1 Introduction

This chapter focuses on a closed-loop identification technique for engine dynamic models in the presence of random disturbances.

Most technical systems operate in a closed loop. Open-loop experimentation is seldom allowed due to the requirements on control quality and safety of operation. Some plants cannot be used in open loop, *e.g.*, integrator-type actuators. Moreover, the plant operating conditions during gathering data for identification must be similar to normal operation, which requires the use of the same feedback. The influence of feedback can result in situations, where the plant model cannot be estimated from experimental data. This problem is referred to as closed-loop identifiability.

To account for the specified difficulties, *identifiability conditions* were introduced. In this chapter existing closed-loop techniques are surveyed, the *closed-loop identifiability* concept is clarified and identifiability conditions are investigated using distribution properties' analysis.

7.2 Concept of Closed-Loop Identifiability

In this chapter, the term *identifiability* means that the experimental data is sufficiently informative for plant model identification. Closed-loop identifiability must be distinguished from the property of *identifiability of the model itself* [1-3], demanding the number of equations being equal to the number of unknown parameters. Closed-loop identifiability embraces the model's identifiability and additionally accounts for excitation properties of the plant input.

Under certain conditions the plant model cannot be identified from closed-loop experimental data, even if the model is identifiable in terms of the classical definition. Hence, this type of identifiability is a necessary, not a sufficient, condition for the plant model identification from its input and output observations.

These two concepts of identifiability can be demonstrated using a simple example. Consider a first-order transfer function:

$$W(s) = \frac{K}{Ts+1} \tag{7.1}$$

The plant itself is undoubtedly observable and identifiable. However, at the steady-state conditions in the absence of external disturbances both the input and output are constant, thus making the model identification not feasible from this particular set of data.

Thus, in addition, strict requirements on the input and output signal properties must exist, guaranteeing successful identification. The condition of *persistent excitation* is often given [4]. The mathematical expression of this condition states that the Fisher's information matrix should be positive definite [5]. However, note that the Fisher's matrix cannot be precisely determined using noisy observations. It is only possible to gain an approximate estimate. Further, the elements of this matrix represent values of the autocorrelation function of the input signal, so the positive definite Fisher's matrix is sometimes interpreted as the presence of a number of nonzero spectral components of the input signal. The minimal amount of the spectral lines depends on the number of unknown coefficients of the plant transfer function.

During closed-loop identification, both exciting properties of disturbances and the influence of feedback must be accounted for.

The measured output $y(t)$ must be the linear reaction of the plant to the measured input $x(t)$. In other words, the influence of the input disturbance $u(t)$ must be more powerful than the output disturbance $v(t)$. If this is true, the plant model can be estimated by direct methods.

Accounting for the models of the closed-loop elements can be partially investigated in terms of classical identifiability. In addition to the model of the plant itself, models of sensors, actuators, A/D and D/A converters must be accounted for. For example, the model of a thermocouple measuring gas temperatures in flowing parts of the engine substantially changes the measured plant dynamics. A thermocouple can be approximately described using a first-order transfer function with the time constant changing with gas speed and pressure. The thermocouple's inertia depends on the gas flow around its sensitive element; therefore gas pressure and shaft speed are accounted for. At some operating conditions the thermocouple's time constant can be greater than the time constant of the engine itself. Similarly, the models of electromechanical, hydraulic and pneumatic actuators of the controller are accounted for. All elements of the closed loop influence the operation of the control system and the results of identification.

When estimating quality of data for identification, some additional factors are considered. These are various disturbances and noises. Usually, their properties are unknown before identification, as well as the points of their application. There is noise that influences measurements only and noise that penetrates into the closed-loop control. In practice, the transfer function of the plant cannot often be identified from the input and output observations in the closed loop. In this situation, the most powerful disturbance is applied to the plant output. Therefore the identifiability problem of the plant model in the closed-loop control is more complicated than in open-loop systems.

In the next sections, existing approaches to the problem of closed-loop identifiability are discussed in detail.

7.3 Existing Approaches to Closed-Loop Identifiability

Dynamic models identified from experimental data can be applied for control design, simulation, condition monitoring, and optimal control purposes. Nowadays analogue controllers have been replaced by digital controllers. These have introduced a number of new noise sources, such as input quantization, prefiltering for enhancing controller response and device driver amplification. Sometimes, this persistent excitation is sufficient for dynamic model identification from observations of the random processes without test signals. Feedback control introduces additional errors in the identified models, creating cross-correlation from the output to the input of the plant. The performance of specialised open-loop or test-signal experiments, however, can be very expensive.

Many authors have examined closed-loop identifiability of dynamic models. Eykhoff [5], Ljung and Glad [6], Söderström and Stoica [7] have carried out theoretical investigations of the identifiability problem. Isermann [8] and Ljung [4] have formulated the identifiability conditions using dynamic models of plant, feedback and random disturbances. In particular, in [4] it is pointed out that direct closed-loop identification can be successful under certain conditions. However, verification of these identifiability conditions requires direct measurement of the disturbances.

Existing approaches can be demonstrated using the example of a closed-loop control system given in Figure 7.1. Two disturbances $u(t)$ and $v(t)$ influence the system's operation. The input disturbance $u(t)$ can be the controller's demand signal, *e.g.*, demanded shaft speed or pressure ratio. The output noise $v(t)$ reflects external environment and internal random processes within the plant. The controller's demand is expected to be constant at the steady-state conditions, so the input noise is zero.

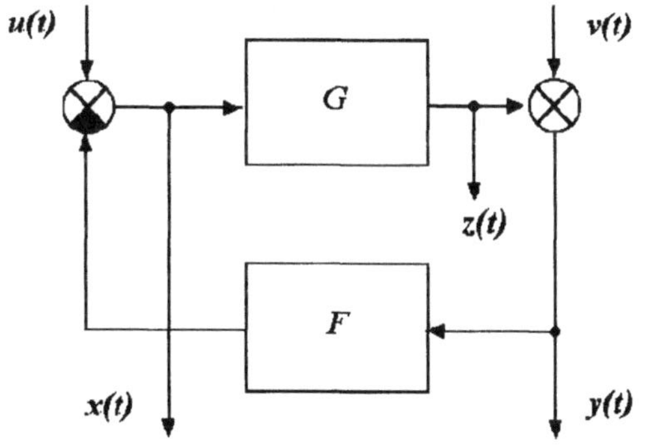

Figure 7.1. Closed-loop control: G is the plant model; F is the feedback model; $u(t)$ and $v(t)$ are the input and output excitation disturbances; $x(t)$, $y(t)$ and $z(t)$ are the input and output measurements

In real-life operating conditions, the demanded value has nonlinear dynamic relations with the control programs, the plant state and the external conditions. The pilot's lever angle (PLA) and the environmental parameters affect the controller's behaviour via the measurement system. As any measurement already includes random errors, accounting for environmental conditions will also alter the demanded value. This produces random processes at the controller's input. Existing approaches to identifiability analysis are based on accounting for models of the closed-loop elements and external disturbances.

Identifiability testing requires knowledge about the random disturbances, their points of application and their models as well as the plant and feedback models. The disadvantage of this technique is the validation difficulty in the absence of direct disturbance measurements. The physical sense of these identifiability conditions consists of the following. The random disturbance applied to the plant input must determine dynamic processes over the closed loop. The measurements of the plant output signal should be determined by the plant linear reaction to the observable input signal.

7.3.1 Parametric Identifiability

In this chapter the term *direct identification* means ignoring the influence of feedback. The signals are processed in the same way as in an open-loop system.

In [8] identifiability conditions are analysed for a discrete system where the output $y(t)$ is measured only. The transfer function of a controller is considered precisely known *a priori*. Random excitation is applied to the plant output only. Indirect identification of the plant model is performed via the ARMA-model estimation from the output signal observations, followed by extracting the plant model from it. The concept of a parametrically identifiable plant is then introduced and the following identifiability conditions are considered:

- the orders of the plant and noise models should be known (note that direct noise measurements are required to build the noise model);
- the orders of the transfer function polynomials of the plant, feedback and noise should meet a system of inequalities. This ensures that all unknown parameters can be found from the equations used in parametric estimation.

Indirect identification requires knowledge of the controller's model. This is true for the digital part of a control system. The models of actuators, sensors and converters are known with some tolerance. Moreover, these models can change depending on operating conditions. Thus, practical application of indirect identification encounters substantial restrictions.

Another example considered in [8] concerns direct identification of the plant model from the input and output measurements. Direct identification with no input disturbance is shown to give the inverse feedback model. However, direct identification with the measurement of the output $z(t)$ without noise produces the plant model estimate. Identifiability conditions for direct closed-loop identification include the restrictions on polynomials' orders introduced above. If the order of the controller's transfer function is not high enough, the advice is to use several control structures in turn, or to use nonlinear control, or to introduce pure delay into the feedback.

This is followed by direct identification in the presence of both excitations $u(t)$ and $v(t)$. Direct parametric identification is stated to be possible, if the input disturbance has a "sufficiently high order." The second identifiability condition is not required here.

7.3.2 Nonparametric Identifiability

Closed-loop related publications often demonstrate the possibility of direct parametrical identification and impossibility of direct nonparametric identification in closed-loop control using correlation and spectral methods.

A classical example is referred to in [5], where direct identification is impossible (see Figure 7.1):

$$u(t) = 0 \text{ and } v(t) \neq 0 \tag{7.2}$$

External excitation is applied to the output and direct identification gives the inverse feedback model:

$$\hat{H} = \frac{S_{yx}}{S_{xx}} = -\frac{1}{F} \tag{7.3}$$

The system is examined in continuous time and is described using continuous transfer functions.

In the work [4] it is also mentioned that spectral analysis cannot be directly applied to closed-loop systems. The plant frequency response estimate by direct spectral methods is stated to converge to the following expression:

$$\hat{H}(\omega) = \frac{S_{yx}(\omega)}{S_{xx}(\omega)} = \frac{S_{uu}(\omega)G(\omega) - S_{vv}(\omega)F^*(\omega)}{S_{uu}(\omega) + S_{vv}(\omega)F^*(\omega)F(\omega)} \tag{7.4}$$

Two extreme cases are suggested for consideration:

- $S_{uu} = 0$;
- $F = 0$.

In the first case the direct spectral estimate corresponds to the inverse feedback model $-1/F$, in the second case to the plant model G. The plant model can be obtained from informative data from a closed-loop system by directly using prediction error methods. Again, parametric methods of direct identification are suggested to be successful for closed-loop systems, as opposed to spectral analysis methods.

Consider the spectral estimate Equation (7.4). From the viewpoint of identification there are two extreme cases:

Case 1. Plant model identification.

$$\text{If } \begin{cases} |S_{uu}(\omega)G(\omega)| >> |S_{vv}(\omega)F(\omega)| \\ |S_{uu}(\omega)| >> |S_{vv}(\omega)F^2(\omega)| \end{cases} \text{ then } \hat{H}(\omega) \approx G(\omega) \tag{7.5}$$

Case 2. Feedback identification.

$$\text{If } \begin{cases} |S_{uu}(\omega)G(\omega)| << |S_{vv}(\omega)F(\omega)| \\ |S_{uu}(\omega)| << |S_{vv}(\omega)F^2(\omega)| \end{cases} \text{ then } \hat{H}(\omega) \approx -\frac{1}{F(\omega)} \quad (7.6)$$

So, there are two extreme situations in closed-loop identification: the plant model G or the feedback model $-1/F$. Similar conclusions are given in [7], where three situations are considered:

- no feedback results in the plant identifiability;
- entrance noise results in the plant identifiability;
- output noise results in the feedback identifiability

In other cases direct application of identification methods is said to give biased estimates. The result of direct closed-loop identification depends on the signals' spectra, frequency responses, the points of the disturbances' application and the points of the signals' measurement. The main condition for the plant identifiability is *persistent excitation*. Then the physical sense of identifiability conditions is that the exciting input noise $u(t)$ must be presented in the input measurements $x(t)$ and also must be more powerful than the output exciting noise $v(t)$. In other words, the *input excitation must be the most powerful* over the closed loop. Also, its spectrum should contain a sufficient number of harmonics for identification of all coefficients of the transfer function. Moreover, the power of these harmonics should be greater than that of other excitations and measurement noise.

In practice, frequency response identification from experimental data is always performed within a given frequency range. In most known works *identifiability* is investigated without such limitations assuming spectral analysis at all frequencies:

$$-\infty < \omega < +\infty$$

However, note that in practice identifiability should be fulfilled over a certain frequency range, where identification takes place.

7.3.3 Common Features of Parametric and Nonparametric Identifiability

Principally, parametric and nonparametric methods for direct identification produce equivalent results. The equations for the estimation of a parametric ARMAX model contain values of correlation functions. Consider ARMAX identification without accounting for the noise component:

$$y_i = a_1 y_{i-1} + \cdots + a_n y_{i-n} + b_1 u_{i-1} + \cdots + b_m u_{i-m} \quad (7.7)$$

During identification, all coefficients a_i and b_i of the transfer function are calculated from the input u_i and output y_i values. For this purpose both parts of the equation are multiplied by u_{i-1} and averaged. The mathematical expectation of such a product is a correlation function. Simple transformations form the following equation:

$$R_{yu}(1) = a_1 R_{yu}(0) + \cdots + a_n R_{yu}(n-1) + b_1 R_{uu}(0) + \cdots + b_m R_{uu}(m-1) \tag{7.8}$$

Similarly, the whole system of $(n + m)$ equations is obtained. Hence, parametric identification is equivalent to estimation of correlation functions followed by the solution of a system of linear equations. Thus, parametric and nonparametric identification methods are closely connected. This conclusion should be taken into account when analysing controversial opinions about closed-loop identifiability.

7.3.4 Summary of Identifiability Concepts

Analysis of existing concepts and approaches has shown that closed-loop identifiability testing consists of two issues:

- input excitation of the plant;
- sufficient power and frequency range of the excitation.

Verification of identifiability conditions requires knowledge about the points of the application of random excitations and their spectral performances. Then restrictions on the orders of transfer functions and forming filters [8] or on the shape of spectral density [4] can then be checked. At present, there is no working method for the identifiability verification. Solution of this problem requires the use of *a priori* information about the system.

The input and output signals of the plant must be measured. In addition, it is desirable to have *a priori* information about the plant, controller, and excitations. As a result, identification represents the improvement of accuracy of a known general model, instead of "black-box" modelling, as it seems to appear.

7.3.5 Example of Closed-Loop Identifiability Analysis

The problem is to analyse identifiability conditions with respect to a fuel metering valve in a closed-loop digital control system of a turbo prop fan engine.

The actuators in this control system are represented by the fuel feed system, which includes the metering valve and fuel pump; see Figure 7.2. Actuators are essentially energy converters and power amplifiers. The operations of energy transformation and amplification inevitably create additional random processes operating as excitations. These are the transform of a numerical output code of the PI controller V_{PI} into the angular position of the metering valve α_{mv}, which in turn is transformed into fuel flow W_{f}. The actuators' models are known only approximately; the models of random excitations in this system were not investigated at the moment of the research.

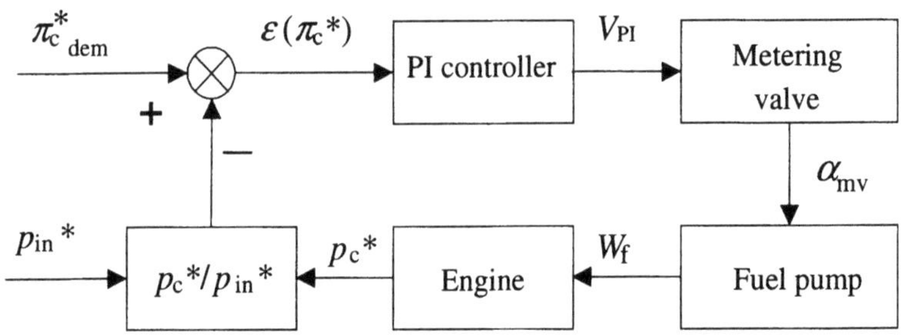

Figure 7.2. Closed-loop control of compressor pressure ratio

Pressures are measured using membrane-type sensors and a separate digital system containing A/D converters and multiplex information channels to transfer the data into the digital controller. Measurement, conversion, and data transfer create additional random errors introduced into the control loop. The models of these processes were unknown at the time of identification.

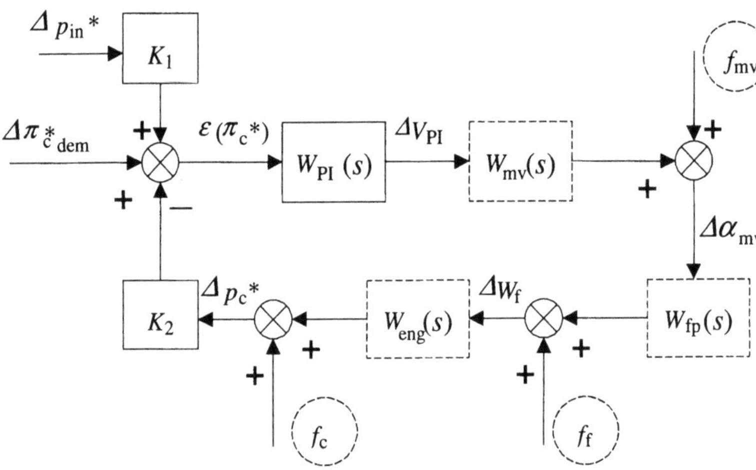

Figure 7.3. Linearised control of compressor pressure ratio: dotted line shows unknown elements

The following excitation factors operate upon the closed-loop control of compressor pressure ratio (see Figure 7.3):

- measurement noise and random disturbances in actuators f_{mv} and f_f;
- noise of input pressure measurement and physical disturbances upon input pressure p_{in};
- noise of compressor pressure measurement and random disturbances within flowing parts f_c;
- random processes within compressor pressure ratio demand $\pi_{c.dem}^*$.

The linear dynamic model "fuel flow to compressor pressure" at the current operating conditions is known only for a general engine. This model is received by linearisation of a performance-based detained nonlinear thermodynamic model. The thermodynamic model was built based on the model of a prototype engine combined with experimentally estimated static models of several demonstration engines.

The system is linearised around steady-state conditions as given in Figure 7.3. The deviation of the compressor pressure ratio $\Delta\pi_c^*$ from a static point is expressed as follows:

$$\Delta\pi_c^* = \Delta\left(\frac{p_c^*}{p_{in}^*}\right) = \frac{\partial\pi_c^*}{\partial p_c^*}\Delta p_c^* + \frac{\partial\pi_c^*}{\partial p_{in}^*}\Delta p_{in}^* = \frac{\Delta p_c^*}{p_{in.st}^*} - \frac{\Delta p_{in}^* p_{c.st}^*}{\left(p_{in.st}^*\right)^2} \tag{7.9}$$

Signals are replaced with their deviations from mean values. The dotted line in Figure 7.3 shows the elements for which models are known with some degree of accuracy. This includes random excitations which are not measured.

The analysis of the simplified block diagram shows that the controller is the only undoubtedly identifiable element of the closed loop. This is due to the fact that the controller's output V_{PI} is a digital signal and it does not contain any noise. Direct identification of other elements will be biased because of the influence of feedback.

Moreover, testing the identifiability conditions in the absence of complete information about the system and disturbances is not feasible. For example, consider identifiability conditions for the frequency response of the fuel metering valve. Direct nonparametric identification is successful under the following conditions:

$$\begin{cases} W_{mv}\left\{\begin{aligned} & S\left(f_{\Delta p_{in}}\right)\left|K_1 W_{PI}\right|^2 + S\left(\Delta\pi_{c.dem}\right)\left|W_{PI}\right|^2 + \\ & S\left(f_c\right)\left|K_2 W_{PI}\right|^2 + S\left(f_f\right)\left|K_2 W_{eng} W_{PI}\right|^2 \end{aligned}\right\} >> S\left(f_{mv}\right)\left|W_{eng} W_{PI} W_{fp} K_2\right| \\ \left\{\begin{aligned} & S\left(f_{\Delta p_{in}}\right)\left|K_1 W_{PI}\right|^2 + S\left(\Delta\pi_{c.dem}\right)\left|W_{PI}\right|^2 + \\ & S\left(f_c\right)\left|K_2 W_{PI}\right|^2 + S\left(f_f\right)\left|K_2 W_{eng} W_{PI}\right|^2 \end{aligned}\right\} >> S\left(f_{mv}\right)\left|W_{eng} W_{PI} W_{fp} K_2\right|^2 \end{cases} \tag{7.10}$$

where $S(\cdot)$ is the power spectral density (reference to frequency $S(\omega)$ is omitted for simplicity).

Testing identifiability in this case would require accurate models of all disturbances and model orders of the closed-loop elements. However, there is only one disturbance available for measurement, which is the pressure ratio demand $\pi_{c.dem}^*$. This example demonstrates that known approaches to closed-loop identifiability deal with mathematical fundamentals, whereas their practical application requires further research.

7.4 Relationship Between Identifiability and Distribution Function

The probability density function contains additional information for the analysis of the system's identifiability. The probability density of a signal passed through a linear system is known to change towards normal distribution [9]. This section shows how this property allows the recognition of nonidentifiability cases.

The output of a linear dynamic system represents a linear combination of the input signal values. This is expressed in the form of the convolution integral [10]:

$$y(t) = \int_{-\infty}^{t} x(\lambda)w(t-\lambda)d\lambda = \int_{0}^{+\infty} x(t-\lambda)w(\lambda)d\lambda \tag{7.11}$$

where $u(t)$ is the input, $y(t)$ is the output, and $w(t)$ is the impulse response or weighting function of the system. Greater time constants of the plant model lead to a greater number of input values to be added into convolution.

According to the central limit theorem of the probability theory [11, 12], adding a large number of independent random values changes the distribution of the sum towards Gaussian. Hence, the distribution of the output signal of a linear system becomes closer to normal distribution compared to the input. The degree of its approximation to the normal law depends on time constants of the plant: the more input values are added, the closer the output is to Gaussian. If the input distribution is closer to normal than that of the output, then there are two possible reasons:

- a powerful non-Gaussian disturbance is applied to the output;
- the influence of the plant nonlinearity.

Exceptions from this rule are some stable distributions. For example, a harmonic sine wave going through a linear system keeps its shape and its distribution does not change.

The normal law of distribution is stable [9]. This means that a composition of normally distributed signals gives a normal law again. Therefore, if the input of a linear system is Gaussian, then the output will be Gaussian too.

The rule about changing output distribution of a linear system towards normal is known. However, in this chapter this property is applied in a new way to test closed-loop identifiability.

Note that approach of the output distribution to Gaussian is a necessary, but not a sufficient, identifiability condition. The rule works similar to testing hypotheses in probability theory: if the condition is not fulfilled, then the hypothesis (about identifiability) should be rejected. The comparative analysis of the input and output distribution allows to elimination of nonidentifiability situations. The degree of approach of a distribution to the normal law can be estimated by known criteria [13], for example, by the Pirson's criterion, or by using asymmetry and excess parameters.

7.5 Identifiability Monitoring via Asymmetry and Excess Analysis

Consider testing the identifiability conditions using the analysis of distribution. Given the processes are stationary and ergodic, the asymmetry and excess parameters can be easily estimated from experimental data. Calculation of asymmetry A and excess E is performed using the following formulae:

$$\mu = \frac{1}{N}\sum_{i=1}^{N} x_i \qquad s = \sqrt{\frac{1}{N-1}\sum_{i=1}^{N}(x_i-\mu)^2} \tag{7.12}$$

$$A = \frac{1}{Ns^3}\sum_{i=1}^{N}(x_i-\mu)^3 \qquad E = \frac{1}{Ns^4}\sum_{i=1}^{N}(x_i-\mu)^4 \tag{7.13}$$

where μ is the mean of a random value, s is the standard deviation and N is the number of samples in the record length.

Variance D of the asymmetry and excess parameters depends on the sample volume N and is determined by the expressions:

$$D(A) = \frac{6(N-1)}{(N+1)(N+3)} \qquad D(E) = \frac{24N(N-2)(N-3)}{(N+1)^2(N+3)(N+5)} \tag{7.14}$$

Deviations $D(A)$ and $D(E)$ allow the estimation of the importance of the deviations A and E from their mathematical expectations. The distribution can be consider normal, if the following inequalities are fulfilled:

$$|A| \le 3\sqrt{D(A)} \qquad |E| \le 5\sqrt{D(E)} \tag{7.15}$$

If even one of these inequalities is not fulfilled, then the hypothesis about normal distribution should be rejected.

Conventional methods for experimental data processing include hypothesis testing of normal distribution. To perform closed-loop identifiability testing, the input and output distributions must be compared. Because of the change of the distribution towards normal, the absolute values of asymmetry and excess should decrease from the input to the output. Since the asymmetry and excess for normal distribution are equal to zero, the identifiability conditions are expressed by the following inequalities:

$$\begin{cases} |A_x| > |A_y| \\ |E_x| > |E_y| \end{cases} \tag{7.16}$$

Note that verification of the proposed conditions Equation (7.16) does not require direct observations of disturbances. Both A and E parameters can be estimated from the experimental data.

7.6 Identifiability Monitoring via Comparison of *a priori* Models with Estimates

The models of the plant and feedback are often known *a priori*, though with some tolerance. Therefore a comparison can be performed between direct identification results and the *a priori* models G and F. If the experimental estimate of the frequency response H is close to the plant model G, then the identifiability conditions can be considered fulfilled. If the estimate H is close to the feedback model $-1/F$, then the identifiability violation should be concluded. Usually, Bode diagrams of the plant and feedback substantially differ both in amplitude and phase in a wide range of frequencies. Therefore, the comparison between the diagrams of the estimated and *a priori* models is not complicated.

The violation of identifiability conditions is an undesirable situation from the viewpoint of identification. However, the information about a nonidentifiability case is of interest for the researcher. This information contains reliable data about the points of application and power spectra of disturbances. Such information is necessary for planning further experimentation with test signals. Such experiments are performed in a series of tests gradually approaching the plant identifiability.

7.7 Identifiability and Instrumental Variable Method

Instrumental variable methods for identification use auxiliary signals correlated with the input and output and not correlated with external disturbances. These methods do not require much information about the system compared to closed-loop identifiability conditions. Most publications on the subject of identifiability describe theoretical properties of identification results, whereas instrumental variable methods are oriented at practical problems in systems identification. Filtering of the input and output can form the instrumental variable. Measurements of external signals operating over the closed loop can also be useful [4, 7, 14].

A practical problem in system identification is the estimation of results' accuracy. Accuracy of the identified models depends on the *signal-to-noise* ratio, where the "useful signal" refers to the cross-correlated component of the measured signals. From the viewpoint of identification, the useful components are those that *go through* the system from the input to the output; they must also appear in the input and output *measurements*. Only these components contain information about dynamic properties of the plant. The essence of identification consists of extracting these properties from the signals' observations.

The accuracy of closed-loop identification increases substantially, when measurements are available for an external signal correlated with the closed-loop dynamics and noncorrelated with the measurement noise. These can be exact measurements of a test signal or the control demand. At the steady-state conditions, the demand, or the reference signal, of the closed-loop control consists of a constant component and a random process imposed upon it. The random component of the demand results from the controller's attempt to compensate environmental fluctuations. The demand is a digital signal within the control system unit, which can be precisely measured. This signal is correlated with the closed-loop dynamic processes and is not correlated with external noise.

This property is used in the instrumental variable method for identification. The method allows compensation of feedback and noise influences on the identified models. Again, there are restrictions on the shape of the input signal, in this case the instrumental variable. Its spectrum must be "rich" enough for accurate identification of the transfer function's coefficients from experimental data. If the existing signal's spectrum does not meet the identifiability requirements, then the researcher has to use test signals with specified spectral structure and of sufficient power, substantially exceeding existing disturbances. The instrumental variable method does not require knowledge of models of all elements of the closed loop, compared with other closed-loop identification techniques.

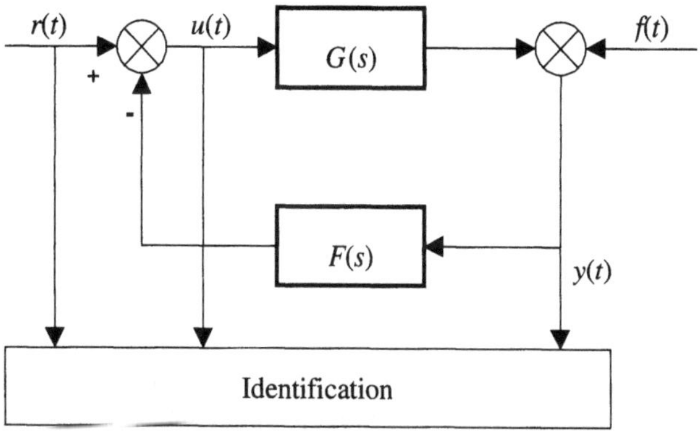

Figure 7.4. Closed loop with demand measured

In most cases the demanded value for the closed-loop output is available for measurement (Figure 7.4). If the demand contains a random part of sufficient power, then the closed-loop dynamics are determined by that signal. This creates the necessary conditions for identification. Power spectrum of the output is determined by the expression:

$$S_{yy} = \left|\frac{G}{1+GF}\right|^2 S_{rr} + \left|\frac{1}{1+GF}\right|^2 S_{ff} \tag{7.17}$$

The models of the plant and feedback are known with some tolerance, so the equivalent transfer function $G/(1 + GF)$ is also approximately known. If the output

$y(t)$ represents the reaction to the measured demand $r(t)$, then identifiability conditions are most likely fulfilled. Otherwise, when the output $y(t)$ is not correlated with the reference signal $r(t)$, then identifiability is almost impossible. This example demonstrates relation between identifiability and controllability: if the output of the plant $y(t)$ is controlled by the demand $r(t)$, then the plant is identifiable as well. If the equivalent transfer function $G/(1 + GF)$ is identified from the observations of the demand $r(t)$ and output $y(t)$, then G should be identifiable from the observations of the input u(t) and output y(t).

There is a spectral analogue for the instrumental variable method [7]. Modified spectral analysis uses measurements of the input and output together with the input disturbance measurements as the instrumental variable:

$$S_{ur} = \frac{1}{1+GF} S_{rr} \tag{7.18}$$

$$S_{yr} = G\frac{1}{1+GF} S_{rr} \tag{7.19}$$

Frequency response is estimated as follows:

$$\hat{H}(\omega) = \frac{\hat{S}_{yr}(\omega)}{\hat{S}_{ur}(\omega)} = G(\omega) \tag{7.20}$$

The estimate Equation (7.20) is unbiased and the knowledge of models of the plant and feedback is not required.

7.8 Concluding Remarks

In this chapter practical issues of identifiability have been considered when using experimental data obtained from closed-loop operation of the real plant. It is highlighted that in some cases it is not possible to identify the system and in other cases a model can only be identified to a certain degree of accuracy. A key consideration for the control engineer in practical applications is his ability to recognise whether it is possible to identify a system from a particular data record. If it is not possible to identify the system then additional information is required and this may involve further active experimentation.

Some new techniques for to verify closed-loop identifiability have been proposed in this chapter. These utilise the probability density function and do not require disturbance measurement making them suitable for practical use. Practical examples of closed-loop identification are considered in Chapters 11 and 15.

Chapter 8
Nonlinear Gas Turbine Modelling

8.1 Introduction

In this chapter several nonlinear model representations are presented along with a general methodology for nonlinear system modelling. Polynomial NARMAX and neural network models are presented in more detail and nonlinear models for the engine are estimated. It is clear that in order to model the global dynamics of the gas turbine a nonlinear model is required.

8.2 Nonlinear System Representation

The identification problem in the time-domain for either linear or nonlinear modelling is to infer relationships between past input–output data and future outputs. If a finite number of past inputs $u(t)$ and outputs $y(t)$ are collected into the vector $\varphi(t)$:

$$\varphi(t) = [y(t-1) \cdots y(t-n_y) \quad u(t-1) \cdots u(t-n_u)]^T \tag{8.1}$$

then the problem is to understand the relationship f between the next output $y(t)$ $y(t)$ and $\varphi(t)$:

$$y(t) \rightarrow f(\varphi(t)) \tag{8.2}$$

To obtain this understanding a set of observed data is required which consists of the input $u(t)$ and output $y(t)$, from which the vector $\varphi(t)$ can be built. The function f can be any function and it is indeed this function which defines the model structure. The case for which f is a linear function has been reviewed in Chapter 5, where model structures such as ARX, ARMAX, OE and BJ were presented. A mature body of work exists for the estimation of linear models in the time-domain, which is considered a relatively simple task by practisioners and indeed, having to estimate a linear model for a system is considered a happy occasion. The case for which f is a nonlinear function presents the most challenging problem for practitioners due to the curse of dimensionality.

8.2.1 Functional Representations

Volterra Series

An analytic response function can be represented by an infinite series called the Volterra series. This is a generalisation of the impulse response function of linear systems and is composed of the convolution integral:

$$y(t) = \int_{-\infty}^{\infty} h(\tau)\, u(t-\tau)\, d\tau \tag{8.3}$$

and a static nonlinearity represented by a Taylor series:

$$y = c_0 + c_1 u + c_2 u^2 + c_3 u^3 + \cdots = \sum_{l=0}^{\infty} c_l u^l \tag{8.4}$$

The Volterra series is then given by:

$$y(t) = h_0 + \sum_{l=1}^{\infty} \int_{-\infty}^{\infty} \cdots \int_{-\infty}^{\infty} h_l(\tau_1, \tau_2, \cdots, \tau_l) \prod_{i=1}^{l} u(t-\tau_i)\, d\tau_i \tag{8.5}$$

which represents a sum of outputs of parallel subsystems called Volterra functionals illustrated graphically in the schematic diagram in Figure 8.1.

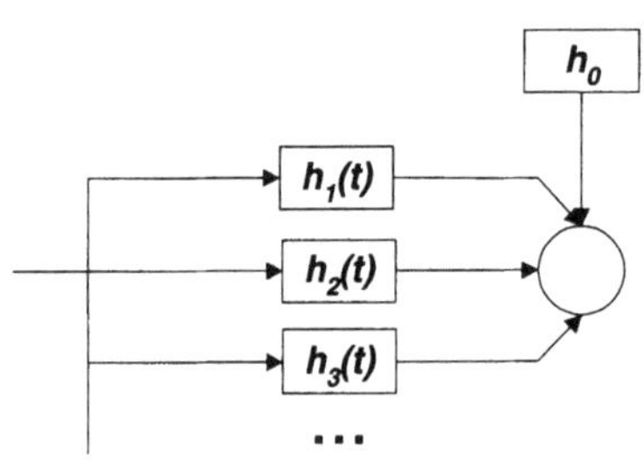

Figure 8.1. Graphical representation of the Volterra series

Nonlinear system identification based on the Volterra representation requires the measurement of the kernels $h_l(\tau_1, \tau_2, ..., \tau_n)$. Several approaches exist in the literature for the estimation of these parameters, the most common based on the extension of correlation methods for linear systems and the use of white Gaussian signals. The estimation of the parameters can be computationally expensive due to the number of parameters that increase exponentially with the degree of the kernel. For example, even for the description of a second-order kernel several hundred parameters may be required. This inherent drawback of the Volterra series, and the fact that nonlinearities such as saturation, backlash, hysteresis and dead zone cannot be modelled, has severely restricted its usefulness for nonlinear system identification.

Wiener Series
A more appropriate alternative for system identification was developed by Wiener who used the Volterra series to construct a new functional series (see [1]):

$$y(t) = \sum_{n=0}^{\infty} \{G_n(k_n, u(t))\} \tag{8.6}$$

In this series a Gram–Schmidt orthogonalisation procedure is used to make the $\{G_n\}$ functionals orthogonal for a white Gaussian excitation and enable their identification in isolation. The identification of the kernels in the Wiener series involves computing multidimensional correlation functions between the Gaussian input and the output, and can be computationally expensive due to the large number of parameters. Again the identification is based on the use of white Gaussian excitation and even though pseudorandom binary signals have been studied as a substitute, it has been shown that they can sometimes lead to disastrous results where even identifiability is lost [2].

8.2.2 Block-Structured Systems

Block-structured systems are systems that can be represented by interconnections of linear dynamic models and static nonlinear elements. The Hammerstein model shown in Figure 8.2 (a) consists of the cascade connection of a static nonlinearity followed by a linear time-invariant system.

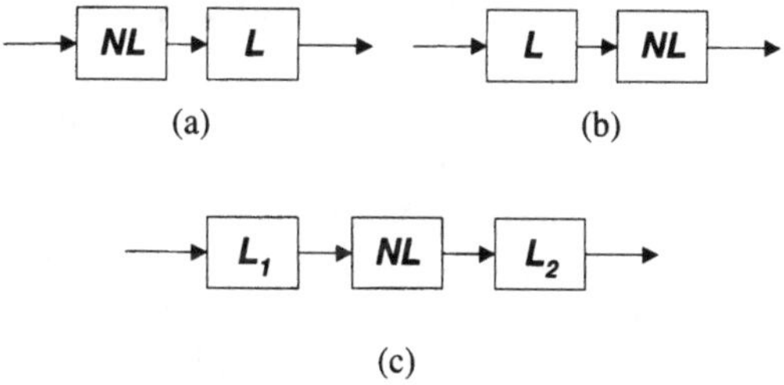

Figure 8.2. Nonlinear model structures: (a) Hammerstein, (b) Wiener and (c) Wiener–Hammerstein or Cascade

It has been first postulated by Narendra and Gallman [3] and has been extensively studied since then. Hammerstein models are usually used to approximate systems where the nonlinearity is caused only by the variation of dc gain with input amplitude. These models have constant dynamic behaviour irrespective of the input amplitude. It thus seems that Hammerstein models are not appropriate for nonlinear gas turbine modelling since the dynamics of the engine change with input amplitude as clearly demonstrated in Figure 6.16. Nevertheless, due to the simplicity of these models and the ease of estimation, Hammerstein engine models are going to be investigated in this chapter.

The Wiener model shown in Figure 8.2 (b) consists of a linear dynamic element in series with a static nonlinear part, and constitutes a reduced form of a general Wiener model described by Schetzen [1]. Unlike Hammerstein models, the nonlinearity in Wiener models is caused by the variation of the system static and dynamic characteristics with input amplitude. Wiener models thus seem to be appropriate candidates for nonlinear gas turbine modelling and they are also going to be investigated in this chapter.

The cascade model shown in Figure 8.2 (c) is the general model for block-structured systems. The theory of separable processes along with Gaussian inputs and cross-correlation techniques were used to decouple the linear dynamics of the system. Billings and Tsang [4] presented a frequency-domain analysis of the models, based on correlation methods and Gaussian inputs. More recently Weiss *et al.* [5] developed specially designed multifrequency signals to measure Volterra kernels and separate the two linear subsystems. This was achieved using combined tests to extract the kernels and estimate the parametric and nonparametric models for the linear subsystems. Similarly, Vandersteen and Schoukens [6] used an excitation, which consists of a large-signal and a small-signal multisine to separate the two linear subsystems. Cascade models are not investigated in this work since their estimation requires special excitation data, which was not available during the engine tests.

The simplicity of block-structured models compared with other identification structures and their relation to traditional linear system representations makes them attractive for nonlinear system modelling. They are not the best for black box identification though, due to the prior information required about the system structure and the restrictions placed on the excitation signals.

8.2.3 The Polynomial NARMAX Approach

Several nonlinear difference equation models that are suitable for identification have been developed. Only the polynomial Nonlinear AutoRegressive Moving Average with eXogenous inputs (NARMAX) model will be considered in this text, due to its application to nonlinear gas turbine modelling which will follow.

One of the most popular structures for linear models presented in Chapter 5 is the autoregressive model with exogenous inputs (ARX):

$$y(t) = -\sum_{i=1}^{n_y} a_i y(t-i) + \sum_{i=1}^{n_u} b_i u(t-i) \tag{8.7}$$

where a_i and b_i are the coefficients of the model and $y(k)$ and $u(k)$ are the output and input vectors of length N, respectively. In the context of system identification and parameter estimation noise terms are usually required in order to produce unbiased estimates. This gives rise to the autoregressive moving average model with exogenous inputs (ARMAX) which was also presented in Chapter 5:

$$y(t) = -\sum_{i=1}^{n_y} a_i y(t-i) + \sum_{i=1}^{n_u} b_i u(t-i) + \sum_{i=1}^{n_e} c_i e(t-i) + e(t) \tag{8.8}$$

A similar representation can be derived if the system is nonlinear. Thus the nonlinear autoregressive moving average model with exogenous inputs (NARMAX) [7] is considered

$$y(t) = F^l \begin{bmatrix} y(t-1),\dots, y(t-n_y), u(t-d),\dots, u(t-d-n_u+1), \\ e(t-1),\dots, e(t-n_e) \end{bmatrix} + e(t) \tag{8.9}$$

where F^l is a nonlinear function which in this work is taken to be a polynomial of degree l (note that for $l = 1$ the resulting model is a linear ARMAX model): $y(t)$, $u(t)$ and $e(t)$ represent the output, input and noise signals respectively; n_y, n_u, and n_e are their associate maximum lags, and $d \in \mathbb{Z}^+$ is the delay measured in sampling intervals, T_s. It has been rigorously proved that a nonlinear discrete system can always be represented as a NARX model (NARMAX model with the noise terms excluded) around an equilibrium point if (i) the response function of the system is finitely realisable and if (ii) a linearised model of the system exists around the equilibrium point. This does not impose severe restrictions on its application, since most practical systems are operated around an operating point. It is noted, that the noise terms are included in the model in order to avoid bias in the parameters and account for uncertainties, possible noise, unmodelled dynamics, *etc.*

Some points about the polynomial NARMAX representation are outlined below to highlight its advantages and justify the selection of this method for nonlinear gas turbine modelling.

- The model has an explicit recursion in the present output $y(t)$. This is an advantage in the identification of nonlinear systems compared with other nonlinear difference equation structures.
- The model is linear in the parameters. This enables the application of well-established parameter estimation techniques readily developed for linear system identification.
- As previously stated the NARMAX representation is an extension of the linear ARMAX model. This makes possible under certain conditions the physical interpretation of the model by computing the Higher-Order Frequency Response Functions (HOFRF).
- There is no restriction in the nature of the excitation in the NARMAX procedure.
- The NARMAX representation includes a family of other nonlinear representations and provides an alternative to block-structured models.

Several successful applications of polynomial NARMAX models are reported in the literature, demonstrating the popularity of NARMAX structures in system identification.

8.2.4 Feedforward Neural Network Models

If the problem of black-box modelling is treated as simply "curve-fitting," neural networks provide an attractive alternative to the previously presented methods. Feedforward neural networks have proved to have excellent function approximation capabilities, thus justifying the enormous amount of research dedicated to the subject in recent years. The two most common types of feedforward neural networks are the *multilayer perceptrons* (MLP) and the *radial-basis functions* (RBF) networks. Both types of neural networks were successfully used in the past not only to model various biological and industrial processes but also to design controllers around them with significant benefits. In this chapter only the MLP network architecture will be used to estimate nonlinear models for the gas turbine. This does not mean in any way that it is the best neural network architecture for nonlinear system identification. On the contrary, the type of neural network used is highly dependent on the specific application and the purpose of the neural network model.

8.2.5 Local Approximations

Local approximations are considered as a simple alternative to nonlinear modelling. They are usually used to model processes which exhibit significantly different dynamics in different regions of operation. In such processes it is likely that a single nonlinear model is not capable of capturing such significant changes. Local methods can provide approximations of higher accuracy than global methods, especially for systems with complex behaviour. It was stressed that the great advantage of local methods (including piecewise-linear approximations) is the flexibility in building the model. One of the first piecewise linear models in the literature was threshold models. The basic idea behind these models was the partition of the *n*-dimensional space Ir^n into a multitude of regions in which a linear model is fitted. Note that the definition of the regions in which a model is valid can be quite difficult and time consuming. Other authors have also dealt with piecewise linear models; see, for example, Billings and Voon [8].

In Johansen and Foss [9], NARMAX models were constructed using linear ARMAX models. This technique is based on decomposing the system operation into a set of operating regimes that are assumed to cover the full range of operation. The next step is to choose a simple (ARMAX) local model for each operating regime and assign a local model validity function for each model, *i.e.*, a function that indicates the relative validity of the local model at a given operating point. Finally, a smooth function is used to interpolate these local models to get a global model.

8.3 Nonlinear System Identification

In order to take the step from general «curve fitting" to system identification the model structure of the dynamical system must be chosen by answering two questions [10]:

- What variables constructed from observed past data, should be chosen as regressors, *i.e.*, as components of $\varphi(t)$ in Equation (8.1).
- What non-linear mapping should $\varphi(t)$ subjected to, *i.e.*, In neural networks, how many hidden layers should be used and how many notes should each layer have?

The first question is answered by reviewing the choice of regressors in the linear case, as presented in Chapter 5. Consider the best-known linear ARMAX model given by

$$A(z)y(t) = B(z)u(t) + C(z)e(t) \tag{8.10}$$

In this model $\varphi(t)$ is made up of past output, input and noise terms. For example, the nonlinear mapping that $\varphi(t)$ is subjected in the NARMAX case discussed in Section 8.1.3, is a polynomial. Following the same regressor choice it is possible to define the neural network dynamic models structures such as:

- NNFIR models, which use $u(t-k)$ as regressors;
- NNARX models, which use $u(t-k)$ and $y(t-k)$ as regressors;
- NNOE models, which use $u(t-k)$ and $\hat{y}_u(t-k\,|\,\theta)$;
- NNARMAX models, which use $u(t-k)$, $y(t-k)$ and $\hat{y}(t-k\,|\,\theta)$;
- NNBJ models, which use all four regressor types.

An example of a NNARMAX model is shown in Figure 8.3. It must be noted here that this is a *recurrent network* since the past prediction errors depend on the model output and consequently establish a feedback. The same applies to the NNOE and NNBJ networks.

Assuming that a model structure has been selected (polynomial NARMAX, NNARX, NNFIR, NNARMAX, *etc.*) the next step in the nonlinear model estimation procedure is to apply the data set to select the «best" model among the candidates contained in the model structure. This stage is called *training*, and it involves the minimisation of a fit criterion, the most common choice being the sum of square errors given by:

$$V_N(\theta, Z_e^N) = \frac{1}{2N}\sum_{t=1}^{N}[y(t) - \hat{y}(t\,|\,\theta)]^2 = \frac{1}{2N}\sum_{t=1}^{N} e^2(t\,|\,\theta) \tag{8.11}$$

where θ are the model parameters (weights in the neural network case), $y(t)$ is the system output, $\hat{y}(t)$ is the model estimate and Z_e^N is a matrix which contains the system output and the regressor matrix $\varphi(t)$ of Equation (8.1):

$$Z_e^N = [y(t) \ \ \varphi(t)] \tag{8.12}$$

known as the estimation or training data. The parameter estimate $\hat{\theta}$ is thus obtained by:

$$\hat{\theta} = \arg\min V_N(\theta, Z_e^N) \tag{8.13}$$

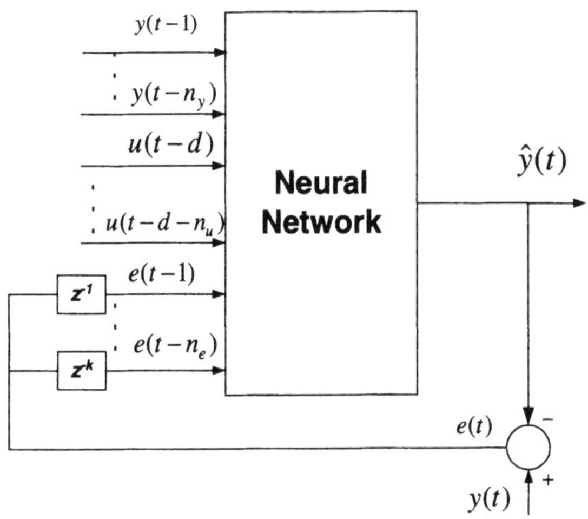

Figure 8.3. The NNARMAX model structure

Generally the estimate in Equation (8.13) is computed using an iterative nonlinear optimisation technique such as the Gauss–Newton or the Levenberg–Marquardt method. The polynomial NARMAX expansion is an exception since the model is linear in the parameters, thus allowing conventional linear estimation techniques to be applied.

An important issue in system identification is the compliance with the principle of parsimony. It would be expected that increasing the size of the NARMAX or neural network model, *i.e.*, increasing the number of regressors, or increasing the number of hidden units and consequently the number of weights, will decrease the criterion in Equation (8.11), thus obtaining a better fit to the estimation data. This does not necessarily mean that the model is able to "generalise," but only that the model is able to adjust on the particular piece of data that it has been trained with. It is indeed remarkable as Ljung [11] notes that if the criterion in Equation (8.11) is evaluated on a validation data set with the same properties as that of the estimation data set, then, asymptotically in N, the expectation of $V_N(\theta, Z_v^N)$ is given by:

$$E\{V_N(\theta, Z_v^N)\} = V_N(\theta, Z_e^N)\left(1 + \frac{dim\,\theta}{N}\right) \tag{8.14}$$

where Z_v^N is a matrix corresponding to Z_e^N evaluated on a validation data set, and the notation *dim* θ mean the number of estimated parameters [12]. It is thus clear that by using validation data to evaluate the sum of square errors as in Equation (8.14) it is possible to penalise each additional parameter that is added to the model. Consequently, in order to select the size of a nonlinear model, models of different sizes are trained on the estimation data and the model which minimises the fit for the validation data in Equation (8.11) is selected as the final model. A general procedure to estimate a black-box nonlinear model of the kind described in this chapter, can be summarised as follows:

Step (1)	Select the model structure, *i.e.*, NNARX.
Step (2)	Split the observed data into an estimation data set and a validation data set.
Step (3)	Select the input and output lag spaces n_u and n_y and the input time delay d.
Step (4)	Using the estimation data set, train a number of nonlinear models, *i.e.*, NNARX (n_u, n_y, d). In the neural network case, for example, start from a simple model structure with one hidden unit (group of weights) and increase the model structure gradually with one hidden unit at a time.
Step (7)	Compute the value of the criterion in Equation (8.11) for each of these models using validation data.
Step (6)	Select as a final model the one that minimises the fit on the validation data.

8.4 Nonlinear Gas Turbine Modeling Using NARMAX Structures

The estimation of NARMAX models consists of two stages, namely, structure selection and parameter estimation. Parameter estimation is a conventional problem and several algorithms are available in the literature, whereas structure selection is a problem which requires more intuition thus presenting the most challenging issue in NARMAX modelling. In this section the algorithms used to estimate a NARMAX model are presented and several issues concerning the crucial issue of structure selection are discussed.

8.4.1 Parameter Estimation

In order to gain a better insight into the NARMAX polynomial structure, Equation (8.9) is expressed in polynomial form:

$$\begin{aligned}
y(t) = & \Bigg[\sum_{i=1}^{n_y} \theta_i \, y(t-i) + \sum_{i=1}^{n_u} \theta_{n_y+i} \, u(t-i) + \sum_{i=1}^{n_y} \sum_{j=1}^{n_y} \theta_{i,j} \, y(t-i) \, y(t-j) \\
& + \sum_{i=1}^{n_y} \sum_{j=1}^{n_u} \theta_{i,n_y+j} \, y(t-i) u(t-j) + \sum_{i=1}^{n_u} \sum_{j=1}^{n_u} \theta_{n_y+i,n_y+j} \, u(t-i) u(t-j) \\
& + \text{higher - order terms up to degree } l \Bigg] \\
& + \Bigg[\sum_{i=1}^{n_y} \sum_{j=0}^{n_e} \theta_{i,0,j} y(t-i) e(t-j) + \sum_{i=1}^{n_u} \sum_{j=0}^{n_e} \theta_{0,i,n_e+j} \, u(t-i) e(t-j) \\
& \sum_{i=1}^{n_y} \sum_{j=1}^{n_u} \sum_{l=0}^{n_e} \theta_{i,j,l} y(t-i) u(t-j) e(t-l) \\
& + \text{all possible combinations of } y(t), u(t) \text{ and } e(t) \text{ up to degree } l \Bigg] \\
& + \Bigg[\sum_{i=0}^{n_e} \theta_i \, e(t-i) + \sum_{i=0}^{n_e} \sum_{j=0}^{n_e} \theta_{i,j} \, e(t-i) e(t-j) \\
& + \text{higher - order terms up to degree } l \Bigg]
\end{aligned} \tag{8.15}$$

which can be rrewritten as

$$y(t) = \Psi_{yu}^T (t-1) \Theta_{yu} + \Psi_{yue}^T (t) \Theta_{yue} + \Psi_e^T (t) \Theta_e \tag{8.16}$$

where Ψ_{yu}^T includes all the output and input terms as well as all the possible combinations up to degree l and time t–1. The parameters of such terms are denoted with Θ_{yu} and the other entries are defined likewise. The problem with Equation (8.16) is that it is unsuitable for estimating the parameter vector $\Theta = [\Theta_{yu}^T + \Theta_{yue}^T + \Theta_e^T]^T$ because the terms $e(t-i)$, $i = 0, 1, \cdots, n_e$ are not known. To overcome this difficulty, Equation (8.16) is written in the prediction error form:

$$\hat{y}(t) = \Psi_{yu}^T (t-1) \hat{\Theta}_{yu} + \Psi_{yu\xi}^T (t) \hat{\Theta}_{yu\xi} + \Psi_\xi^T (t) \hat{\Theta}_\xi + \xi(t) \tag{8.17}$$

where $\xi(t)$ is the residual at time t and is defined as:

$$\xi(t) = y(t) - \hat{y}(t) \tag{8.18}$$

Finally, Equation (8.16) can be written in the concise form:

$$\hat{y}(t) = \left[\Psi_{yu}^T(t-1) + \Psi_{yu\xi}^T(t) + \Psi_{\xi}^T(t)\right] \begin{pmatrix} \hat{\Theta}_{yu} \\ \hat{\Theta}_{yu\xi} \\ \hat{\Theta}_{\xi} \end{pmatrix} + \xi(t) \tag{8.19}$$

$$\hat{y}(t) = \Psi^T(t-1)\hat{\Theta} + \xi(t)$$

The parameter vector $\hat{\Theta}$ can be estimated by minimising the cost function:

$$V_{LS} = \left\| y(t) - \Psi^T(t-1)\hat{\Theta} \right\| \tag{8.20}$$

where $\|\bullet\|$ is Euclidean norm. A similar cost function which is similar to the quadratic cost function in Equation (8.20) can be defined as:

$$V_{PE}(\hat{\Theta}) = \log_e \det Q(\hat{\Theta}) \tag{8.21}$$

where $Q(\hat{\Theta})$ is the sample covariance matrix of the residuals and is defined as:

$$Q(\hat{\Theta}) = \frac{1}{N}\sum_{t-1}^{N} \xi(t)\xi^T(t) \tag{8.22}$$

In Chen *et al.* [13], three decompositions were investigated for the solution of this least squares problem, namely, (i) Cholesky, (ii) orthogonal and (iii) singular value. The second option is the most widely used since it enables the simultaneous solution to the structure selection problem as it will be shown later in this chapter. The orthogonal estimation algorithm will be used in this chapter for both parameter estimation and structure selection when estimating NARMAX models for the gas turbine.

Consider a linear-in-the-parameters polynomial model:

$$y(t) = \sum_{i=0}^{n_\theta} \theta_i p_i(t) + \xi(t) \tag{8.23}$$

where $p_i(t)$ represent the various terms (regressors) in the polynomial and n_θ the number of coefficients in the model. An auxiliary model can then be built by transforming this equation into:

$$y(t) = \sum_{i=0}^{n_\theta} g_i \omega_i(t) + \xi(t) \tag{8.24}$$

where g_i are the model coefficients and ω_i are the model terms constructed to be orthogonal over the data records as follows:

$$\omega_0(t) = p_0(t) = 1$$

where:

$$\omega_m(t) = p_m(t) - \sum_{i=0}^{m-1} a_{im}\omega_i(t) \qquad m = 1, 2, \ldots, n_\theta \tag{8.25}$$

$$a_{im} = \sum_{t=1}^{N} p_m(t)\omega_i(t) \Big/ \sum_{t=1}^{N} \omega_i(t)^2 \qquad 0 \le i \le m-1 \tag{8.26}$$

and in this case the following orthogonality property holds:

$$\frac{1}{N}\sum_{t=1}^{N} \omega_i(t)\omega_{j+1}(t) = 0 \qquad i = 1, 2, \cdots, j \tag{8.27}$$

The second step consists of estimating the coefficients g_i and transforming them back to the system parameters θ_i. The parameters of the auxiliary model g_i can be estimated as [14]:

$$\hat{g}_i = \frac{1}{N}\sum_{t=1}^{N} \frac{y(t)\omega_i(t)}{\omega_i^2(t)} \qquad i = 1, 2, \cdots, n_\theta \tag{8.28}$$

provided that $\omega_i^2(k) \neq 0$. The coefficients $\hat{\theta}_i$ can be obtained from g_i according to the formula:

$$\hat{\theta}_i = \sum_{i=m}^{n_\theta} \hat{g}_i \upsilon_i \qquad m = 1, 2, \ldots, n_\theta \qquad \text{where} \tag{8.29}$$

$$\upsilon_m = 1$$

$$\upsilon_i = -\sum_{r=m}^{i-1} a_{ri}\upsilon_r \qquad i = m+1, \cdots, n_\theta \qquad \text{and} \tag{8.30}$$

$$a_{ij} = \frac{1}{N}\sum_{t=1}^{N} \frac{\omega_i(t)p_j(t)}{\omega_i^2(t)} \qquad i = 1, 2, \cdots, j-1 \qquad j = 2, \cdots, n_\theta \tag{8.31}$$

Matrix formulae have also been developed for the above algorithm, and four different methods for achieving orthogonal decompositions were compared in Chen *et al.* [13], namely, (i) classical Gram–Schmidt, (ii) modified Gram–Schmidt, (iii) Householder transformation and (iv) Givens method. Throughout this work the modified Gram–Schmidt method is preferred since it is more accurate than the others.

8.4.2 Structure Selection

A clear disadvantage of polynomial models is the enormous number of terms a general nonlinear polynomial may have. The number of terms in a polynomial grows very rapidly even for relatively low values for l, n_y, n_u and n_e. In particular, it can be seen from Equation (8.15) that the maximum number of terms that can be considered in a NARMAX model is given by:

$$n_\theta = M + 1 \qquad \text{where} \tag{8.32}$$

$$M = \sum_{i=1}^{l} n_i$$
$$n_i = \frac{n_{i-1}(n_y + n_u + n_e + i - 1)}{i} \qquad n_0 = 1 \tag{8.33}$$

In practical systems, it seldom happens that a full NARMAX expansion is required to characterise the nonlinear mechanism. A full expansion would lead to numerical ill-conditioning and consequently discrepancy in the parameter estimates due to the large number of terms in the model. On the other hand in the simple example considered, there are ten terms in the full expansion and 1024 (2^{10}) models to choose from. This implies that the problem of selecting the most important of all possible terms is too difficult to be solved by trial and error. The problem of determining the structure of nonlinear systems thus involves two major aspects:

- the specification of the dynamic order n_y, n_u, n_e, the input time delay d and the degree of nonlinearity l;
- the selection of the most significant regressors from the set specified by the full NARMAX expansion.

It must be noted here that even if the dynamic order, input time delay and degree of nonlinearity are accurately estimated, failure to select the correct model structure – in other words, to comply with the second aspect – leads to either an overparameterised or an underparameterised model with undesirable consequences. An underparameterised model will obviously omit important terms which are essential to model the system dynamics leading to an incomplete model. An overparameterised model will contain unnecessary terms, and the global behaviour of such a model can be quite different from the behaviour of the underlying system.

The methods described in the literature use information criteria to develop a model structure which is optimal in a statistical sense. However, as Aguirre [15] points out, there is no guarantee that the statistically optimal model will be dynamically optimal as well. During the same work, it was concluded that these methods provide models which are close to the dynamical optimum models, but difficult to apply due to the appearance of local minima, which make the selection procedure very complicated since it is not usually obvious which one to choose. A popular alternative solution to the structure selection problem is provided by the *error reduction ratio* (ERR) [14]. The popularity of this method for structure selection of NARMAX models is mainly due to the fact that (i) it does not require the estimation of a complete model to determine the significance of each candidate term and its contribution to the output, (ii) the ERR test is derived as a by-product of the orthogonal estimation algorithm.

Consider the auxiliary model in Equation (8.24). Multiplying this model by itself and taking the time average gives:

$$\frac{1}{N}\sum_{t=1}^{N} y^2(t) = \frac{1}{N}\sum_{t=1}^{N}\left(\sum_{i=0}^{n_\theta} g_i^2 \omega_i^2(t)\right) + \sum_{t=1}^{N} \xi^2(t) \tag{8.34}$$

assuming that $\xi(t)$ is a zero mean independent random sequence and the orthogonality property of Equation (8.27) holds. Thus:

$$\frac{\frac{1}{N}\sum_{t=1}^{N}\left(\sum_{i=0}^{n_\theta} g_i^2 \omega_i^2(t)\right)}{\frac{1}{N}\sum_{t=1}^{N} y^2(t)} \le 1 \tag{8.35}$$

provided that the measured outputs are not all zero for $t \in [1, N]$. The equal sign holds only if the estimated variance of the residuals is zero, which in turn means that there are no estimation errors and the system under consideration is deterministic. If the above equation is written as:

$$\sum_{i=1}^{n_\theta} \frac{\frac{1}{N}\sum_{t=1}^{N} g_i^2 \omega_i^2(t)}{\frac{1}{N}\sum_{t=1}^{N} y^2(t)} \le 1 \quad \text{or} \tag{8.36}$$

$$\sum_{i=1}^{n_\theta} ERR_i \le 1 \quad \text{with} \tag{8.37}$$

$$ERR_i = \frac{\frac{1}{N}\sum_{t=1}^{N} g_i^2 \omega_i^2(t)}{\frac{1}{N}\sum_{t=1}^{N} y^2(t)} \tag{8.38}$$

The maximum *mean squared prediction error* (MSPE) is achieved when no terms are included in the model and in this case equals to $\Sigma y^2(t)/N$. Thus from Equation (8.36) the reduction in the mean square error due to the inclusion of the

ith term $g_i\omega_i(k)$, in the auxiliary model of Equation (8.24) is $\Sigma g_i^2\omega_i^2(t)/N$. The ERR thus represents the reduction in mean squared error as a fraction of the total MSPE. It is thus reasonable that a good model should, in addition to other requirements, have a ratio defined by Equation (8.36) near unity. A modified version of the ERR was proposed by Liu [16] to avoid the problem of estimating incorrect dynamic models when a constant term dominates the second-order moment of the output about zero, and is given by:

$$ERR_i = \frac{\frac{1}{N}\sum_{t=1}^{N} g_i^2\omega_i^2(t)}{\frac{1}{N}\sum_{t=1}^{N} y^2(t) - \left(\frac{1}{N}\sum_{t=1}^{N} y(t)\right)^2} \tag{8.39}$$

The ERR Algorithm

A usual practice to assess the model terms is by using a forward-regression algorithm which selects at each step the term with the highest ERR. The ERR test thus provides the means of ordering all the candidate terms according to a hierarchy which depends on the relative importance of each term. It must be noted here that the ERR test described above is first performed to estimate the deterministic part of the model that is, the terms which do not contain $\xi(t)$ (noise terms), also known as the *process model.* The same algorithm is then applied in the same way to select the noise terms. However, the following question arises: How many terms should be included in the model? A practical way of addressing this problem is by means of information criteria such as the previously mentioned FPE, AIC, BIC and LILC given by:

$$\begin{aligned}
FPE &= N\log_e[\sigma^2(n_p)] + N\log_e\frac{N+n_p}{N-n_p} \\
AIC(a) &= N\log_e[\sigma^2(n_p)] + an_p \\
BIC &= N\log_e[\sigma^2(n_p)] + n_p\log_e N \\
LILC &= N\log_e[\sigma^2(n_p)] + 2n_p\log_e\log_e N
\end{aligned} \tag{8.40}$$

The basic algorithm to perform the ERR test and select the deterministic part of a polynomial NARMAX model is given below:

Step (1) Select values for l, n_y, n_u, n_e, in Equation (8.15).

Step (2) Construct all possible regressors using $u(t)$ and $y(t)$ according to Equation (8.15).

Step (3) Consider all deterministic terms $p_i(t)$ $(i = 1,2,\cdots,n_\theta)$ as candidates for $\omega_0(t)$ in Equation (8.25) and determine the ERR_i

for each term using Equations (8.26), (8.28) and (8.39). Select the term with the largest ERR_i.

Step (4) Remove the selected term from the set of deterministic terms and go to Step (3). Note that n_θ is reduced by one. Proceed to Step (5) only when all terms are ordered.

Step (5) Start from a model with one term and use the information criteria of Equation (8.40) on a validation data set. Increase the model terms by one according to the hierarchy established by the ERR test and select the model or models which minimise these criteria.

After the deterministic part of the model is estimated, the forward-regression algorithm can be used to estimate a noise model as follows:

Step (1) Using the estimated process model calculate $\xi(t)$ using Equation (8.24).

Step (2) Using $\xi(t)$ apply the forward regression algorithm to the noise model and determine the noise term which has the maximum ERR_i.

Step (3) Add the selected noise term to the model and recompute $\xi(t)$ using Equation (8.24). Apply the orthogonal estimator to each of the current noise terms and recompute $\xi(t)$ using Equation (8.24) until the parameters converge to constant values. Go to Step (2) and repeat until $\xi(t)$ converges to a white sequence.

The ERR algorithms described in this section have been used successfully in various applications for the estimation of polynomial NARMAX structures. This does not mean that the ERR will always give a parsimonious model structure. On the contrary, if care is not taken the ERR test can result in a model which is overparameterised, incorrect, and which exhibits dynamical characteristics different to that of the real system.

8.4.3 Model Validation

Model validation is the most essential part of the identification process. An engineer should never deliver a product without a statement about its precision. Unfortunately, much less attention has been paid to the derivation of model validation and quality assessment tools than to the development of model derivation algorithms. This is because until the early 1980s most of the research in system identification focused on questions of convergence and efficiency, with the underlying assumption that the system belonged in the model set. It is only recently that system identification has been seen as an approximation theory, and it has become essential to be able to quantify the error between the model and the underlying system. The topic of estimating the bias and variance error of identified models is now well covered for both frequency- and time-domain linear models, as was shown in Chapter 5. The issue of model validation is even more critical in nonlinear system identification, due to the enormous number of possibilities that exist concerning the model structure and the different dynamical regimes introduced by the inaccurate selection of the model structure. In what follows an overview of the available validation tools for nonlinear system identification is presented, with special emphasis paid to the issue of cross-validation.

Statistical validation. The classical approach to validating identified NARMAX models is by the use of high-order correlation functions [17] to detect the presence of unmodelled terms in the residuals of nonlinear discrete models. Thus if a model of a system is adequate the following conditions should hold:

$$\begin{aligned}
&\phi_{\xi\xi}(\tau) = E\{\xi(t-\tau)\xi(t)\} = \delta(\tau) && \textit{i.e.}\text{, an impulse} \\
&\phi_{\xi u}(\tau) = E\{u(t-\tau)\xi(t)\} = 0 && \forall\tau \\
&\phi_{\xi(\xi u)}(\tau) = E\{\xi(t)\xi(t-\tau)u(t-\tau)\} = 0 && \tau \geq 0 \\
&\phi_{u^{2'}\xi}(\tau) = E\{[u^2(t) - E\{u^2(t)\}]\,\xi(t-\tau)\} = 0 && \forall\tau \\
&\phi_{u^{2'}\xi^2}(\tau) = E\{[\,u^2(t) - E\{u^2(t)\}]\;\xi^2(t-\tau)\} = 0 && \forall\tau
\end{aligned} \tag{8.41}$$

where $\delta(\tau)$ is the Kronecker delta, $\xi(t)$ are the residuals defined in Equation (8.18), $u^{2'}(t) = u^2(t) - E\{u^2(t)\}$ and $u(t)$ is the input to the system. The rationale of Equation (8.49) is that for a model to be statistically valid there can be no predictable terms in the residuals. Because the data length N can never be infinite, in practice the mathematical expectations are estimated and instead of checking if the cross-correlation functions are zero, confidence bands are plotted on the graphs to reveal if the correlation between the examined variables is significant or not. The 95% confidence bands are approximately $\pm 1.96/\sqrt{N}$ and one or more points of the function lying outside the respective confidence band will indicate significant correlation.

It must be stressed here that the above correlation tests only provide statistical information. A model which is statistically valid may turn out to be dynamically inappropriate [15]; in other words, the correlation tests do not necessarily confirm that a particular model has captured or indeed is able to reproduce the dynamics of the original system. It is for this reason that the above statistical tests are not going to be considered important in the identification of nonlinear models for the gas turbine dealt with in this chapter.

Cross-validation. The most powerful approach to the validation of an estimated model is to verify if the model is able to closely predict measured data. It must be stressed here that the measure of the model ability to predict is not the variance of the one-step-ahead prediction errors in Equation (8.20), but the variance of the long-term prediction errors obtained by simulating the model. A naïve approach to this is to validate the estimated models on the data that the model was estimated, known as *estimation* or *learning data* [15]. A model is always bound to predict closely to the estimation data, since it was estimated on that data. This does not mean that the model is capable of representing the system dynamics, but only that the model is able to adjust to the estimation data. It is thus necessary, in order to check the ability of the model to "generalise," to validate the estimated models on an independent set of data, called the *validation data*. This is usually referred to as *cross-validation* and it is the most valuable tool not only for the validation of the estimated model but also for the estimation of the model itself.

Recently Ljung [11] has stressed the need to compute the value of a criterion for a model quality, using validation data. According to the author the final model should be the one that minimises the fit for the validation data. Referring back to the previous section where the structure selection of NARMAX models was discussed, it can be seen that after the ordering of the candidate terms using the ERR, a cut-off point is provided by the use of the statistical criteria in Equation (8.40). If these criteria are evaluated on the estimation data, then the estimated model will, as mentioned before, be able to adjust only to the estimation data. On the other hand, if the criteria are evaluated on the validation data the estimated model will be able to "generalise." It is thus possible to incorporate cross-validation into the structure selection procedure in order to ensure that the estimated model is able to predict on validation data and capture the system dynamics.

A different way of using cross-validation to assess the performance of the estimated nonlinear model is to compare the static behaviour of the model against measured system static behaviour. This implies that data or knowledge of the system static behaviour is available. If this is the case, then *a priori* knowledge can be incorporated in the model estimation procedure as in Aguirre *et al.* [18]. The static nonlinearity of an estimated NARMAX model can be obtained either through simulation or with the simple technique recently proposed by Aguirre *et al.* [18]. A complementary approach is to linearise the estimated nonlinear model at different operating points and compare the resulting linear models to the modes estimated using linear modelling techniques. This in turn implies that data for linear model estimation is available as in the gas turbine case.

8.4.4 A Proposed Identification Scheme

Considering the techniques and the issues discussed in the previous sections, an identification scheme is proposed for the estimation of a NARMAX model. The proposed scheme is shown diagrammatically in Figure 8.4 where it can be seen that the ERR criterion is used as a constructive technique to build a model, and then cluster analysis is used as an eliminating technique to refine the model structure in order to arrive at the "best" model. The procedure starts by applying the forward-regression orthogonal algorithm to build a model according to the ERR criterion. The procedure is stopped by the AIC criterion, and then the model cluster coefficients are examined. Spurious clusters are deleted from the model structure, and the forward regression algorithm is executed again. Compensation terms are eliminated in accordance with their relevance to the model according to the ERR criterion. After the model structure is refined by cluster analysis and the removal of compensation terms, the model is validated using cross-validation. If any of these tests fail then the term with the lowest ERR should be deleted from the model. The proposed identification scheme uses cross-validation as an essential tool for the estimation of the structure of the NARMAX model.

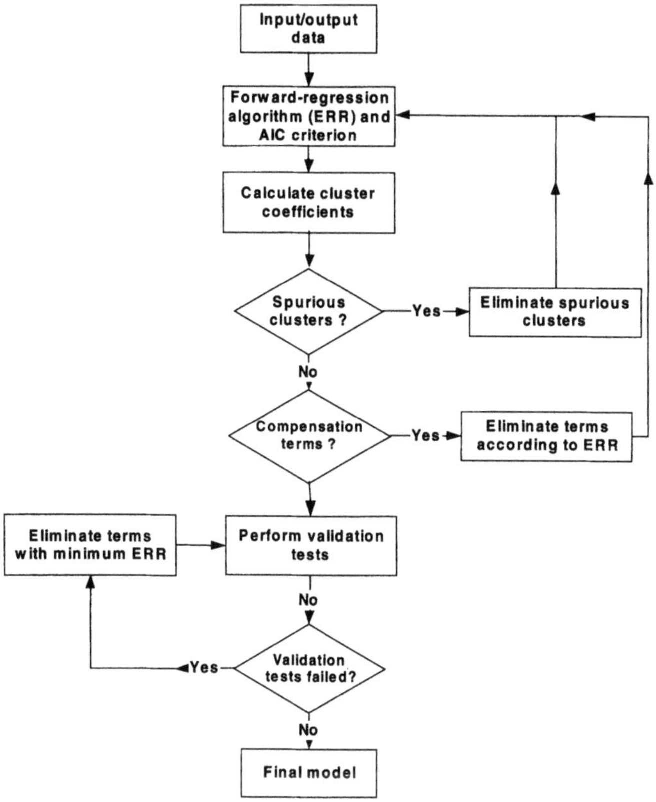

Figure 8.4. Proposed methodology for NARMAX modelling

8.4.5 Gas Turbine Modelling

Using the proposed methodology and a concatenated data set of small signal IRMLBS tests covering an operating range between 55 and 85% N_H, the NARMAX model of Equation (8.50) was estimated for the HP shaft:

$$y(t) = 0.7260y(t-1) + 0.2885y(t-2) + 0.0066u(t-1) - \\ -3.31\times10^{-5}u(t-1)y(t-2) - 0.7202 - 2.56\times10^{-4}y(t-1)y(t-2) \tag{8.42}$$

The model was cross-validated by calculating the mean square errors (MSE) on engine data gathered at different operating points using Equation (8.11). It must be stressed here that the long-term predictions were used to calculate the MSEs, *i.e.*, pure simulation and not the one-step predictions. Table 8.1 shows a comparison of the MSE values achieved with the estimated HP shaft NARMAX model of Equation (8.42) and the MSE values achieved with linear models estimated at different operating points, on the small-signal validation data.

Table 8.1. Comparison of linear and nonlinear models, HP shaft

Test type	NARMAX model	Linear models (% N_H)			
		55	65	75	85
Multisine (1) 55% N_H	0.0243	0.0085	0.0547	0.1287	0.2179
IRMLBS 65% N_H	0.0311	0.1068	0.0101	0.0433	0.1284
Multisine (1) 75% N_H	0.087	0.8140	0.1204	0.0075	0.1017
IRMLBS 85% N_H	0.0124	1.4123	0.4310	0.1017	0.0020

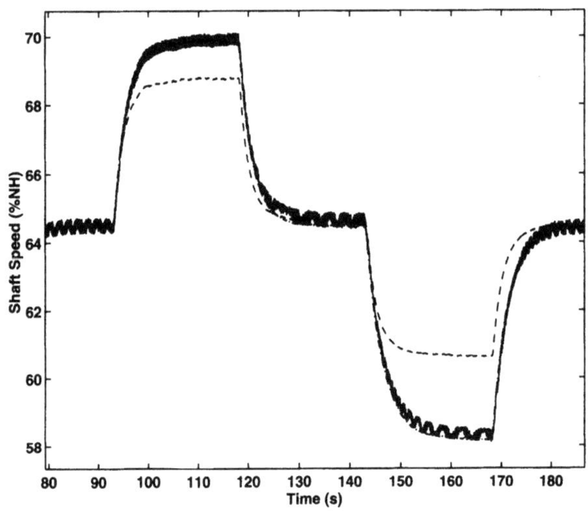

Figure 8.5. Output of three-level HP shaft test. Measured output (solid), linear ARMAX model (dashed), nonlinear model output (dash-dot)

It is clear from Table 8.1 that the linear models perform better than the nonlinear model only at the operating point that they were estimated. At other operating points the linear models fail to capture the engine dynamics as revealed by the high MSE values in the table. This means that the linear models provide a better local approximation than the estimated nonlinear NARMAX model, whereas the NARMAX model provides a good global approximation of the engine dynamics at the different operating points. This was something to be expected, since the nonlinear model tries to approximate the variation in dynamics with different operating points whereas the estimated linear models dealt only with local data not influenced by nonlinearity as discussed in Chapter 6. The same pattern is observed when a comparison is made between the LP shaft Wiener model and local linear LP shaft models. The results confirm that the estimated nonlinear models are able to represent the dynamics of the engine at the different operating points, thus enabling the use of a single nonlinear model in the place of the family of linear models.

The improvement achieved using a nonlinear model can also be demonstrated on high-amplitude validation data. Figure 8.5 shows a comparison of the performance of the estimated with the NARMAX model against a linear ARMAX model estimated on a high-amplitude data set. It can be clearly seen that the nonlinear model is capable of dealing with large deviations in speed whereas the linear model even though it was estimated using this high-amplitude data set, fail to capture the engine's high-amplitude dynamics.

8.5 Nonlinear Gas Turbine Modeling Using Neural Network Models

The use of neural networks for the identification of nonlinear systems has been advocated in earlier in this chapter. Here, multilayer perceptron (MLP) neural networks are trained on the gas turbine estimation data in order to provide global nonlinear models for both the HP and LP shafts. The selection of the size of the estimated model is based on the performance of the model on the validation data sets. The estimated HP and LP shaft models are validated using the validation data sets and by comparing their static behaviour with the static behaviour of the engine derived from measured data.

8.5.1 Multilayer Perceptron Neural Networks

The basic neural network element (the neuron) is a processing element that takes a number of inputs, applies some weights and sums them up, and feeds the result to an activation function. The most popular activation functions used in system identification are the sigmoid function and the closely related hyperbolic tangent function (tanh):

$$f_{\text{sigmoid}}(x) = \frac{1}{1+e^{-x}} \tag{8.43}$$

$$f_{\tanh}(x) = 2f_{\text{sigmoid}}(x) - 1 \tag{8.44}$$

so that it makes no difference which of these two functions is used.

A feedforward multilayer-perceptron network is constructed by ordering the units into layers, with each layer taking as input other external inputs or outputs of units in previous layers. An example of a two-layer feedforward network is shown in Figure 8.6 where it can be seen that the second layer produces the output, thus referred to as the *output layer*. The first layer is called a *hidden layer* since it is hidden between the external inputs and the output layer. The network is described by the magnitude of the weights and biases and should be determined by *training* the network on the estimation data. The estimation of the weights is usually a conventional estimation problem and several algorithms are available for this purpose.

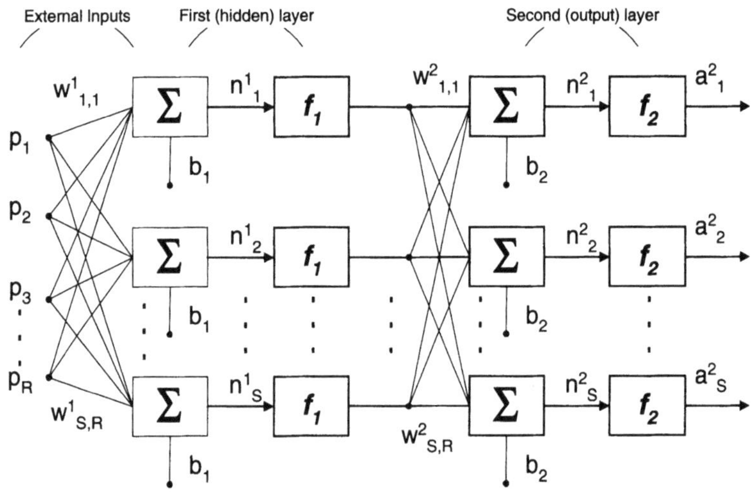

Figure 8.6. Two-layer feedforward neural network

Assuming that a model structure has been selected as discussed previously (NNARX, NNFIR, NNARMAX, *etc.*) the next step in the neural network estimation procedure is to apply the data set to select the «best" model among the candidates contained in the model structure. The procedure described in the previous section is then followed in order to ensure that a parsimonious model is estimated. Consequently, in order to select the size of the neural network model, models of different sizes are trained on the estimation data and the model which minimises the fit for the validation data in Equation (8.11) is selected as the final model. The whole procedure for estimating a feedforward multilayer perceptron neural network model of the kind described in this chapter, is based on the general methodology presented previously and can be summarised as follows:

Step (1)	Select the neural network model structure, *i.e.*, NNARX.
Step (2)	Split the observed data into an estimation data set and a validation data set.
Step (3)	Select the input and output lag spaces n_u and n_y and the input time delay d.
Step (4)	Using the estimation data set, train a number of neural network models, *i.e.*, NNARX(n_u, n_y, d). Start from a simple model structure with one hidden unit and increase the model structure gradually with one hidden unit at a time.
Step (5)	Compute the value of the criterion in Equation (8.11) for each of these models using validation data.
Step (6)	Select as a final model the one that minimises the fit on the validation data.

All estimators and algorithms required for the estimation of a multilayer perceptron feedforward neural network are implemented in the Neural Network System Identification Toolbox for MATLAB written by Nørgaard *et al.* [19].

8.5.2 Gas Turbine Modelling Using Multilayer Perceptron Neural Networks

In order to estimate a neural network model for the engine as described in the previous section, it is necessary to choose (i) the model structure, *i.e.*, NNARX, NNARMAX, (ii) the input n_u, output n_y and noise n_e lag spaces, and the input time delay d and (iii) how many hidden units and consequently how many parameters (weights) should be used in order to achieve a parsimonious model. In this work several neural network structures were investigated on the gas turbine data, and it was found that the NNARX and NNARMAX structures provided the best results. The lag spaces where chosen according to *a priori* knowledge, so n_u, n_y and n_e were set equal to 2, and the delay was set to one sampling period since this provided the best results after trying different delays. An attempt was then made to select a parsimonious model using the procedure described previously that is, by training several neural networks of different sizes and selecting the one which minimises the MSE criterion of Equation (8.11) on the validation data sets.

It must be stressed here that since the validation data used for this purpose consists of four different small-signal tests the criterion of Equation (8.11) will be different on each of these data. It is thus necessary to select the neural network which provides a reasonable compromise between performance on all operating points and number of parameters. It must be also noted that the estimated values of the weights are strongly dependent on their initial values, since a nonlinear optimisation technique is used to train the neural networks. It is thus impossible to claim that the neural networks presented in the remainder of this chapter are perfectly optimised on the gas turbine data. The use of the presented neural networks should be seen in the context of black-box models with high approximation accuracy, defined by the performance of these networks on the engine data.

Figures 8.7 and 8.8 show the variation of the criterion in Equation (8.11) evaluated at the four small-signal validation data sets, with the number of hidden units for different NNARX and NNARMAX models trained on the HP shaft estimation data. It can be seen in the NNARX case that a neural network with six hidden units is the one that minimises the MSE criterion on the validation data sets considered, whereas in the NNARMAX case a neural network with three hidden units is required.

Table 8.2 shows the MSE values obtained from the two neural network model structures on the validation data. Looking back at Table 8.1, it can be seen that both neural network models in Table 8.2 give comparable and in some cases better performance than the previously estimated NARX models on the small signal tests.

This behaviour is not replicated on the high-amplitude tests since the neural network models are found to be less accurate than the estimated NARX models when tested with these signals. In Figure 8.9 a comparison of the static behaviour of the two models and the static behaviour of the engine derived from static tests is made. It can be seen that the two models have almost identical static characteristics and they are both capable of approximating the static behaviour of the HP shaft.

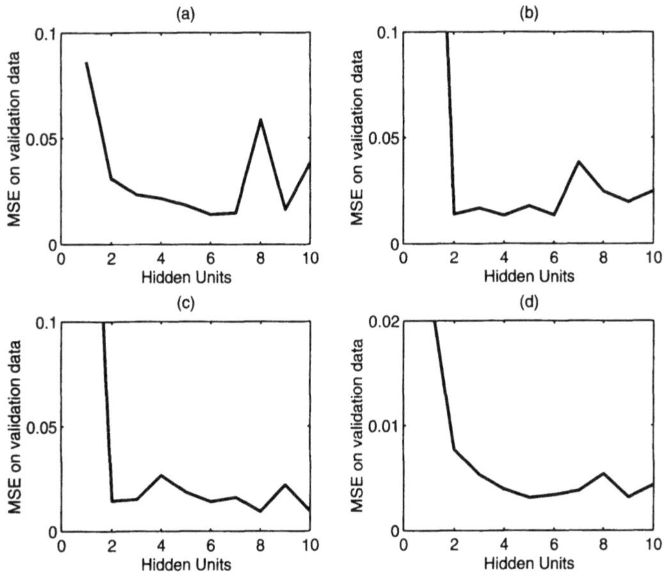

Figure 8.7. Ten different NNARX HP shaft models showing the sum of squares of the long-term prediction errors on validation data: (a) multisine (1) at 55% N_H, (b) IRMLBS at 65% N_H, (c) multisine (1) at 75% N_H and (d) IRMLBS at 85% N_H

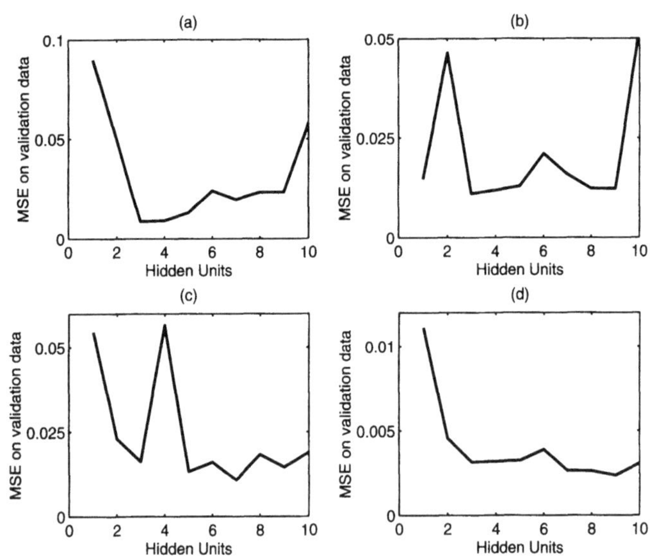

Figure 8.8. Ten different NNARMAX HP shaft models showing the sum of squares of the long-term prediction errors on validation data: (a) multisine (1) at 55% N_H, (b) IRMLBS at 65% N_H, (c) multisine (1) at 75% N_H and (d) IRMLBS at 85% N_H

Table 8.2. Values for the estimated HP shaft models

Test type	NNARX	NNARMAX
Multisine (1) 55% N_H	0.0142	0.0086
IRMLBS 65% N_H	0.0135	0.0111
Multisine (1) 75% N_H	0.0141	0.0163
IRMLBS 85% N_H	0.0034	0.0032
Three-level periodic 58-70% N_H	0.0524	0.0684
Triangular wave 55-85% N_H	0.1327	0.1761
Triangular + IRMLBS 55-85% N_H	0.0873	0.1110

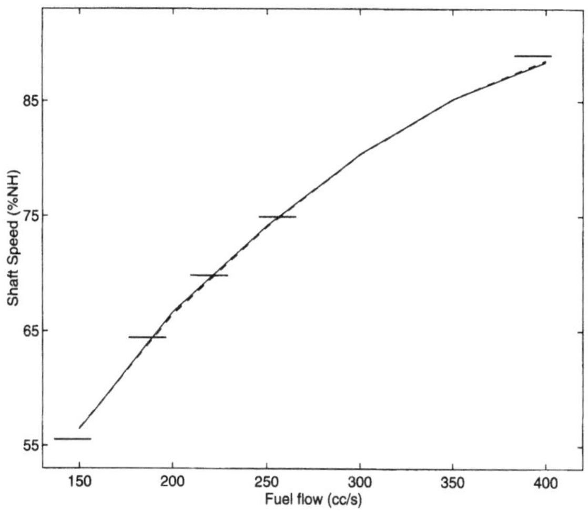

Figure 8.9. Static relationship between HP shaft speed and fuel flow (solid bars), NNARX model (solid), NNARMAX model (dashed)

8.6 Concluding Remarks

This chapter dealt with the application of modern nonlinear system identification techniques to model the dynamic relationship between the fuel flow and the shaft speed dynamics of a Rolls-Royce Spey engine. Several nonlinear model representations were presented and special attention was paid to polynomial NARMAX models and multilayer perceptron neural network models. A concatenated data set of small signal IRMLBS tests covering an operating range of 55-85% N_H was used to estimate polynomial and neural network models for the engine. The performance of the estimated models was demonstrated on a range of engine tests, and it was seen that nonlinear models are essential to model the engine dynamics in a global sense. It was also clear that the linear models estimated in Chapter 6 are only valid for a restricted operating range since they can only capture the local dynamics of the engine.

PART III. NEW PERSPECTIVES IN MODELLING, IDENTIFICATION, CONDITION MONITORING AND CONTROL

Chapter 9
Nonlinear Model Structure Selection Using Evolutionary Optimisation Methods

9.1 Introduction

In previous chapters, it was shown how linear and nonlinear models of a gas turbine engine could be obtained using identification methods. *A priori* knowledge of the engine dynamics was used to estimate engine model parameters when the model structure is selected.

In this chapter this approach is extended in two ways:

- an evolutionary computing approach is used to aid selection of an appropriate structure for gas turbine engine model, and
- these models are assessed against a set of objectives, using a multiobjective optimisation approach.

These extensions are realised through genetic programming in a multiobjective fashion to generate global models, which describe the nonlinear dynamic behaviour of the gas turbine engine.

The introduction of the validation stage of system identification into the multiobjective tool is also explored in this chapter, in order to direct the identification process to a set of global models of the system.

9.2 Genetic Algorithms and Genetic Programming

Evolutionary algorithms (EAs) use computational models of evolutionary processes in the design and implementation of computer-based problem solving. A general definition and classification of these evolutionary techniques is given in Bäck [1]. Bäck defines an EA as a search and optimisation algorithm, inspired by the process of natural evolution, which maintains a population of structures that evolve according to rules of selection and other operators such as recombination and mutation.

Three main evolutionary algorithms methods exist: evolution strategies [2], evolutionary programming [3] and genetic algorithms [4]. Genetic programming

(GP) introduced by Koza [5] is a branch of GAs. In this chapter, genetic algorithms and genetic programming will be described in detail.

9.2.1 Genetic Algorithms

Genetic algorithms (GAs) are probably the most popular of the evolutionary algorithms; the general structure of a GA is described in Figure 9.1. The simple GA consists of a population of *fixed-length* individuals, each one being a candidate solution. Two different representation levels are considered for each individual: *genotypic* and *phenotypic*. An individual representation, commonly binary (such as is depicted in Figure 9.2), expresses its *genotype*, while the value of each individual, in the problem domain over which the objective function is defined, is denoted as the *phenotype*.

```
procedure GA {
  initialise(Chrom);
  while not finished do {
    Cost = objv_fun(decode(Chrom));
    Sel = select(Chrom, Cost);
    Sel = recombine(Sel);
    Chrom = mutate(Sel);
    Gen = Gen + 1;
  }
}
```

Figure 9.1. A genetic algorithm

A simple GA works by creating a random initial population of fixed-length chromosomes, Chrom (see Figure 9.1). Each iteration (generation), the population evolves by means of the use of selection, recombination and mutation, which are the main genetic operators in GAs. Individuals are chosen, based on their fitness measure, to act as parents of offspring which will constitute the new generation.

Once a population has been produced it may be evaluated using an objective function or functions that characterise an individual's performance in the problem domain. Where the encoding of chromosomes uses a mapping from the decision variables to some other alphabet, *e.g.*, real-values encoded as binary strings, it will be necessary to decode the chromosomes before the objective function may be evaluated and a cost vector, Cost, assigned to the population. The objective function(s) is also used as the basis for selection and determines how well an individual performs in its environment. A fitness value is then derived from the raw performance measure given by the objective function(s) and is used to bias the selection process towards promising areas of the search space. Highly fit individuals will be assigned a higher probability of being selected for reproduction than individuals with a lower fitness value. Therefore, the average performance of individuals can be expected to increase as the fitter individuals are more likely to be selected for reproduction and the lower fitness individuals get discarded. Note that individuals may be selected more than once at any generation (iteration) of the EA and that the temporary vector of selected individuals, Sel, may therefore contain more than one copy of any individual in the original population.

Selected individuals are then reproduced, usually in pairs, through the application of genetic operators and these new individuals may then overwrite their parents in the vector, Sel. These operators are applied to pairs of individuals with a given probability and result in new offspring that contain material exchanged from their parents. The offspring from recombination are then further perturbed by mutation. These new individuals then make up the next generation, Chrom. These processes of evaluation, selection, recombination and mutation are then repeated until some termination.

9.2.2 Genetic Algorithm Operators

The ***selection*** process may be implemented in a number of ways [6]. Its simplest form has been described by Goldberg [7] as the *roulette wheel* where each current individual in the population has a roulette wheel slot sized in proportion to its fitness. The N_{sel} individuals are chosen by N_{sel} trials where the probability of each individual being selected in each trial remains constant.

Recombination (or crossover) is the main operator in genetic algorithms, and is commonly included with a high probability. There are numerous ways to implement recombination at both genotypic-level and phenotypic-level (real-value chromosome representation). The crossover operator illustrated in Figure 9.2 is the simplest one and is known as "single-point" crossover, where two parent individuals exchange a portion of their chromosomes to produce two offspring.

Mutation works by altering the binary information by randomly selecting positions as shown in Figure 9.2. In this case of binary representation, mutation swaps 0's with 1's, and vice versa. As with crossover (or recombination), different implementations of this operator have been described.

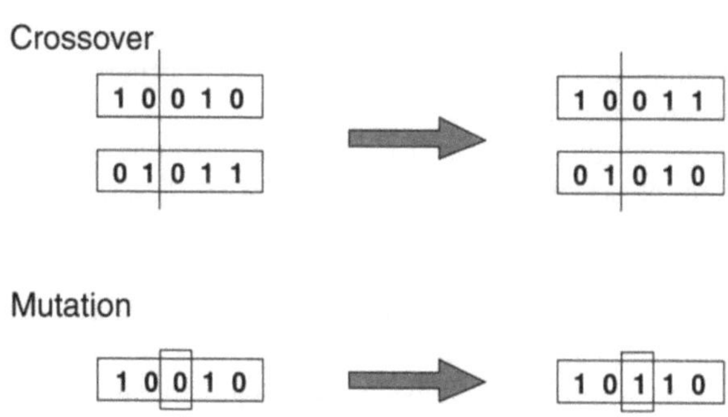

Figure 9.2. Single-point crossover and mutation for binary representation

Since Holland's original proposal, the application domain has been extended. It has motivated the development of alternative representations and new genetic operators allowing the GA to be used in a variety of problem domains. In spite of the fact that binary encoding has been shown to be closer to natural evolution and the genetic representation of living organisms, there has emerged the need for more complex and powerful representations of the GA.

Searching for alternative individual encodings has given rise to the concept of genetic programming [5]. GP is an extension of GAs that uses a more flexible representation, allowing dynamic encoding. This novel paradigm is introduced in more detail in the next section.

The population-based nature of GAs (and GP) enables multiobjective optimisation to be addressed in a satisfactory way. We will use this later to assess our models against a variety of criteria.

9.2.3 Genetic Programming

The adaptive search algorithm called genetic programming (GP) was proposed by Koza [5]. GP is an evolution-based search model that is a subclass of genetic algorithms. Koza introduced a more complex representation based on *computer programs*. Although finding algorithms or programs is more difficult than finding a single solution, it is more useful since generalised solutions work for an entire class of tasks. The fact that many problems from a wide range of domains can be expressed as the discovery of computer programming problems makes this field an interesting area for study. We will demonstrate below that this approach enables us to search effectively over potential structures for the polynomial representation of dynamic gas turbine engine models.

From this point, GP can be defined as a GA designed to evolve populations of hierarchically structured computer programs according to their performance on a previously specified fitness criterion. The main difference between GP and its predecessor GA is the fact that GP genotypes, or individuals, are programs which are not fixed in length or size. The maximum depth height of the parse tree of the program is specified *a priori* to constrain the search space but all solutions up to and including this maximum are considered. (Thus when we investigate model structures for the Spey engine, unlike the approach adopted in Chapter 8 we are not limited to quadratic polynomial model structures.) When genetic operators operate over the population of computer programs, the new genotypes differ from their parents in structure (size, shape and contents).

Each hierarchical genotype consists, then, of functions that can be composed recursively from the set of N_F functions from $F = \{f_1, f_2,..., f_{NF}\}$, and the set of N_T terminals from $T = \{a_1, a_2,..., a_{NT}\}$. The **function set** can consist of any arithmetic, Boolean, mathematical, or any other more complex functions (routines). The **terminal set** basically contains variables or constant values. To illustrate the hierarchical encoding used for GP, Figure 9.3 gives a simple example where the operations +, –, *, % (protected division), abs, log, exp, sin and cos belong to the function set and the variables A, B and C constitute the terminal set.

It is seen from Figure 9.3 that the number of arguments (*arity*) taken for each of the functions {+, –, *, %, abs, log, exp, sin, cos} are {2, 2, 2, 2, 1, 1, 1, 1, 1}, respectively. Consistency in this respect must be applied in order to create valid hierarchical structures and to select valid substructures for crossing over and mutating.

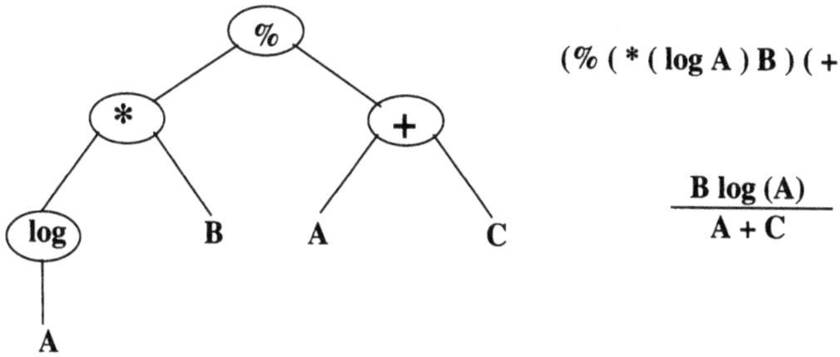

Figure 9.3. A mathematical expression represented hierarchically by its parse tree

The GP population then consists of hierarchical structures, each of them created by randomly generating a rooted, point-labelled tree with ordered branches. The root of any structure has to be an element from the function set (an operation) in order to generate hierarchical individuals avoiding structures consisting of a single terminal. When a terminal is chosen to be the label for any point in the tree, this node becomes an end-point and the generating process is terminated for that point.

This generating tree process can be implemented in different ways. Koza [5] defines two alternatives: the "*full*" method and the "*grow*" method. The grow method is used in this work, where the initial population consists of structures that have varying lengths. The length of a path between the root and the endpoint can vary up to the specified maximum depth.

9.2.4 Genetic Programming Operators

As for the conventional GA, selection and crossover are considered the main genetic operators, mutation being a secondary operator.

Selection in GP works in a similar way to that in a GA, being one of the foundations of the survival of the fittest. It is an asexual operator that selects an individual structure according to some selection method based on the fitness measures. The selected individual is then copied without any alteration to the new population.

One of the main differences between GP and the traditional implementation of GA is the fact that ***GP crossover*** does not preserve any kind of context in the chromosome. This is due to the fact that the standard crossover defined by Koza [5] exchanges subtrees which are chosen at random in both parents. Koza has pointed out that random subtree crossover maintains diversity in the population because crossing two identical structures, generally, will create different offspring. This is because the crossover points are, in general, different in the two parents.

Crossover, a sexual operator, works by first selecting a pair of structures from the current population. Then, a node rooted from each parent is randomly selected. These nodes become the roots for the substructures lying below the crossover point. In the next step, the substructures are exchanged between the parents producing two new structures which are usually of different sizes to their parents. Figure 9.4 illustrates the crossover operation over a function set and terminal set defined as for Figure 9.3. Note that for GP-crossover, the crossover point can be either a terminal or an internal point. In this example, the crossover point in both parents are internal nodes. This means that function nodes are chosen as roots for the substructures to be exchanged. Based upon the parent tree structures of Figure 9.4, a second case shows the crossover process where a terminal node and an internal node, as the root of the substructure, are chosen in the first and second parents, respectively (see Figure 9.4). When an internal node is selected, the number of arguments taken by the associated function must be considered in order to exchange a valid substructure.

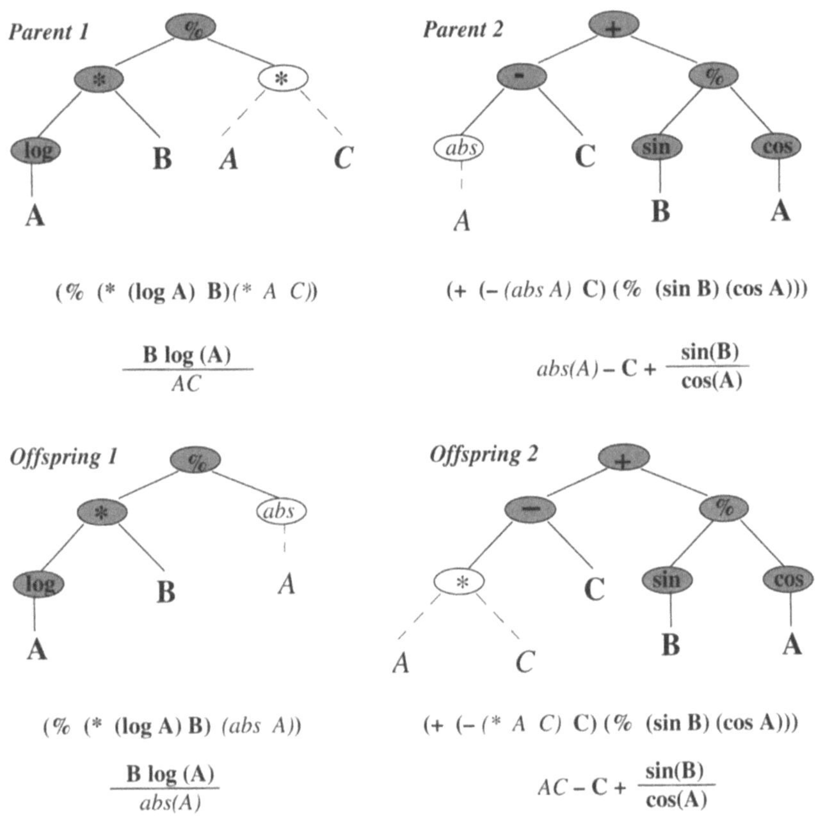

Figure 9.4. GP-crossover operation

Mutation is considered a secondary operator. It operates by randomly selecting a node, which can be either a terminal or internal point, and replacing the associated substructure with a randomly generated subtree up to a maximum size. In a conventional GA, the mutation operator introduces a certain degree of diversity into the population which is being beneficial. In contrast, the GP-crossover operation is the mechanism for diversification in the GP population. This fact is the justification given by Koza [5] for using a 0% mutation probability. Hence, convergence of the population is unlikely in genetic programming.

9.3 Multiobjective Optimisation

Consider the following multiobjective optimisation (MO) design problem:

$$\begin{aligned} &\min \mathbf{F}(\mathbf{p}) \\ &\mathbf{p} \in \Omega \end{aligned} \tag{9.1}$$

where $\mathbf{p} = [p_1, p_2, \ldots, p_q]$, Ω defines the set of q free variables, $\mathbf{p}$, subject to any constraints and $\mathbf{F}(\mathbf{p}) = [f_1(\mathbf{p}), f_2(\mathbf{p}), \ldots, f_n(\mathbf{p})]$ are the design objectives to be minimised.

Clearly, for this set of functions, $\mathbf{F}(\mathbf{p})$, it can be seen that there is no one ideal "optimal" solution, rather a set of Pareto-optimal solutions for which an improvement in one of the design objectives will lead to a degradation in one or more of the remaining objectives. In Figure 9.5 there are two objectives, f_1 and f_2, to be simultaneously minimised. These objectives are competing with one another such that there is no single solution. Candidate solution point A has a lower value of f_2, but a higher value of f_1, than candidate solution point B. Thus, it is not possible to state that one point on the trade-off curve shown in Figure 9.5 is better or worse than another. Such solutions are known as Pareto-optimal solutions (alternatively as *noninferior* or *nondominated* solutions) to the multiobjective optimisation problem.

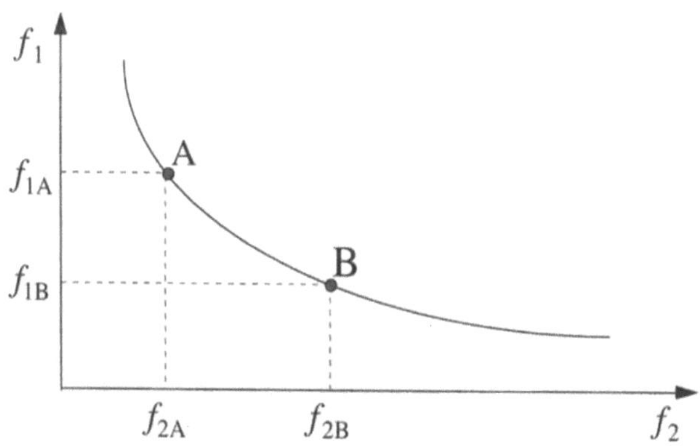

Figure 9.5. Pareto-optimal set of solutions for 2-objective problem

Hitherto, members of the Pareto-optimal solution set have been sought through solution of an appropriately formulated nonlinear programming (NP) problem. A number of approaches are currently employed including the ε-constraint, weighted-sum and goal attainment methods [8]. However, such approaches require precise expression of a usually not well understood set of weights and goals. If the trade-off surface between the design objectives is to be better understood, repeated application of such methods is necessary. In addition, NP methods cannot handle multimodality and discontinuities in function space well and can thus only be expected to produce local solutions.

The population-based nature of genetic algorithms (GAs) enables the evolution of a Pareto-optimal set of solutions and Pareto-ranking multiobjective genetic algorithm approach is described in the next Section.

9.4 Pareto-ranking Method

The multiobjective genetic algorithm approach proposed by Fonseca and Fleming [9] uses a rank-based fitness assignment, where the rank of a certain individual x_i at generation t is related to the number of individuals $p_i(t)$ in the current population by which it is dominated. This is given by:

$$\text{rank}(x_i, t) = p_i(t) \tag{9.2}$$

All nondominated individuals are assigned rank 0 and remaining individuals are penalised according to Equation (9.2).

Fitness is assigned by interpolating from the best individual (rank = 0) to the worst, and then the fitness assigned to individuals with the same rank is averaged where the global population fitness is kept constant. However, such fitness assignment tends to produce premature convergence due to the fact that all nondominated (best rank) points are considered equally fit (Figure 9.6).

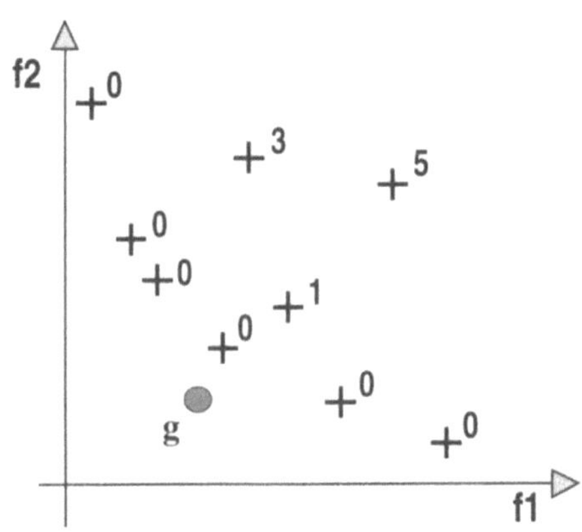

Figure 9.6. Pareto-ranking method

In order to overcome this deficiency, Fonseca and Fleming have used a niche induction method to promote the distribution of the population over the Pareto-optimal front in order to maintain diversity. This is achieved by a method of fitness sharing which encourages the reproduction of isolated individuals and favours diversification.

Preference information can also be introduced in the form of a goal vector, **g**, which provides a means of evolving only a specific region of the search space. This allows the decisionmaker to focus on a region of the Pareto front by providing external information to the selection algorithm.

In many practical applications, design problems may be represented by a set of competing requirements (or objectives), which can be formulated as a multiobjective optimisation problem. There is seldom a unique solution to such problems since objectives may be in conflict with one another. Rather, a set of solutions emerges where, in the absence of information concerning the importance of each objective relative to the others, each solution is deemed equivalent to the others in the set. Thus, the concept of multiobjective optimisation is defined as the problem of finding the vector of decision variables $\vec{x}$ that optimises the n components vector function:

$$\vec{f}(\vec{x}) = \left(f_1(\vec{x}),\dots,f_n(\vec{x})\right) \tag{9.3}$$

producing a set of equally efficient alternative solutions known as the nondominated or Pareto-optimal set .

9.5 Example of Engine Model Structure Selection Using Genetic Programming

In this section, multiobjective genetic programming (MOGP) is applied to sets of actual input–output observation data from a gas turbine engine system. This case study utilises real engine data records, from a Rolls-Royce Spey engine, collected by the Defence Evaluation and Research Agency, Pyestock, U.K.

Most previous research on system identification of aircraft engines has been based on *linear* frequency- and time-domain identification and has not been directly applicable to *nonlinear* modelling. Some approaches which deal with nonlinear modelling are based upon the identification of local linear models. The basis of these approaches is a decomposition of the system's full range of operation into a number of possible overlapping operating regions. In each operating region, a simple local model is applied. These local models are then combined in some way to yield a global model. However, an important question arises: How many local models are required to cover the operating range of the system? It is clear, therefore, that there is a compromise to be made between the number and size of local models and their complexity.

The identification approach presented in this paper deals with the identification of *nonlinear* models that can cope with the entire range of working conditions. This global model is then used to model the relationship between the fuel flow consumption and the spool speed of a gas turbine engine. The identification is based upon a set of multisine excitation signals [10] at different operating conditions, applied to the engine system which is described in the next section.

9.5.1. Description of the Engine System

A schematic of the measurement system is shown in Fig 6.1 The reheat system is inoperative during the tests and the compressor bleed valve is closed. The angle of the inlet guide vanes and the reheat nozzle area are fixed at their low speed positions for the duration of the tests. The engine speed control is operated in open-loop and a perturbed fuel demand signal fed to the fuel feed system, which regulates the fuel flow to the engine by means of a stepper valve.

The fuel feed system exhibits both linear and nonlinear dynamics, which affect the actual fuel flow applied to the gas turbine. It is important to eliminate these effects from the estimated engine models, and this is achieved by measuring the actual fuel flow downstream of the fuel feed, using a turbine fuel meter. The speed of the low-pressure (LP) shaft is measured by counting the rotations of the turbine blades and the speed of the high-pressure (HP) shaft by measuring the rotation of a gear linked to the shaft itself.

The shaft speeds are the primary outputs of a gas turbine, from which the internal engine pressures and thrust can be calculated. The dynamic relationship between these shaft speeds and the measured fuel flow is the purpose of this model identification example.

9.5.2. Result Analysis

Using the MOGP identification approach we are able to include a set of objectives which assess the performance of a candidate model at individual points over the operating range of the engine. A series of MOGP approaches is detailed below.

The initial approach did not take into account the validation criteria, dealing only with model complexity and performance. These were considered as "soft" objectives (that is, not "hard" constraints) and had the same level of priority. With this approach only linear models emerged to model the relationship between fuel flow demand and the HP shaft speed. These results are graphically represented in Figure 9.7, where the eleven objectives are ranged along the *x-axis* and the performance achieved for each objective is indicated in the *y direction*. In this "parallel coordinates representation," each line in the graph represents a potential solution to the design problem, indicating the achieved objective values for that solution. All solutions are both nondominant and satisfy the prescribed *goals* as represented by the "×" marks. Five models emerged which exhibited similar performance in terms of residual and long-term prediction error measures. These models (indicated by individual lines) satisfy all the objectives and belong to the ARX model (linear) expressed by

$$y(k)=\sum_{i=1}^{n_y} a_i y(k-i)+\sum_{i=1}^{n_u} b_i u(k-i)+c \tag{9.4}$$

where $n_y = n_u = 2$ (maximum lags), and c is a constant value.

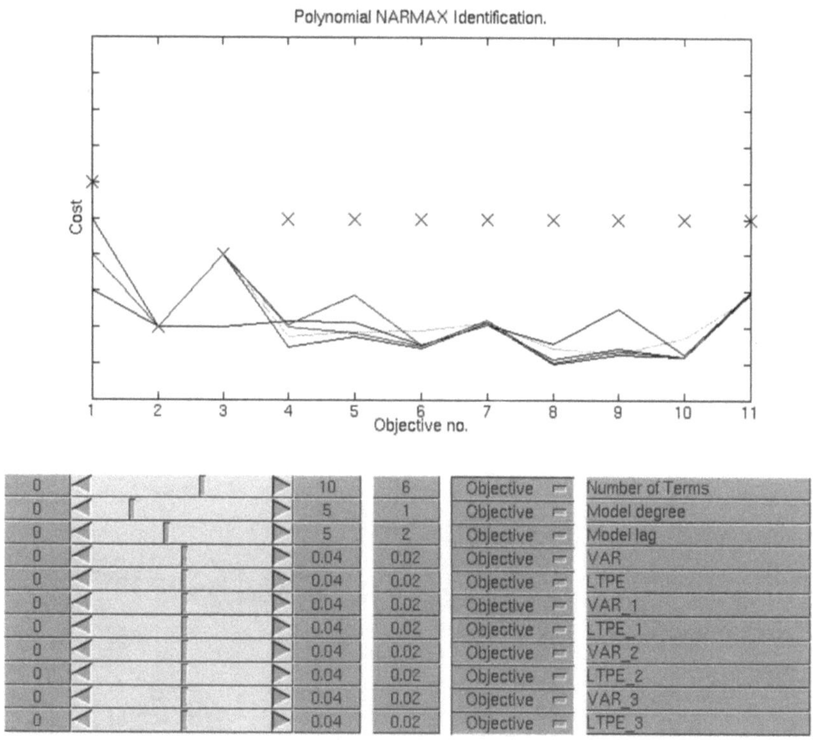

Figure 9.7. Multiobjective genetic programming framework, multisine input signals (VAR: residual variance; LTPE: long-term prediction error)

Table 9.1. Fuel flow-HP shaft speed linear model structures

Model	a_1	a_2	b_1	b_2	c
1	▲		▲		▲
2		▲	▲	▲	▲
3	▲		▲	▲	▲
4	▲	▲	▲		▲
5	▲	▲	▲	▲	▲

The structure of each model is specified in Table 9.1. Note that while the MOGP process obtained only linear models in the nondominant set, the formulation permitted nonlinear model descriptions.

Linear Validation

Since linear engine models were obtained from the previous experiment, the first step in model validation is to test the associated correlation functions. Therefore, in a second approach to the identification of this system, a validation stage was included, based on the evaluation of the autocorrelation of the residuals (*ACF*), the crosscorrelation between the residuals and input (*CCF*). This functions are given as

$$\Phi_{\varepsilon\varepsilon}(\tau) = E[\varepsilon(k-\tau)\varepsilon(k)] = \delta(\tau)$$
$$\Phi_{u\varepsilon}(\tau) = E[u(k-\tau)\varepsilon(k)] = 0 \quad \forall\tau \tag{9.5}$$

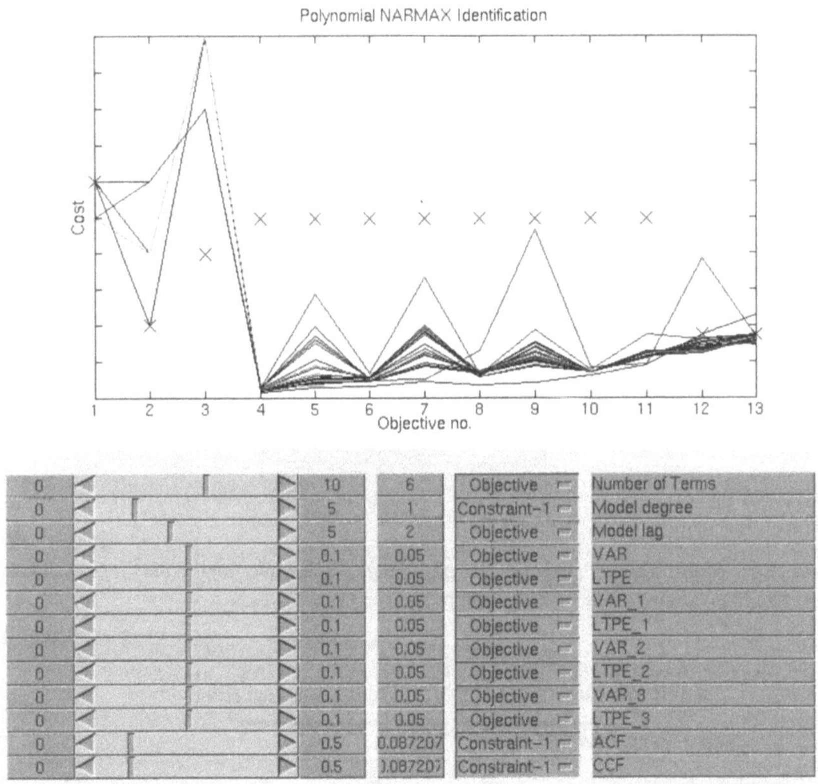

Figure 9.8. Model complexity, performance and linear statistical validation

The correlation objective functions were cast as constraints. The target value to be attained was given by the 95% confidence limit. Scalar measures of correlation were selected to be

$$\begin{aligned} CCF &= \max\{|\Phi_{u\varepsilon}(\tau)|\} \\ ACF &= \max\{|\Phi_{\varepsilon\varepsilon}(\tau)|\} \end{aligned} \tag{9.6}$$

where $\Phi_{\varepsilon\varepsilon}(\tau)$ is set equal to 0, for $\tau = 0$, otherwise ACF would always be δ.

The identification then sought valid models in the space of all possibilities. In this experiment, this method provided a set of nonlinear, instead of linear, models which satisfy the validation requirements. Note that in Fig 9.8 there are two linear models but these do not meet the constraints related to the validation criteria, Objectives 12 and 13. The optimisation produced these linear models because Objective 2 (see Table 9.2) was assigned the same level of priority as the correlation functions. Note that models with a nonlinear degree of two and three possess a few more terms than the linear models generated in the previous approach (see Objective 1).

Table 9.2. Description of the objectives considered in the MOGP-identification procedure

Attribute	Objective	Description
Model complexity	Model size (1)	Number of process and noise terms
	Model degree (2)	Maximum order term
	Model lag (3)	Maximum lagged input, output and noise terms
Model performance	Residual variance (4,6,8,10)	Variance of the predictive error between the one step ahead prediction and the measured outputs
	Long-term prediction error (5,7,9,11)	Variance of the LTPE
Model validation	ACF (12) CCF (13)	Autocorrelation, crosscorrelation functions

Validation of Nonlinear Engine Models

Although the ACF and CCF constraints were satisfied, these two conditions are not sufficient in order to provide unbiased nonlinear models. For the purpose of nonlinear system validation, high-order correlation tests [11] must also be evaluated, and these are described by:

$$\begin{aligned}
&\Phi_{\varepsilon(\varepsilon u)}(\tau) = E[\,\varepsilon(k)\,\varepsilon(k-\tau)\,u(k-\tau)\,] = 0 \quad \tau \geq 0 \\
&\Phi_{u^2\varepsilon}(\tau) = E[\{u^2(k) - E[u^2(k)]\,\}\varepsilon(k-\tau)] = 0 \quad \forall\tau \\
&\Phi_{u^2\varepsilon^2}(\tau) = E[\{u^2(k) - E[u^2(k)]\}\varepsilon^2(k-\tau)] = 0 \quad \forall\tau
\end{aligned} \tag{9.7}$$

The objectives considered for the next application of MOGP now relate to model structure, performance and *nonlinear* validation aspects (as summarised in Table 9.2). This objective function vector is defined as:

$$F = [NT, DEG, LAG, VAR_i, LTPE_i, ACF, CCF, HOC_j] \tag{9.8}$$

where *NT* is the number of model terms, *DEG* is the degree of nonlinearity, *LAG* is the maximum model lag, VAR_i and $LTPE_i$ correspond to the residuals variance and long-term prediction error, respectively, and *i* identifies the test signal used.

The validation stage was based on the evaluation of the autocorrelation of the residuals (*ACF*), the crosscorrelation between the residuals and input (*CCF*), and higher-order correlation functions (HOC_j). The HOC_j functions, as defined in Equation (9.7), are used to determine whether the correct nonlinear terms are detected in the model.

By evaluating the higher-order correlation tests, the number of nondominated (or preferred) models tended to be more selective because of the number of restrictions (constraints) that had to be satisfied. The final set of valid nonlinear models generated by means of this new approach is described in Table 9.3. As in the previous approach, Objective 2 (model degree) and the higher-order correlation functions were considered as "hard" objectives, *i.e.*, constraints.

Table 9.3. Quadratic polynomial models

Term\model	1	2	3
C	▲	▲	▲
$y(k-1)$	▲	▲	▲
$y(k-2)$	▲	▲	▲
$u(k-1)$			▲
$u(k-2)$			▲
$y(k-1)^2$			▲
$y(k-2)^2$	▲		
$y(k-1)u(k-1)$	▲	▲	
$y(k-2)u(k-2)$	▲	▲	

As can be observed from these results, even though the terminal set $\mathcal{T}$ in the GP method included past values of the residuals ($\{e(k-1),\ldots, e(k-n_e)\}$), no NARMAX structures arose. Instead, only NARX model structures emerged.

Validation Using Global Simulation

In this section, the previous set of quadratic models is validated on different data sets from the ones used for identification. Ramp testing signals were used to exercise the nonlinear models over a wide operating range. Figures 9.9 to 9.11 show the measured and the long-term predictive outputs of models 1 to 3, respectively, over a triangular wave of period 100 s with amplitude such as to cause variations of the *NH* from 65% to 85%.

Models 1 and 3 exhibit a better response than model 2. Model 3 predicts better than model 1, which shows a small offset of *NH* (Figure 9.12). For the three-level periodic sequence (Figure 9.13), again Model 2 gives an inferior predictive response.

Based upon the nonlinear structure of model 3, three-level (Figure 9.13) and large transient (Figure 9.14) data were additionally used for modelling the engine relationship between the measured fuel flow and the high-pressure shaft speed. As a result of this study, this model is selected as the most appropriate; the structure and parameters of this quadratic NARX engine model are given in Equation (9.9):

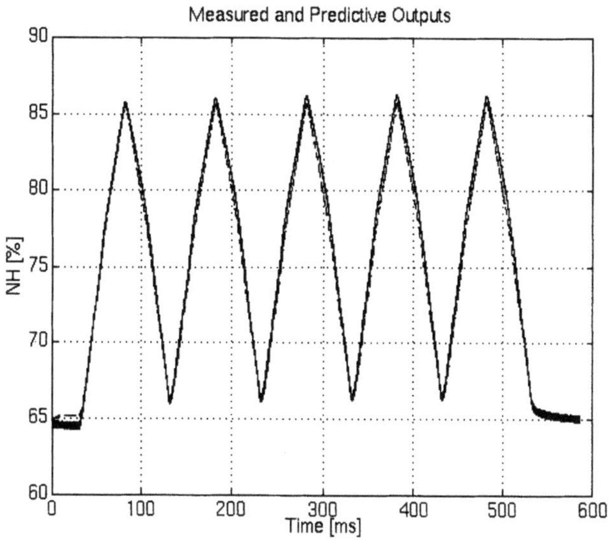

Figure 9.9. Predictive output of a triangular wave using model 1: solid line (measured), dashed line (predictive)

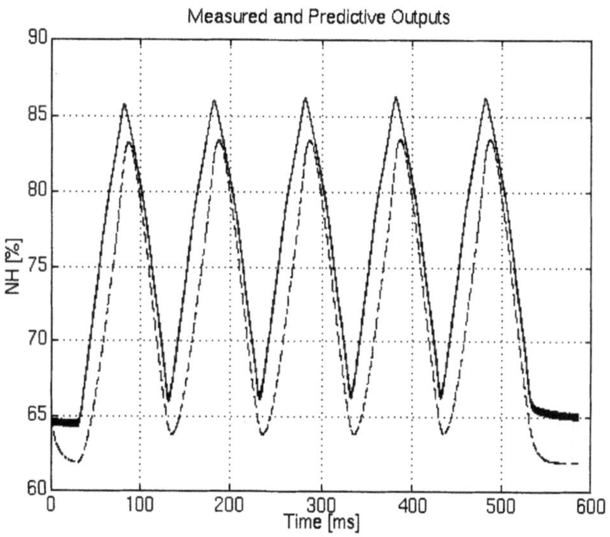

Figure 9.10. Predictive output of a triangular wave using model 2: solid line (measured), dashed line (predictive)

$$\begin{aligned} y(k) = & -1.16236 \cdot 10^{0} \quad + 7.96090 \cdot 10^{-1}\, y(k-1) \\ & + 2.39363 \cdot 10^{-1}\, y(k-2) + 4.33425 \cdot 10^{-3}\, u(k-1) \\ & - 5.90401 \cdot 10^{-4}\, u(k-2) - 4.37337 \cdot 10^{-4}\, y(k-1)^{2} \end{aligned} \tag{9.9}$$

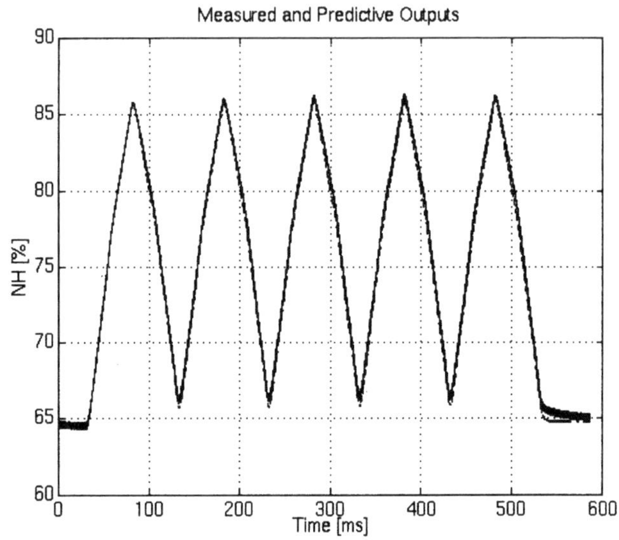

Figure 9.11. Predictive output of a triangular wave using model 3: solid line (measured), dashed line (predictive)

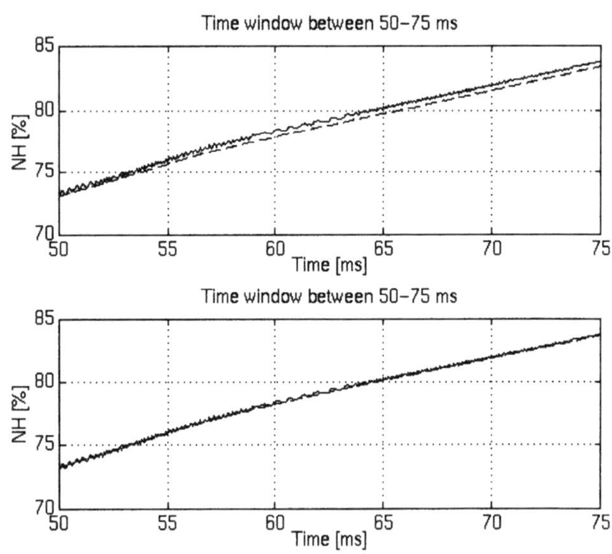

Figure 9.12. Predictive output of a triangular wave using model 1 (top) and model 3 (bottom): solid line (measured), dashed line (predictive)

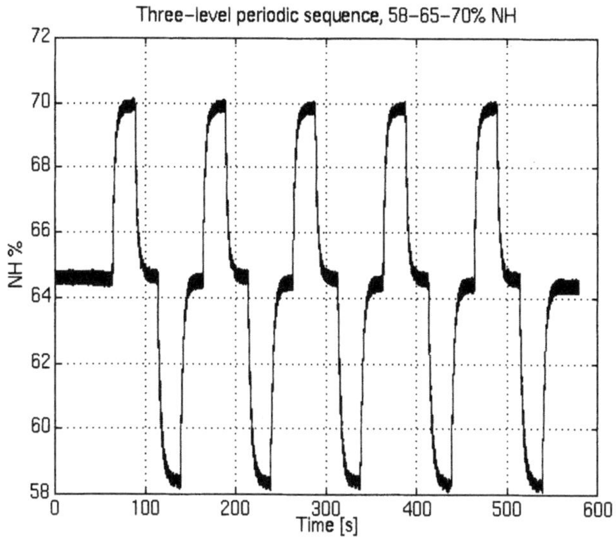

Figure 9.13. Three-level periodic sequences of period 100s, each period consisting of quarter-periods at levels 65%, 58%, 65% and 70% of the NH shaft speed: solid line (measured), dashed line (predictive)

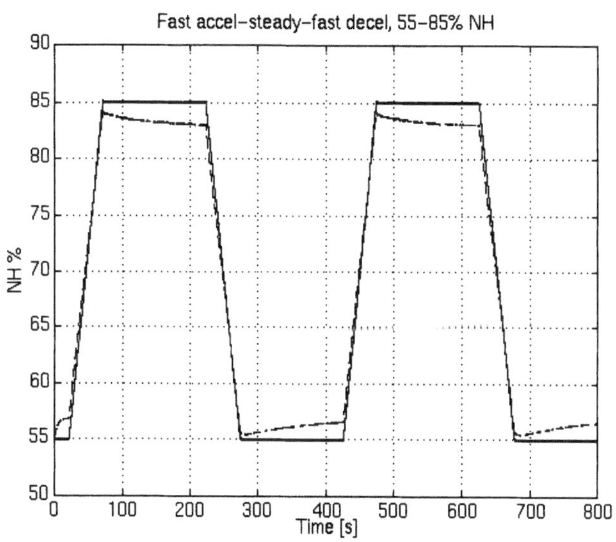

Figure 9.14. Fast acceleration/deceleration signal (55% to 85% *NH*): solid line (measured), dashed line (predictive)

9.6 Concluding Remarks

An evolutionary algorithm based upon the NARMAX representation has been introduced in this chapter as an alternative approach for nonlinear system identification problems. Genetic programming has proved to be a powerful tool for formulating and solving complex system identification problems, in particular for determining system structure.

Additionally, the incorporation of a multiobjective approach has enabled the separate consideration of different objectives related to model complexity and model performance. The validation process can be also included in this multiobjective framework.

In comparison with *conventional* identification techniques multiobjective evolutionary identification methods provide a family of candidate models which satisfy diverse objectives. This enables the modeller to select the appropriate model depending on specific circumstances.

In the search for a valid engine model, the multiobjective GP identification method has been demonstrated to perform effectively. The set-up of statistical validation criteria as "hard" constraints instead of being considered as "soft" objectives, directed the search process to reveal valid models.

Fitness function evaluation is time consuming when using this approach, in view of the multiple objectives involved. Compared with "conventional" methods, more time is spent running the identification method. While this is a shortcoming of this approach, it is offset by the ability of this evolutionary-based identification method to produce valid nonlinear models which confer a high level of credibility on the model. The ability to cope with different situations, as shown by the different test responses for the gas turbine engine example, demonstrates the flexibility of the approach.

Chapter 10
System Identification Using Frequency Response Techniques with Optimal Spectral Resolution

10.1 Introduction

This chapter describes an approach to the problem of dynamic model identification of aero engines using frequency response techniques. A problem with frequency response identification is obtaining an optimal spectral resolution. A visual-analysis-based optimization technique for identification is proposed providing a compromise between the bias and variance of the spectral estimate. The accuracy of the identification is provided by the optimal choice of spectral resolution using *a priori* information about the engine dynamic model.

Most modern aircraft engines are provided with a digital controller. This enables the measurements made for engine control to be also passed to a data-logging device. The data obtained from this source have been shown [1] to supply the information needed for identification.

Identification using experimental data represents an inverse problem, which is ill-conditioned [2-4]. The experimental estimates are not stable and precise for the following reasons:

- control processes in the engine are not strictly stationary and ergodic;
- the engine model is nonlinear;
- there is an influence from unknown random noise and feedback;
- linear and nonlinear model identification are performed separately;
- black-box approaches lead to unsatisfactory results.

Nonparametric identification of the frequency response function (FRF) by means of spectral analysis is usually considered well-investigated. Statistical methods for spectral estimation have been the object of study by Otnes and Enochson [5], Marple [6], Bendat and Piersol [7], and Ljung [4, 8]. However, for the present the choice of the identification parameters (record length, correlation window size, number of averages, *etc.*) is very subjective and weakly formed [9, 10]. In practice, trial and error is used. The choices made are largely empirical and require the experience of a skilled control engineer. Nevertheless, the optimal choice of the identification parameters can be carried out using *a priori* information about the system. This information can be obtained from thermodynamic models and previous results of data processing.

10.2 Problem Formulation

Given a closed-loop control as shown in Figure 10.1, let the plant (gas turbine engine) and feedback (controller) models G and F be obtained by the linearisation of the nonlinear thermodynamic model and be known *a priori* for an average system with some tolerance. The unobservable random disturbances $u(t)$ and $v(t)$ are applied to the measured outputs of the plant and feedback $x(t)$ and $y(t)$. Suppose the amplitude of random processes $x(t)$ and $y(t)$ makes it possible to neglect the influence of plant and control model nonlinearity. The objective is then plant model estimation using the input and output signal observations $x(t)$ and $y(t)$.

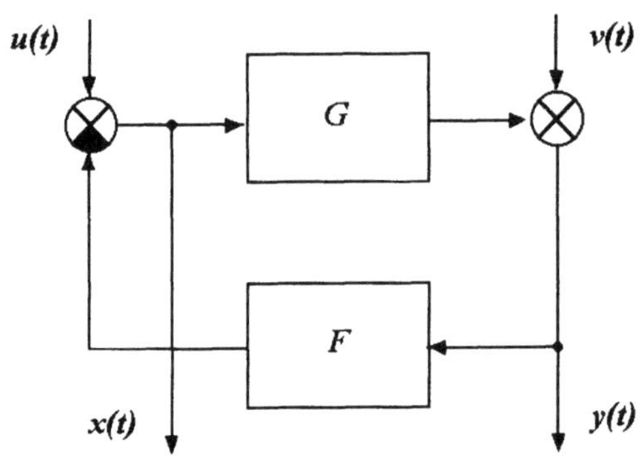

Figure 10.1. Closed-loop control system

In this chapter, a direct identification method is considered in the absence of feedback. The next chapter will explore the identifiability conditions accounting for the operation of the whole closed loop. A simple open-loop system is shown in Figure 10.2.

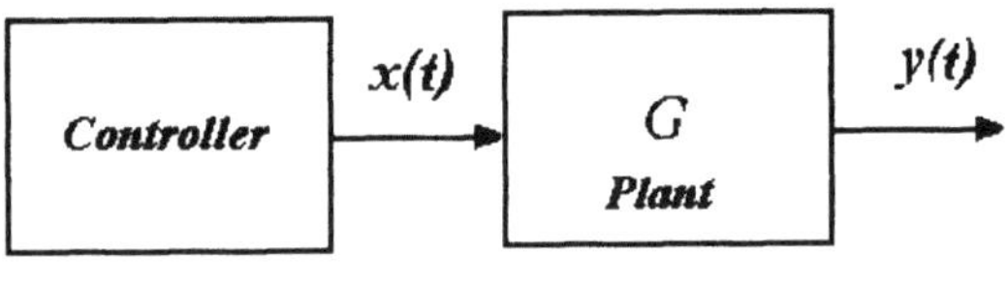

Figure 10.2. Open-loop system

10.3 Spectral Estimation

Recent work [1, 11, 12] performed the analysis of experimental data from a number of aero engine control systems. This showed that spectra of control and output signals (W_f, n_{HP}, n_{LP}, p_c^*, T_t^*, *etc.*) are often "rich" enough for identification of linear dynamic models using methods of correlation and spectral analysis.

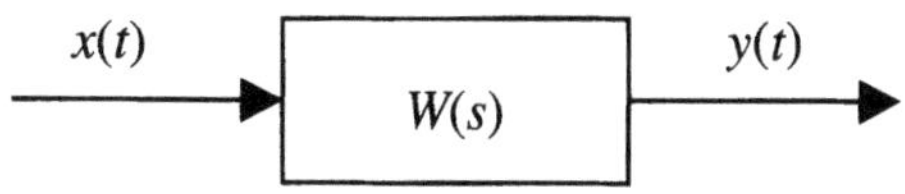

Figure 10.3. SISO linear system

These methods use known relationships, which can be demonstrated on the following example. Consider a linear system with the transfer function $W(s)$, input $u(t)$, and output $y(t)$ shown in Figure 10.3.

Note that the transfer function $W(s)$ represents an analytical function of a complex variable s. This relationship can also form a frequency response function (FRF) $W(\omega)$ for a set of frequencies ω FRF identification using spectral analysis consists of two stages. Firstly, the input, output, and cross spectra are estimated and then, secondly, FRF and coherence estimates are calculated. FRF is estimated as the ratio of the cross spectrum to the input spectrum:

$$\hat{H}(\omega) = \frac{\hat{S}_{yx}(\omega)}{\hat{S}_{xx}(\omega)} \tag{10.1}$$

There are three main nonparametric spectral estimation methods: band filtering, periodogram and correlogram [6, 7]. All these methods produce asymptotically equivalent results. Each method has advantages and disadvantages. Accordingly, the application areas of the methods for spectral analysis differ slightly.

The filtering method consists of band filtering of a signal, squaring and averaging:

$$S_{xx}(\omega) = 1/B \; \mathrm{E}[\, x^2(t,\omega,B) \,] \tag{10.2}$$

$$\mathrm{Re}[S_{yx}(\omega)] = 1/B \; \mathrm{E}[\, x(t,\omega,B \,)\, y(t,\omega,B \,)] \tag{10.3}$$

$$\mathrm{Im}[S_{yx}(\omega)] = 1/B \; \mathrm{E}[\, x(t,\omega,B \,)\, y_q(t,\omega,B \,)] \tag{10.4}$$

where E[.] is the operation of averaging;

$x(t,\omega,B)$ is the result of band filtering of the signal $x(t)$;

ω is the central frequency of the filter;

B is the pass band of the filter;

$y_q(t,\omega,B)$ is the signal $y(t,\omega,B)$ shifted by 90° phase.

The Fourier transform method (periodogram) consists of a Fourier transform of the initial realisation (data record) followed by averaging:

$$S_{xx}(\omega) = 1/T_s \, \mathrm{E}[\, |X(\omega)|^2 \,] \tag{10.5}$$

$$S_{yx}(\omega) = 1/T_s \, \mathrm{E}[\, Y(\omega)\, X^*(\omega) \,] \tag{10.6}$$

where T_s is the sampling interval;

$X(\omega)$ is a fast Fourier transform (FFT) of $x(t)$;

$X^*(\omega)$ is a complex conjugate of $X(\omega)$.

The Blackman–Tukey method (correlogram) consists of a Fourier transform of the appropriate correlation functions $R(t)$ multiplied by the window $U(t)$:

$$\begin{aligned} S_{xx}(\omega) &= \int_{-\tau_m}^{+\tau_m} R_{xx}(t)U(t)e^{-j\omega t}\,dt \\ S_{yy}(\omega) &= \int_{-\tau_m}^{+\tau_m} R_{yy}(t)U(t)e^{-j\omega t}\,dt \\ S_{yx}(\omega) &= \int_{-\tau_m}^{+\tau_m} R_{yx}(t)U(t)e^{-j\omega t}\,dt \end{aligned} \tag{10.7}$$

In this work, the *correlogram method* was chosen for processing experimental data for the following reasons. The filtering method requires development of digital filters for each frequency. The periodogram method does not allow calculation of spectra at any arbitrary frequency. The advantage of the correlogram method is that a Fourier transform is applied only once. Hence, the fast version of the transform (FFT) is not required. Spectral density can be estimated for any number of frequencies, enough for visual analysis of results. Moreover, correlation functions are calculated as an intermediate step, thus giving some additional information on the signal properties. Visual analysis of correlation functions allows estimation of the presence of a nonzero average, pure time delay, periodical properties of the signal, and the character of the signal as a whole (broadband or narrow-band noise). The specified properties are much more difficult to analyse from the plot of spectral density, because of the frequency-domain representation and because of the effect of spectral windowing (smoothing).

In investigating nonlinear dynamic systems by linear methods, the following approach is adopted. Dynamic processes in a nonlinear system at the steady-state conditions are considered a sum of static (mean value or the nearest point of a static model) and dynamic components (random deviation from the static point). If the amplitude of deviation is rather small, then the nonlinear properties of the plant can be neglected, and a linear model can describe the process with sufficient accuracy.

In this work, the term *correlation function* is used. However, in various sources this characteristic is referred to as *correlation* or *covariation* function. When analysing a centred random process these terms are equivalent. The term *correlation* function prevails in Russian books. It is used with reference to the mixed second-order central moment. This means centring of the random value, or subtracting the mean value [13-17]. Most western authors prefer to distinguish correlation and covariation functions. *Correlation* functions usually designate the initial moments, whereas *covariation* functions refer to centred moments, or characteristics of a stationary process with zero mean [4, 7, 18-20].

A possible source for the difference in opinions is from the viewpoint of random processes in a nonlinear system. When a nonzero-mean random signal goes through a linear system, both random and constant components of the signal follow the known relationship of the convolution integral with the weighting function. However, real physical objects are seldom completely linear. When considering nonlinear systems, a linear relation can be accepted only between dynamic components of the input and output processes. Constant components of signals are linked through the nonlinear static characteristic. Usually, there is no linear relation between mean values.

However, small changes of constant components (average values at the steady-state conditions) correspond to a linear model and hence determine the "direct current" factor of the transfer function, or DC gain. Thus, there is an alternative method to estimate the gain of a dynamic model. The use of a set of independent methods of identification of the same model allows increase in the results' accuracy and reliability in processing experimental data. In other terms, this approach can be referred to as *data fusion* or *model fusion*. In the subsequent sections this will be discussed in more detail.

10.4 Spectral Analysis Calculations

Using the correlogram method, the calculations are performed in the following sequence. Firstly, visual analysis of the input and output signals is done, and sharp mistakes and apparent problems with the data are handled. Once the process' stationarity is checked then a number of analysis techniques are applied. These are described in the following sections.

10.4.1 Mean Value

Before performing spectral analysis, a mean value should be subtracted from the data in order to obtain deviations from the static point. The mean is estimated as an average value over the record length:

$$\begin{aligned} &\mu_u = E[u(t)] \qquad 0 < t < T_r \\ &\Delta u(t) = u(t) - \mu_u \end{aligned} \tag{10.8}$$

where μ_u is the average;

T_r is the record length;

$\Delta u(t)$ is the deviation from the average.

In the text below, the deviation symbol Δ is omitted, and all further calculations deal with deviations, if not stated.

Some authors recommend using a narrow-band filter instead of removing the average. The removal of a linear trend is sometimes more preferable, depending on a particular situation. The main problem in the choice of centring the process is the researcher's ability to identify the valuable component in the data. The information useful for the linear dynamic model identification is contained in the cross-correlated components of the input and output signals. If the trend is related to the static model, then it should be excluded. The useful trend is part of a low-frequency component passing through the linear model.

If the average is not removed, then the results of spectral analysis are distorted. The correlation function's estimate appears to rise by the square of the signal mean. The spectral estimates will contain an impulse delta-function at zero frequency. Resulting from spectral smoothing, this impulse would deform the spectrum in the low-frequency range.

Discrete-time processes are usually represented by regular observations of the signals' values, sampled by time and magnitude. Consider observations of the input and output at the time moments iT_s as the following time series:

$$u(iT_s),\ y(iT_s),\ i = 1,\dots,N$$

where T_s is the sampling time;

$T = NT_s$ is the record length.

In the next sections the parameter's estimates are considered. Accounting for the time quantisation, the basic calculation formula is the following. The mean value estimate is:

$$\mu_x = \frac{1}{N}\sum_{i=1}^{N} x(iT_s) \tag{10.9}$$

10.4.2 Correlation Functions

The estimates of the (auto) correlation $R_{xx}(kT_s)$ and $R_{yy}(kT_s)$ and cross-correlation $R_{yx}(kT_s)$ functions are calculated using the following formulae:

$$R_{uu}(kT_s) = \frac{1}{N-k}\sum_{i=1}^{N-k} u(iT_s)u((i+k)T_s) \qquad k = -K,\dots,K \tag{10.10}$$

$$R_{yy}(kT_s) = \frac{1}{N-k} \sum_{i=1}^{N-k} y(iT_s)\, y((i+k)T_s) \qquad k = -K, \ldots, K \tag{10.11}$$

$$R_{yu}(kT_s) = \frac{1}{N-k} \sum_{i=1}^{N-k} u(iT_s)\, y((i+k)T_s) \qquad k = -K, \ldots, K \tag{10.12}$$

10.4.3 Spectral Density

In estimating the autocorrelation functions, only nonnegative delays are used, as these functions are symmetric by the axis of ordinates. Autospectral and cross-spectral power densities are found as Fourier transforms of the appropriate correlation functions. These functions are real, so their spectra are real too and have the property of symmetry. The phase of autospectral density is zero. The spectral properties are expressed below:

$$\begin{aligned}
&|S_{yx}(\omega)| = |S_{yx}(-\omega)| \\
&\angle S_{yx}(\omega) = -\angle S_{yx}(-\omega) \\
&S_{yx}(\omega) = S_{yx}{}^{*}(-\omega) \\
&S_{xx}(\omega) = S_{xx}(-\omega) \\
&S_{yy}(\omega) = S_{yy}(-\omega) \\
&\angle S_{xx}(\omega) = 0 \\
&\angle Syy(\omega) = 0
\end{aligned} \tag{10.13}$$

10.4.4 Spectral Windowing

Due to the specified property of the spectra, it is enough to calculate the appropriate spectral density only for positive frequencies [7]. Further, a single-sided spectral density is discussed, which is defined in the range of frequencies $\omega \geq 0$. In estimating the spectral density $S(\omega)$, a number of values of the appropriate correlation function $R(t)$ are used within the interval:

$$-\tau_m \leq t \leq \tau_m$$

In fact, the function $R(t)$ is nonzero outside the specified interval. Therefore, the correlation function estimate Equation (10.10) can be represented as an "infinite" correlation function $R_\infty(t)$ defined at the interval $-\infty < t < +\infty$ and multiplied by the rectangular window $u_{rec}(t)$. The index ∞ in $R_\infty(t)$ means a sampling estimate of the autocorrelation function from the infinite record ($T_r = \infty$):

$$R(t) = R_\infty(t)\, u_{rec}(t) \tag{10.14}$$

$$u_{\text{rec}}(t)=\begin{cases}0,\ t<-\tau_m\\ 1,\ -\tau_m\le t\le\tau_m\\ 0,\ t>\tau_m\end{cases} \tag{10.15}$$

A Fourier transform of the product in the time-domain is equivalent to a convolution of Fourier transforms in the frequency-domain: the spectrum of the "infinite" correlation function $S_\infty(\omega)$ and the spectral window $U_{\text{rec}}(\omega)$. The spectral window $U_{\text{rec}}(\omega)$ represents a Fourier transform of the correlation window $u_{\text{rec}}(t)$:

$$S(\omega)=S_\infty(\omega)*U_{\text{rec}}(\omega) \tag{10.16}$$
$$F[R(t)]=F[R_\infty(t)\times u_{\text{rec}}(t)]\ =F[R_\infty(t)]*F[u_{\text{rec}}(t)]$$

where $*$ is the symbol of convolution;

$F[.]$ is a Fourier transform.

$$U_{\text{rec}}(t)=2\tau_m\left(\frac{\sin\omega\tau_m}{\omega\tau_m}\right) \tag{10.17),}$$

$$\hat{S}_{xx}(\omega)=\int_{-\infty}^{\infty}S_{xx}(\omega)U(\omega-\xi)d\xi\approx\frac{1}{B}\int_{\omega-B/2}^{\omega+B/2}S_{xx}(\xi)d\xi$$

The function $U_{\text{rec}}(\omega)$ in Equation (10.17) has large side lobes. Energy of other harmonics can leak within the spectral estimate through the side lobes of the spectral window, and as a result the spectral estimate Equation (10.16) is biased. Moreover, the estimation bias can lead to the occurrence of negative spectrum estimates at some frequencies. To minimise the spectral leakage effect, a Hanning window will be used, which has much smaller level of side lobes and can be easily realised computationally. Both the rectangular and Hanning windows are represented in Figure 10.4, with the corresponding spectral windows.

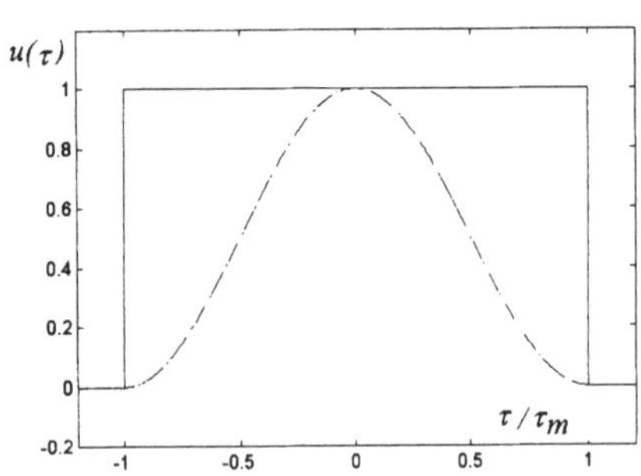

(a)

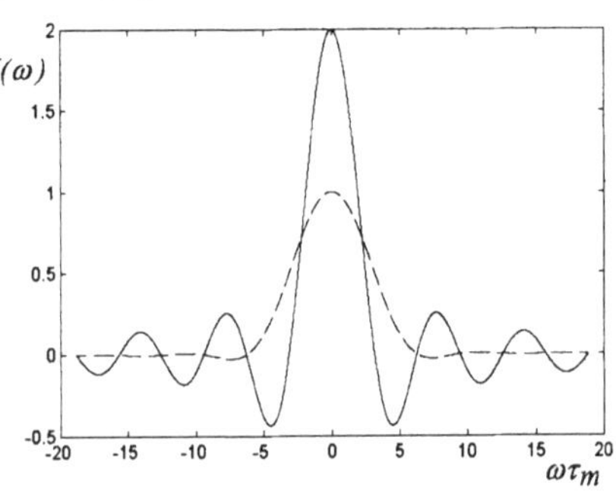

(b)

Figure 10.4. Rectangular (solid) and Hanning (dashed) windows in time- (a) and frequency-domains (b)

The shape of the Hanning window is defined by a biased cosine:

$$R_{xx}^{\text{han}}(\tau) = R_{\infty}(\tau)u_{\text{han}}(\tau) \tag{10.18}$$

$$u_{\text{han}}(\tau) = \begin{cases} 0.5(1+\cos(\pi\tau/\tau_m)), |\tau| \le \tau_m \\ 0, \ |\tau| > \tau_m \end{cases} \tag{10.19}$$

The "true" spectrum $S_{\infty}(\omega)$ of a signal cannot be measured precisely. Any method for spectral estimation inevitably leads to the occurrence of errors. Spectral bias (systematic error of estimation) is caused mainly by the effect of smoothing. Any spectral window has a finite size. A spectral estimate using a Hanning window Equation (10.19) represents a convolution of the "infinite" spectrum $S_{\infty}(\omega)$ with a spectral equivalent of a correlation window — the spectral window $U_{\text{han}}(\omega)$:

$$S_{xx}^{\text{han}}(\omega) = S_{xx}^{\infty}(\omega) * U_{\text{han}}(\omega) \tag{10.20}$$

$$U_{\text{han}}(\omega) = 0.5U_{\text{rec}}(\omega) + 0.25U_{\text{rec}}(\omega-\pi/\tau) + 0.25U_{\text{rec}}(\omega+\pi/\tau) \tag{10.21}$$

Power spectral density of the input is estimated by the formula

$$S_{xx}(j\omega) = 2T_s \sum_{k=-K}^{K} R_{xx}(kT_0)u_{\text{han}}(kT_0)e^{-j\omega kT_s} \qquad 0 \le \omega \le \pi/T_s \tag{10.22}$$

Only single-sided spectral densities are used, defined for nonnegative frequencies. Therefore, the formulae for spectral estimation are multiplied by 2. Then Parseval's theorem [7] is kept valid: the total amount of energy in the frequency-domain is equal to the total energy in the time-domain:

$$\int_{-\infty}^{+\infty} u^2(t)dt = \int_{-\infty}^{+\infty} |U(f)|^2 df \tag{10.23}$$

or in the discrete form:

$$T_s \sum_{i=1}^{N} u^2(iT_s) = \Delta f \sum_{k=0}^{K} |U(k\Delta f)|^2 \tag{10.24}$$

This corresponds to the average square of a process:

$$\frac{1}{N}\sum_{i=1}^{N} u^2(iT_s) = \Delta f \sum_{k=0}^{K} S_{uu}(k\Delta f) = \sigma^2 \tag{10.25}$$

Thus, the integral of the power spectral density is equal to the deviation of the process with a zero mean. Hence, power spectral density of the input is determined by the formula:

$$S_{yy}(j\omega) = 2T_s \sum_{k=-K}^{K} R_{yy}(kT_s) u_{\text{han}}(kT_s) e^{-j\omega k T_s} \qquad 0 \le \omega \le \pi/T_s \tag{10.26}$$

The cross-spectrum is estimated using the formula:

$$S_{yx}(j\omega) = 2T_s \sum_{k=-K}^{K} R_{yx}(kT_s) u_{\text{han}}(kT_s) e^{-j\omega k T_s} \qquad 0 \le \omega \le \pi/T_s \tag{10.27}$$

10.4.5 Frequency Response and Coherence

After the power spectral densities have been defined, frequency response is calculated as a complex-value function:

$$H(\omega) = \frac{S_{yx}(\omega)}{S_{xx}(\omega)} \tag{10.28}$$

which is usually represented in the form of magnitude and phase:

$$A(\omega) = |H(\omega)| \qquad \varphi(\omega) = \angle H(\omega) \tag{10.29}$$

The coherence function is estimated in the following way:

$$\gamma^2(\omega) = \frac{|S_{yu}(\omega)|^2}{S_{uu}(\omega) S_{yy}(\omega)} \qquad 0 \le \omega \le \pi/T_s \tag{10.30}$$

The calculation formulae deal with the angular frequency ω. This is determined by the fact that in visual analysis of the FRF time constants are easily determined graphically. The inverse abscissa of a bending point at the asymptotic logarithmic plot gives the time constant's estimate.

10.5 Errors of FRF Estimation by Spectral Methods

Mean square error of the spectral estimation ε^2 consists of a variance and bias square:

$$\varepsilon^2 = \varepsilon_b^2 + \varepsilon_r^2 \tag{10.31}$$

The finite record length T_r results in the variance ε_r^2. The bias ε_b is the effect of windowing.

10.5.1 Spectral Estimation Bias

Consider the analysis of the spectral estimation errors made by Bendat and Piersol [7]. The bias in spectral analysis can be estimated by the following method. Consider a power spectral density estimate $S_{xx}(\omega)$ obtained by the correlogram method. The index "xx" is omitted to simplify equations. The correlogram method for obtaining $S_{xx}(\omega)$ consists of a Fourier transform of a correlation function estimate $R(t)$, multiplied by a windowing function $u(t)$. The product in the time domain corresponds to the convolution of the true spectrum $S(\omega)$ and spectral window $U(\omega)$ in the frequency-domain:

$$\hat{S}(\omega) = \int_{-\infty}^{+\infty} \hat{R}(\tau)u(\tau)e^{-j\omega\tau}d\tau = \int_{-\infty}^{+\infty} S(\xi)U(\omega-\xi)d\xi \tag{10.32}$$

The spectral window $U(\omega)$ generally has a complicated form with a central "bell" and a number of side lobes. As a first approximation, a rectangular window can replace it. The window width B will represent an equivalent frequency bandwidth, which corresponds to the width of the spectral window $U(\omega)$ at the half-peak level:

$$U(\omega) = \begin{cases} 0, \ \omega < \omega_0 - B/2 \\ 1/B, \ \omega_0 - B/2 \le \omega \le \omega_0 + B/2 \\ 0, \ \omega > \omega_0 + B/2 \end{cases} \tag{10.33}$$

where ω_0 is the central frequency of the window.

Note that the spectral window is symmetric in relation to the central frequency ω_0:

$$|U(\omega_0 - \xi)| = |U(\omega_0 + \xi)| \tag{10.34}$$

Finally, the expression for the biased power spectral density estimate has the following form:

$$\hat{S}(\omega) = \int_{-\infty}^{+\infty} S(\xi)U(\omega-\xi)d\xi \approx \int_{\omega-B/2}^{\omega+B/2} S(\xi)d\xi \tag{10.35}$$

Thus, when calculating a spectrum at a frequency ω_0, an average of a true spectrum is actually estimated in a range of frequencies from $(\omega_0 - B/2)$ to

$(\omega_0 + B/2)$. Therefore, the application of a correlation window is referred to as averaging, smoothing, or windowing. Only an average spectrum can be estimated.

Expanding Equation (10.35) in a Taylor series about the point ω and rejecting the terms of higher than second degree, an estimate of bias can be obtained. In [7] the normalised bias error of the spectrum estimate is defined as the first approximation by the following expression:

$$\varepsilon_b[S(\omega)] \approx \frac{S^{(2)}(\omega)}{24S(\omega)\tau_m^2} \tag{10.36}$$

where ε_b is the normalised bias; $S^{(2)}(\omega)$ is the second derivative of spectral density by frequency (determining the curvature of the function); $1/\tau_m$ is the spectral resolution connected with the finite correlation window size τ_m.

The analysis of Equation (10.36) shows that the bias error rises with the degree of the spectrum's curvature. Thus, the bias results in smoothing the complex shape of the spectrum: maxima are underestimated, and minima are overestimated. Spectral bias is caused by the fact that the spectrum is not constant in the considered frequency range.

For a more complicated shape of spectral window, Equation (10.35) represents a first approximation producing qualitatively correct results. Note that exact measurement of power spectral density is physically impossible. This statement applies to any density, including power spectral density [5]. When measuring density, the quantity of substance, energy or power contained in a finite volume is divided by that volume. In the case of spectral estimation, power is divided by the frequency interval. Instead of point estimation, an average is measured for a chosen interval. One can see a parallel to measuring a derivative. A derivative estimation is a complicated ill-posed inverse problem, greatly influenced by noise.

An increase in stability of ill-posed problem resolution requires the application of regularisation methods. Regularisation employs additional *a priori* information. Regularisation in spectral analysis would require some additional information about spectral estimation properties. In this chapter, such information represents prior knowledge of the mechanism of bias resulting from smoothing.

When an *a priori* dynamic model is available for the system under investigation, a numerical simulation of the identification process can be performed. This would produce preliminary estimates of the necessary window parameters. The analysis of estimation errors in this chapter is greatly simplified. Actually, the operations of convolution are applied to complex-value functions. Nevertheless, the general trend of distortion of spectral estimates can be demonstrated based on that analysis.

10.5.2 Spectral Estimation Variance

Variance is the second component of the estimation error. Under certain reasonable conditions [5], spectral estimates have a distribution χ^2 with n degrees of freedom, where n is determined by the expression:

$$n = 2\,BT_r \tag{10.37}$$

where B is the effective spectral resolution (the width of the spectral window at the half-power level); T_r is the record length.

The χ^2 distribution for $n \geq 30$ can be approximated by a Gaussian distribution. In this case, the normalised standard deviation of the estimated parameter is defined as:

$$\varepsilon_r = \frac{\sqrt{\mathrm{var}[\hat{S}_{xx}(\omega)]}}{|S_{xx}(\omega)|} = \frac{1}{\sqrt{BT_r}} \tag{10.38}$$

Random deviation of an estimate is connected to the limited interval of observation (finite record length). The error of estimation of a correlation function $R_{xx}(\tau)$ grows with the lag increase, because a lesser amount of the data participates in the estimate. Therefore, the approach of the correlation window size τ_m to the record length T_r makes the spectral estimate $S_{xx}(\omega)$ statistically unstable because of accounting for unstable elements of $R_{xx}(\tau)$.

Spectral resolution of a Hanning window is:

$$B = 2\pi / \tau_m \tag{10.39}$$

Then the total normalised standard deviation as a first approximation can be represented as:

$$\varepsilon^2(S_{xx}(\omega)) = \varepsilon_r^2 + \varepsilon_b^2 \approx \frac{\tau_m}{T_r} + \left(\frac{S_{xx}^{(2)}}{24\,S_{xx}(\omega)}\right)^2 \frac{1}{\tau_m^4} \tag{10.40}$$

The choice of the window size is a compromise between the variance and bias of the estimation: a greater window τ_m (improved resolution) causes a decrease in bias and an increase in variance. Thus the form of a spectral curve becomes closer to the "true" spectrum; simultaneously, more false deviations appear.

A plot of the estimation error Equation (10.40) as a function of the window size is shown in Figure 10.5. One can see the existence of the optimal window size τ_{opt}, ensuring the minimal square error for a fixed frequency. Note that the optimum correlation window τ_{opt} can vary at each frequency.

In practice, the error ε^2 cannot be estimated directly from experimental data. In FRF identification from experimental data, the "true" model is unknown (or is known with some tolerance). Therefore, drawing an experimental diagram like Figure 10.5 and choosing the optimum window from it for any particular situation is not feasible.

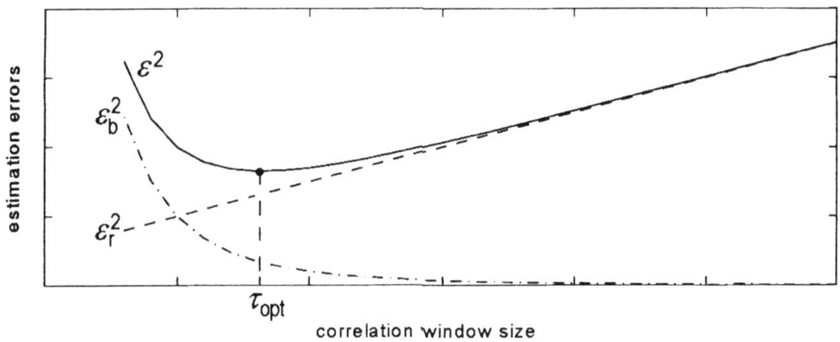

Figure 10.5. Spectral estimation errors: mean square error (solid), variance (dashed) and bias square (dash-dotted)

The choice of the window size in spectral analysis applications represents a complicated problem. Most authors like Marple [6] and Ljung and Glad [8] give a typical recommendation: "to try out a variety of different values of window size and find the balance between true details and random fluctuations."

A more exact criterion can be proposed using an *a priori* plant model. A thermodynamic model of the engine can be built based on physical laws. For instance, two-shaft engine dynamics at the steady-state conditions are described by a second-order model. Therefore, the approximate values of the coefficients are known before the experiment and the utilisation of this information can improve the identification results. A novel approach for the window size choice is discussed below bringing about a compromise between the bias and variance of the estimated model.

Let the plant input be white noise $S_{xx}(\omega) = 1,\ 0 < \omega < \infty$. Then the input spectrum estimate has no bias. In this case, the bias of the FRF estimate is determined by the cross-spectrum estimate bias caused by windowing. The first approximation of the correlation window in the frequency-domain is the rectangular window:

$$H(\omega) = S_{yx}(\omega) = \int_{-\infty}^{\infty} S_{yx}(\xi) U(\omega - \xi) d\xi \approx \frac{1}{B} \int_{\omega - B/2}^{\omega + B/2} S_{yx}(\xi) d\xi = \frac{1}{B} \int_{\omega - B/2}^{\omega + B/2} H(\xi) d\xi \quad (10.41)$$

The use of *a priori* information for the plant model allows the prediction of the biased FRF estimate as a function of the window size; see Figure 10.6. The amplitude estimate increases at the beginning. This behaviour is the result of bias. The shape of the experimental estimate is close to the prediction in this part of the plot. The choice represents the optimal window size, which ensures the minimal estimation error. The optimal window size τ_{opt} providing the minimal estimation error is chosen from the comparison of those graphs from experimental data and *a priori* prediction. Further increasing of the window size greater than the optimal value makes the estimate error grow because of the variance rise. The shape of the phase estimate plot is similar to that of amplitude. So, the optimum window can be found as the beginning of the saturation part of the curve.

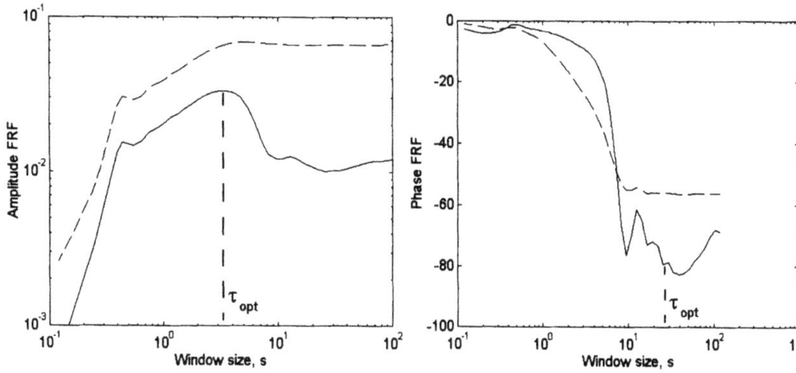

Figure 10.6. FRF amplitude (left) and phase (right) estimates: real data (solid) and *a priori* prediction (dashed)

The expression for the normalised standard deviation of a cross-spectrum estimate $S_{yu}(\omega)$ is similar to the formula for a power spectrum error. Additionally, the coherence function value is taken into account:

$$\varepsilon_r^2\left[\hat{S}_{yu}(\omega)\right] = \frac{\tau_m}{\gamma^2(\omega)T_r} \tag{10.42}$$

The normalised variance of the FRF amplitude estimate depends on the same factors — coherence, record length and window size (spectral resolution):

$$\varepsilon_r\left[\hat{H}(\omega)\right] = \sqrt{\frac{[1-\gamma^2(\omega)]\,\tau_m}{\gamma^2(\omega)T_r}} \tag{10.43}$$

The variance of the FRF phase estimate expressed in radians is similar to the previous expression:

$$\varepsilon_r\left[\hat{\varphi}(\omega)\right] \approx \sqrt{\frac{[1-\gamma^2(\omega)]\,\tau_m}{\gamma^2(\omega)T_r}} \tag{10.44}$$

Expressions for the biased cross-spectrum and FRF estimates will be more complicated and less transparent, compared to the power spectrum estimate with a rectangular spectral window. In this case, the convolution operation is performed over complex numbers, not real. However, the general character of the estimates' behaviour is similar to power spectrum: the bias decreases with a widening correlation window and grows with the second derivative's increase. Moreover, the

FRF amplitude estimate's bias will depend mainly on the shape of FRF and to a lesser degree on the spectrum's shape.

10.6 Optimal Spectral Resolution

Predicting the biased FRF and calculating the estimate errors allows choice of optimal smoothing. The selection of the optimal window size (or, more generally, spectral resolution) requires the use of additional information. When identifying, an *a priori* dynamic model is always known from physical laws and previous experiments. The estimate bias can be calculated from the *a priori* model. Three possible methods for the bias calculation can be suggested: a Taylor series expansion, simulation and integration of Equation (10.35).

A method for estimating the spectrum bias has been proposed [7] using expansion of Equation (10.35) in a Taylor series. This method can produce acceptable results only for relatively small windows. More exact prediction of the biased spectral estimate can be obtained by integrating Equation (10.41) at every frequency. The third possibility consists of simulating an *a priori* model response to the real input $x(t)$ followed by FRF identification with various window sizes. After the bias and deviation have been calculated, the relationship between the FRF estimate and window can be plotted.

The common procedure for FRF estimation deals with the same window size for all frequencies [4]. However, an extended research [21-23] has shown that the optimal values of the window size are not necessarily identical for different frequencies. The proposed technique makes the experimental model more precise at the expense of the complication of the identification procedure.

The optimal choice of the correlation window size is illustrated in Figures 10.7 and 10.8. A general rule for the window choice can be formulated such that the optimal window size is less for high frequencies. The overall relationship between frequency and window size can often be approximated by the formula:

$$\log \tau_m = A - B \log \omega \qquad (10.45)$$

This is the equation of a straight line in the logarithmic plot scales. The parameters A and B are chosen to match the contour plots Figures 10.7 and 10.8.

Based on the analysis of identification errors, the following procedure for optimal FRF identification is proposed:

- choose the record length T_r much greater than T;
- estimate the FRF and coherence for a range of τ_m;
- plot the graph $\hat{H} = f(\omega, \tau_m)$;
- choose the optimal window size $\tau_{opt} = f(\omega)$;
- calculate optimal FRF and coherence estimates for each frequency;
- estimate confidence intervals;

- compare the estimate with the *a priori* model;
- make decision on model quality.

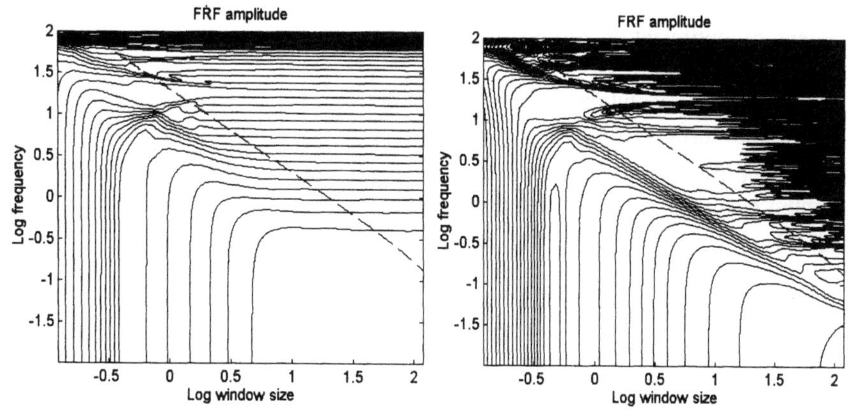

Figure 10.7. Optimal window size (dashed) for amplitude FRF: simulated prediction (left) and estimate (right)

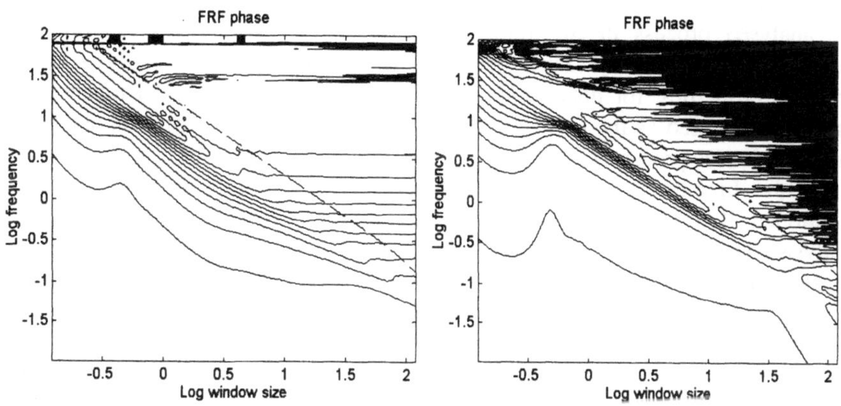

Figure 10.8. Optimal window size (dashed) for phase FRF: simulated prediction (left) and estimate (right)

10.7 Graphical Interpretation

Consider the diagrams for the optimum spectral resolution choice shown in Figure 10.7 and 10.8. In the case of a correlogram method for spectral estimation, this applies to the choice of the correlation window size. A logarithmic scale is more suitable for this purpose than linear scale diagrams. When analysing the diagram, the researcher should visually distinguish the point where the bias-caused changes diminish. Such an approach is usually used to determine the moment of optimization termination. In the linear scale, this plot represents an exponential curve. In the logarithmic coordinates, asymptotically linear parts form the curve. On such a diagram, the sections of almost straight lines (inclined and horizontal) can be easily recognised. The beginning of a horizontal line is also the required moment of optimization termination. In our case, this is the optimum spectral resolution.

Note that the resolution varies for different frequencies. In a three-dimensional view of the surface $H = f(\omega, \tau_m)$, a "ridge" can be allocated which becomes a straight line in the logarithmic scale. On the contour plot, the straight line of the optimum resolution goes through the "corner" points, which are clearly seen in the diagrams.

10.8 Optimum Spectral Resolution and Wavelets

In recent years, wavelet transforms have become popular in different application areas including time series analysis [24-26]. The term *wavelet* means simply "a small wave." A set of orthogonal functions (wavelets) is used for decomposition of the analysed process in two dimensions: time and scale, which can often be interpreted as frequency. Such decomposition is used for the analysis of the process' properties, for data compression and filtering. The wavelet transform is usually considered as the opposite to a Fourier transform in terms of applicability to nonstationary processes.

In the literature on wavelet transforms, the uncertainty principle is often mentioned. It can be expressed as the constant *time-frequency window* in spectral analysis. Therefore, the product of the time and frequency windows should be constant in any method of spectral analysis. So, the time resolution is inversely proportional to the resolution in the frequency-domain. Moreover, a classical Morlet wavelet transform can be considered to be a Fourier transform of a non-stationary time series with a variable spectral window size, again with the constant time-frequency window.

A family of functions is derived from one mother wavelet. This function represents a cosine with a bell-shaped window applied upon it. The window is close to Gaussian and is limited in length. A whole family of orthogonal functions is formed from the mother wavelet by shifting the starting point in time and by changing the time scale. The time scaling changes the spectral and time windows simultaneously. The time shift extracts the current part of the initial data. Note that the time scaling of a single mother wavelet results in a constant product of the time and frequency windows. Moreover, spectral resolution determined by the time scale is set individually for each frequency. This is why a wavelet transform is often referred to as multi-resolution analysis.

Note that multiresolution analysis is mostly used for nonstationary processes. Moreover, advances in the multiresolution technique of wavelet analysis have not produced appropriate improvements in conventional spectral analysis. Existing commercial and freely distributed software for spectral analysis uses constant spectral resolution for all frequencies analysed [9, 10]. Typical recommendations on the resolution choice are formulated as general advice such as "find a compromise between variance and bias." Also, it is suggested that several spectral estimations with different resolution (again, constant for all frequencies) are performed.

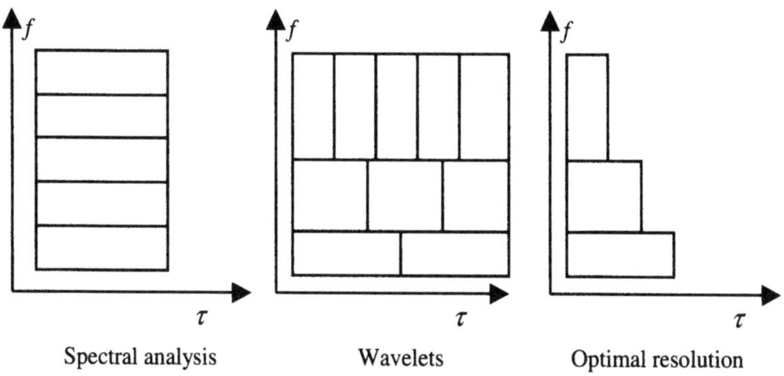

Figure 10.9. Comparison of optimal resolution and wavelets

Coming back to stationary processes, the multiresolution principle from wavelet analysis can be applied to a Fourier transform of correlation functions in spectral analysis (Blackman–Tukey method); see Figure 10.9. The whole record length of the data available is used for analysis here. However, the correlation window is not imposed on the initial data, but on the correlation function, where information on all the data is "compressed." Therefore, the application of a variable time window for stationary and quasistationary processes results in just one spectral estimate at each frequency. In comparison, a wavelet transform gives a number of spectral estimates distributed over the time of the process observation. This constitutes the major difference in analysis of stationary and nonstationary time series: one estimate of a constant parameter *versus* a series of the parameter's estimates changing in time.

10.9 Concluding Remarks

In this chapter frequency response techniques for systems identification have been discussed. It is highlighted that the choice of the window size is important to obtain optimal results with respect to bias and variance. An optimization technique for the window size choice has been derived for dynamic model identification. It is based on *a priori* model information and provides high precision of the FRF estimates. This technique uses the normal operation data and does not require additional tests and equipment.

A comparison is also made between conventional spectral analysis and wavelet transforms highlighting areas of commonality and difference in the choice of spectral resolution.

In Chapters 11 and 15 the frequency response techniques developed in this chapter and optimal windowing selection are applied to real-life examples of turbo jet and turbo prop fan engines demonstrating the power of the technique.

Chapter 11
Turbo Prop Fan Engine Identification: Practical Issues

11.1 Introduction

In this chapter, the identification techniques considered in previous chapters are applied to a real-life power plant. Dynamic models of a turbo prop fan aero engine are identified. Practical issues of closed-loop identifiability are also demonstrated.

11.2 Turbo Prop Fan Identification

An example of dynamic model identification of a turbo prop fan engine is considered demonstrating viability of the method proposed in Chapter 9. The results of the experimental data processing show the principal possibility of closed-loop identification of an aero engine model at steady-state conditions without test signals.

Dynamic model identification of a turbo prop fan engine confirms the viability of the existing general model derived from a performance-based thermodynamic model. In addition, this enables the validation of the FRF identification technique described in Chapters 7 and 10.

11.2.1 Description of Experiment

Recorded data from normal engine operation has been exploited for dynamic model identification. This would usually require test inputs to be applied to excite sufficiently the engine dynamics. However, in this work a novel approach is adopted utilising existing random disturbances created by the controller and operating environment during normal engine operation as the test source. An essential advantage of this passive identification approach is that special engine test-runs and monitoring equipment are not required. Optimal resolution provided the compromise between the bias and variance of spectral estimates.

The optimization technique proposed in Chapter 9 for the spectral estimation was applied to the experimental data from the aircraft engine test bed. The turbo prop fan engine was developed for a medium-size transport aircraft. A dual-lane digital controller controls the engine. The data was transmitted to a personal computer directly from the DAC using a serial interface as described in [1]. This scheme excludes the use of additional data acquisition equipment. The engine and controller parameters were logged at steady-state conditions.

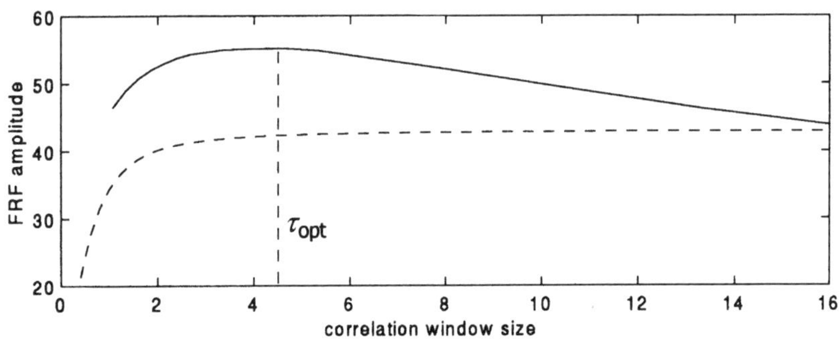

Figure 11.1. Optimal correlation window at fixed frequency: prediction (dashed) and experimental estimate (solid)

FRF estimation of dynamic models was performed using the MATLAB environment [2]. Optimal spectral resolution determined by the correlation window was chosen based on *a priori* prediction [3], as shown in Figure 11.1. Fragments of random processes were chosen at steady-state conditions. The record length was 26.7 s, or 1000 sampling points; and the sampling interval was 26.7 ms.

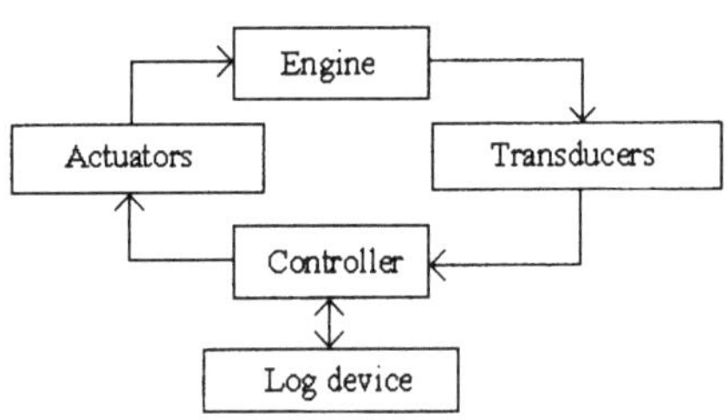

Figure 11.2. Data acquisition during engine tests

During the engine tests, experimental data was transferred to a logging device from the control computer. Figure 11.2 shows a general schematic of the data acquisition. The logging device represents a portable personal computer with a serial interface. This enables data to be gathered during on-ground and in-flight tests. Along with measured engine parameters, some control signals are also transmitted and logged. However, this does not require additional equipment because the digital controller performs all measurements.

11.2.2 Plant Decomposition

The gas turbine engine represents a multivariable plant with five major control inputs and tens of outputs. The order of the multiinput-and-multioutput (MIMO) model should not be less than four. For simplicity of modelling in the controller design, the model is divided into four relatively independent parts: the two-stage gas-producing turbine, LP and HP variable stator vanes (VSV) and two prop fans. The propellers are connected with the free power turbine via a differential reduction gear. The prop fan dynamics are closely interdependent and can be described by a second-order MIMO model.

Figure 11.3 demonstrates decomposition of the engine model for control design purposes. Besides the engine model, the following actuators were also investigated because of their contribution to the plant dynamics:

- the hydro-mechanic fuel feed system (fuel metering valve and fuel pomp);
- the step drives and hydromechanic actuators of the prop fans;
- the pneumatic actuators driving VSV.

Some of them are considered as part of the plant model because of insufficient number of measured signals.

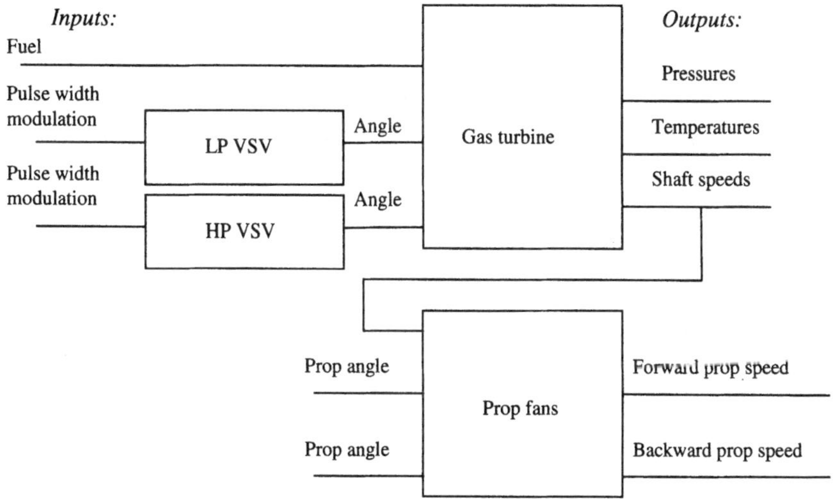

Figure 11.3. Decomposition of engine model

11.2.3 Identification of Fuel Metering System

Consider the closed loop of compressor pressure ratio control shown in Figure 7.2. The control loop consists of a linear proportional integrate (PI) controller, metering valve (mv), fuel pump (fp) and gas turbine. This type of control is activated at steady-state conditions as part of thrust control. The demanded value of compressor pressure ratio is maintained via fuel feed. Identification of dynamic models is performed using an open-loop technique assuming that identifiability conditions are fulfilled. No test signal was applied to the system, and identifiability was maintained only by ambient noise. The fuel metering valve was modelled during control design with a first-order transfer function:

$$W(s) = \frac{K}{Ts+1} \tag{11.1}$$

where the input is the control signal V_{PI} fed by the PI controller, and the output is the angular position of the metering valve α_{mv}.

In this example, the operating point of maximum take-off thrust was examined. Disturbances in this mode achieve maximum amplitude, providing best identifiability of dynamic models, whilst at the same time magnitude of random processes does not exceed the linearity zone, allowing the use of linear modelling techniques (relative STD is around one per cent).

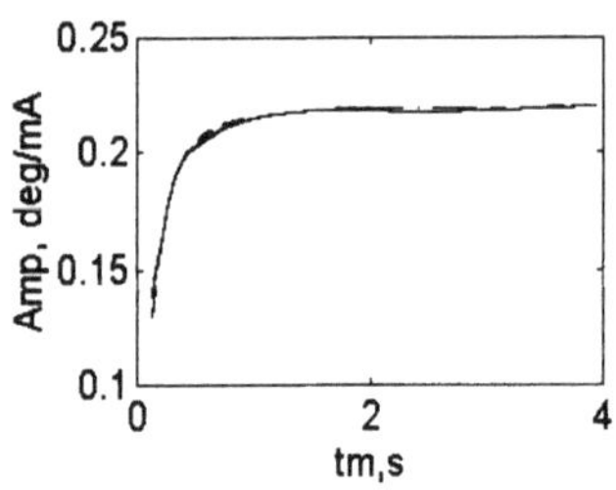

Figure 11.4. Choice of resolution

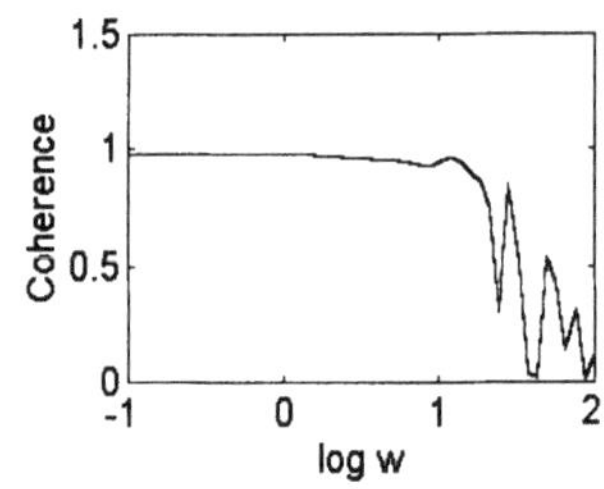

Figure 11.5. Coherence function

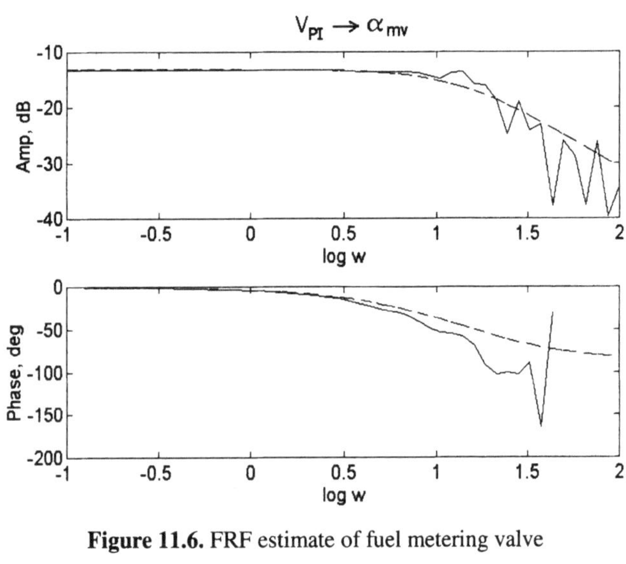

Figure 11.6. FRF estimate of fuel metering valve

The choice of optimal resolution is illustrated in Figure 11.4. The FRF amplitude *Amp* at low frequencies achieves its steady state with the correlation window t_m of 1.5 s. Further increase of the window size does not improve the estimate. The coherence function shown in Figure 11.5 is nearly one at the frequencies less than 10 rad/s. This is caused by low-pass filtering of the metering valve; see Equation (11.1). Also, high coherence proves that the input–output relationship is linear and the ambient excitation is sufficient for identification as compared with possible additive noise.

Estimates of FRF amplitude and phase are presented in Figure 11.6 with solid lines. Dashed lines show the existing model Equation (11.1) of the valve. Within the frequency range of high coherence, the estimates are similar to the model. Random deviations at high frequencies rise because the output signal contains mostly low frequency harmonics. The low-pass filtering plant Equation (11.1) suppresses high-frequency harmonics.

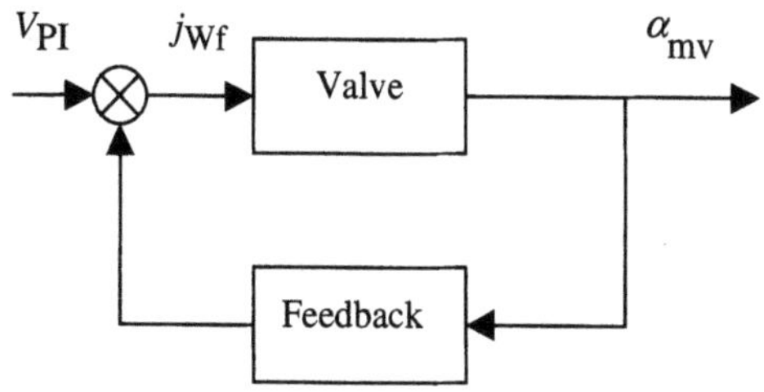

Figure 11.7. Local feedback of fuel metering valve

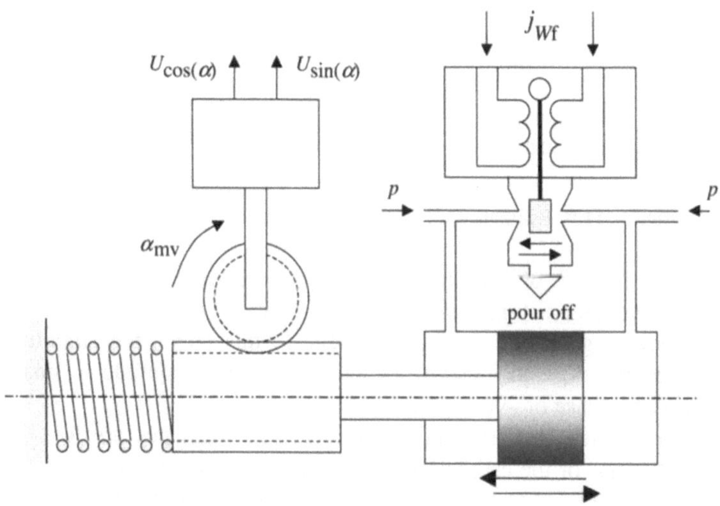

Figure 11.8. General schematic of fuel metering device

The valve considered above operates with a local feedback; see Figure 11.7. The PI controller output V_{PI} is a digital signal. The control error determines direct electric current j_{Wf} being the actual physical signal fed to the actuator, see Figure 11.8. It controls fuel pressure p moving the piston connected with a spring.

The sensor of angular position α_{mv} measures the piston's horizontal coordinate. The dynamics of the valve itself are described by a first-order nonperiodic model with a large time constant, or approximately by an integrate transfer function:

$$W(s) = \frac{K}{Ts+1} \approx \frac{K/T}{s} \tag{11.2}$$

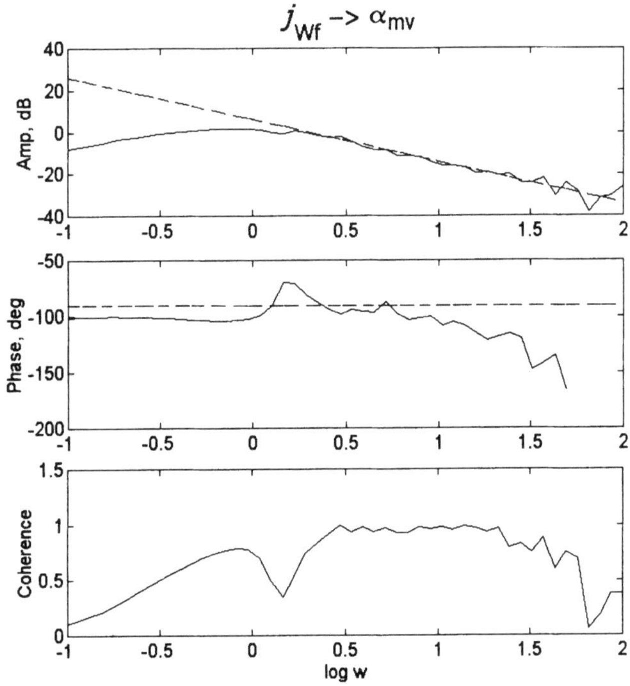

Figure 11.9. Estimates of FRF and coherence of metering valve

Results of estimation of FRF and coherence are presented in Figure 11.9. The dashed line shows the existing *a priori* model Equation (11.2). Coherence is high in the frequency range of 0.7 through 40 rad/s. In this range, the amplitude and phase estimates are close to those of an integrator with amplitude inclination of –20dB/decade and phase of –90 degrees.

At low frequencies, the amplitude and coherence estimates are low. This is explained by the bias of the estimation method. Ideally, the output of an integrator can contain infinite energy at zero frequency. This can be observed with an infinite record length only. In addition, before spectral analysis, mean values are subtracted from experimental data. This leads to elimination of energy at very low frequencies, resulting in zero spectral estimates. Moreover, the effect of windowing produces nonzero estimates at low frequencies due to energy leakage from neighbouring harmonics. This distortion of FRF and coherence estimates is typical for integrator identification and can be obtained in simulation with a finite record length. This phenomenon helps to easily distinguish the plant structure in spectral identification; compare Figures 11.6 and 11.9.

11.2.4 Identification of Shaft Speed Dynamics

In the example considered, the gas turbine engine operates under closed-loop control of compressor pressure ratio, which is one of gas turbine outputs. Hence, the feedback does not directly influence all other outputs including shaft speeds. This allows the feedback operation to be ignored, as far as LP and HP shaft speed dynamics are concerned. Therefore, direct methods can be applied to dynamic model identification.

Shaft speed dynamics are described by a second-order linear model accounting for the interaction between the high-pressure and low-pressure rotors of the engine. An *a priori* model was obtained by thermodynamic model linearisation:

$$G(s) = \frac{K(T_1 s + 1)}{(T_2 s + 1)(T_3 s + 1)} \tag{11.3}$$

This model accounts for inertial accumulation of mechanical energy and interaction of the shafts only. The thermodynamic model has a higher order, as described in Chapter 5. The identification process is illustrated in Figures 11.10-11.15. Figure 11.14 shows the FRF estimate of the low-pressure (LP) turbine rotor dynamics compared with the model Equation (11.3).

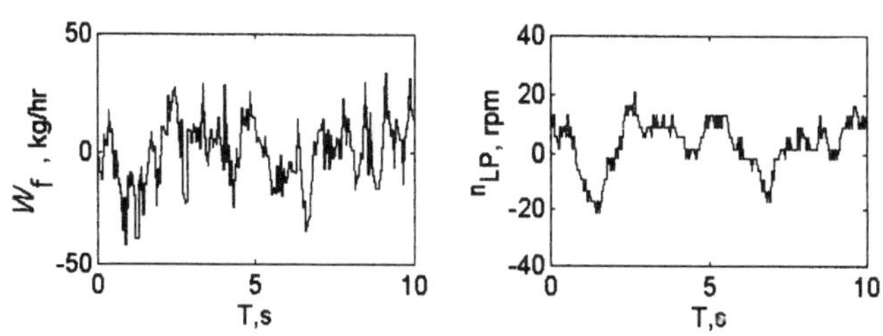

Figure 11.10. Input–output random processes of LP shaft speed

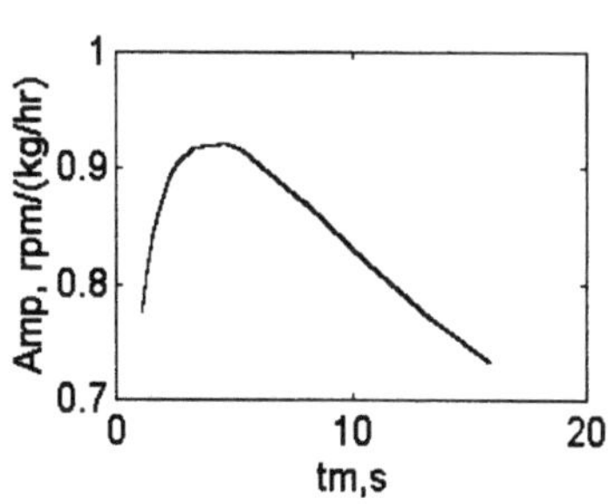

Figure 11.11. Choice of resolution

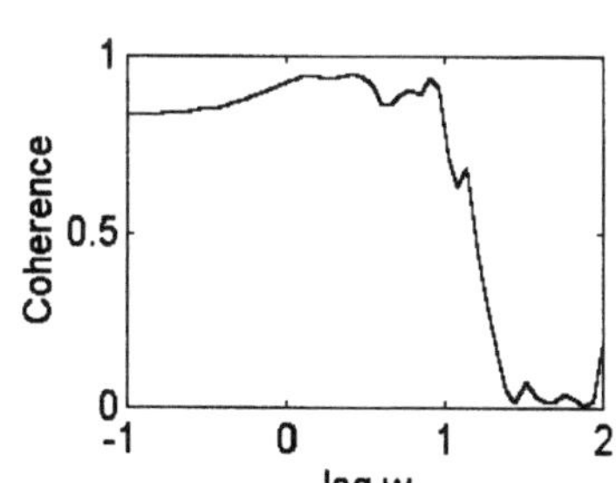

Figure 11.12. Coherence function

The transfer function from the fuel flow to shaft speed Equation (11.3) represents a low-pass filter. The bandwidth of this model is limited by 20 rad/s approximately. There are high coherence value and precise FRF estimates in this frequency range; see Figure 11.11. Similarly, HP shaft speed dynamics are identified.

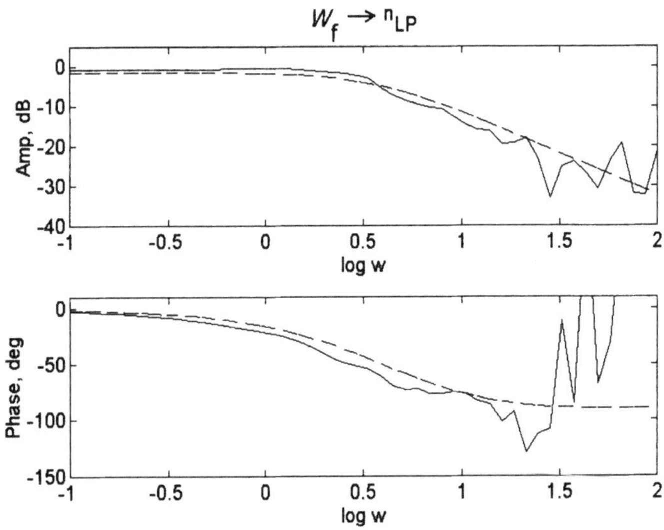

Figure 11.13. FRF of LP shaft speed: *a priori* thermodynamic model (dashed) and estimate (solid)

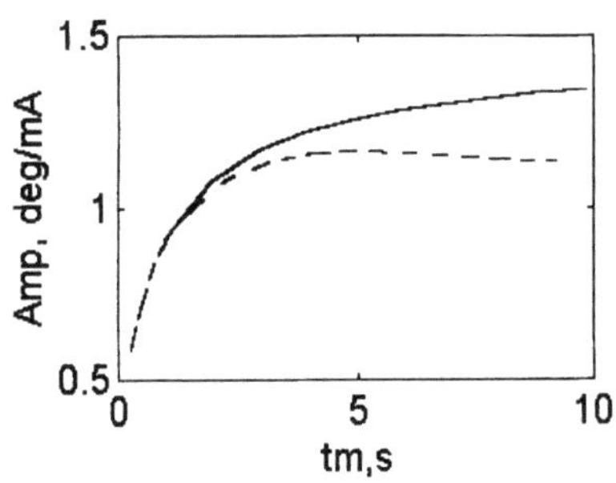

Figure 11.14. Choice of resolution: estimate (solid) and prediction (dashed)

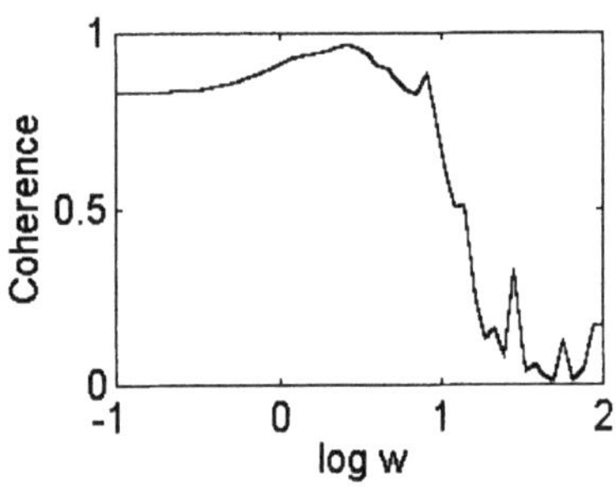

Figure 11.15. Coherence function

11.2.5 Identification of Compressor Pressure Dynamics

A dynamic model of compressor pressure reaction to fuel flow was identified similarly to the shaft speed dynamics; see the previous section. This part of the plant operates under closed-loop control of compressor pressure ratio; see Figures 7.2 and 7.3. The linear model of compressor pressure dynamics has the following structure:

$$W(s) = K\frac{(Ts+1)(Ts+1)}{(Ts+1)(Ts+1)} \tag{11.4}$$

The identified dynamic model is close to the linearised performance-based *a priori* model. Identification of elements of the fuel metering system was also successful. This indirectly witnesses the situation of closed-loop identifiability. In other words, the most powerful excitations apply to the plant input. These are the compressor pressure ratio demand and measured input air pressure. Both signals contain random processes stemming from ambient and measurement noise, which creates necessary excitation conditions for identification. The noise of operation of the fuel metering system does not influence the closed-loop operation.

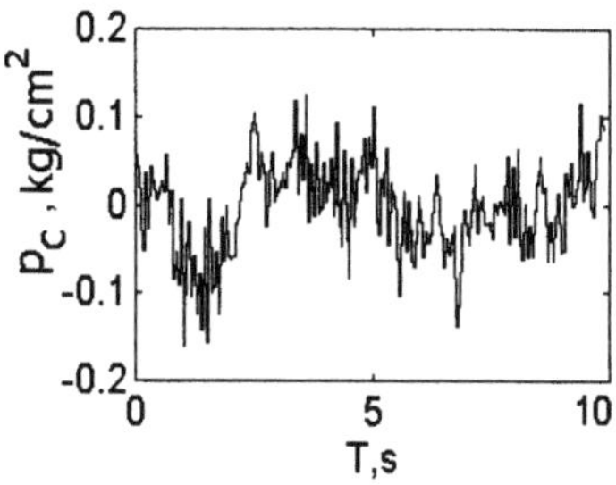

Figure 11.16. Compressor pressure

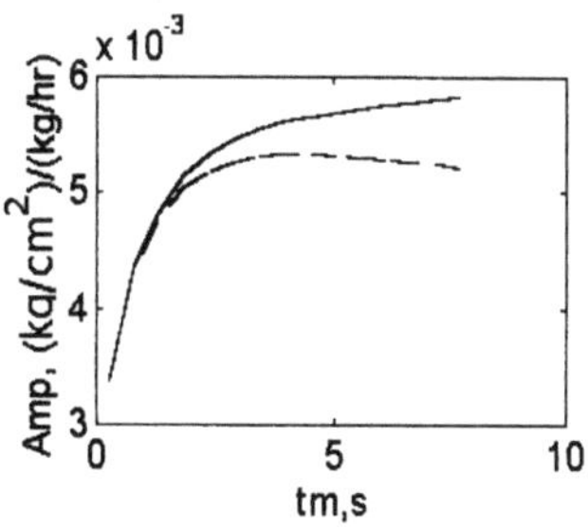

Figure 11.17. Choice of resolution: estimate (solid) and prediction (dashed)

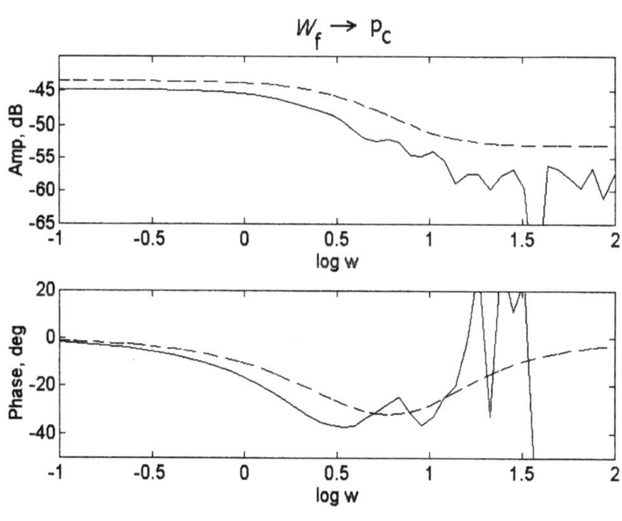

Figure 11.18. FRF of compressor pressure: *a priori* model (dashed) and estimate (solid)

Based on this conclusion, an improvement was introduced into the controller. A low-pass filter was applied to measured input air pressure before supplying it to the control loop. Consequently, STD of compressor pressure at steady-state conditions decreased by almost three times, allowing the controller to meet the system requirements. This example demonstrates a possible way that identification can affect control quality via analysis of random phenomena.

11.2.6 Identification of Turbine Temperature Dynamic Model

Turbine temperature is measured with a thermocouple. This adds the thermocouple dynamics to the plant model, when using temperature measurements with no compensation. The thermocouple is modelled with a first-order nonperiodic transfer function.

The time constant T depends on the operating point of the engine. Inertia of the thermocouple is determined by airflow speed around, which depends on shaft speed. According to the existing model, $T \approx 0.6$ s at the considered operating point of maximum take-off thrust.

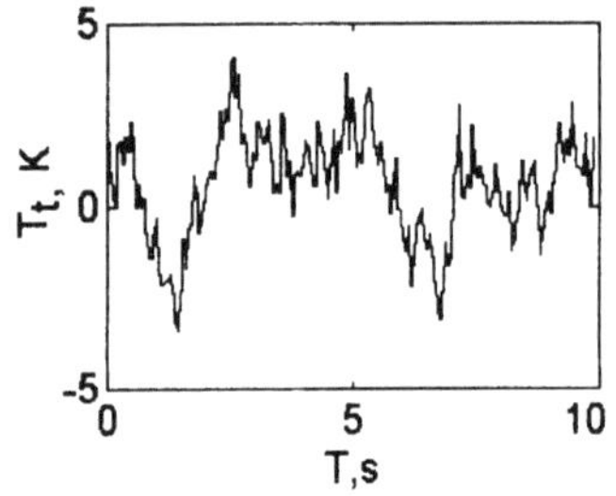

Figure 11.19. Input–output random processes of turbine temperature

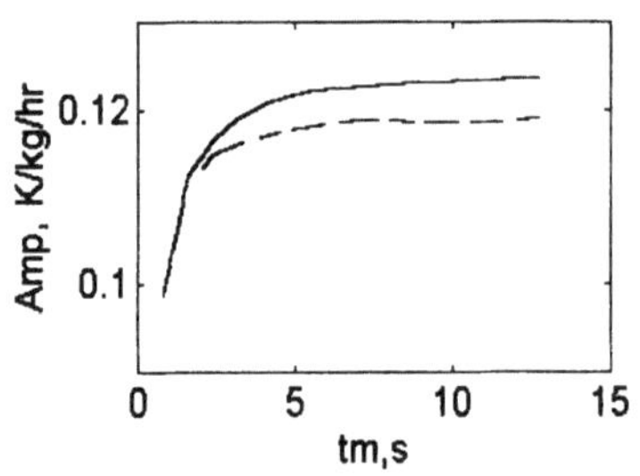

Figure 11.20. Choice of resolution: estimate (solid) and prediction (dashed)

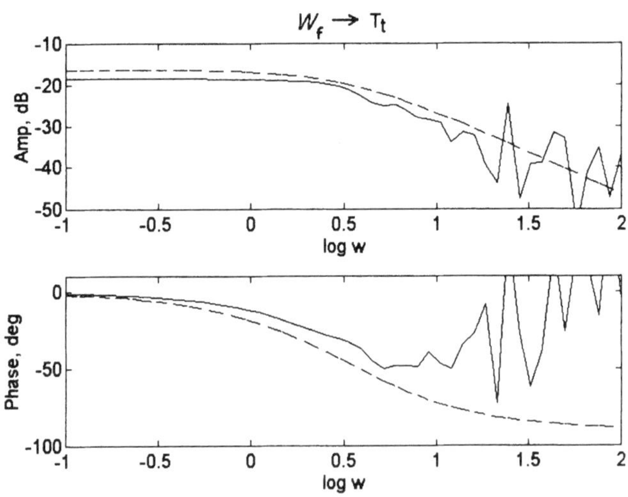

Figure 11.21. FRF of turbine temperature: *a priori* model (dashed) and estimate (solid)

The experimental estimates of FRF are presented in Figure 11.21 compared with the existing model accounting for the plant and thermocouple dynamics. This model includes a second-order model type of Equation (11.3) and a first-order model Equation (11.1). The plot of the identified FRF corresponds to the shape of a first-order transfer function in the range 0.1 through 100 rad/s in Figure 11.21. In this frequency range, the dynamics of measured turbine temperature are determined mostly by the thermocouple dynamics.

11.2.7 Identification of VSV Actuators in LP and HP Compressors

Both LP and HP stages of the compressor are equipped with variable stator vanes (VSV). The angular position of VSV is demanded as a function of the corresponding shaft speed. This creates optimal geometry for the airflow and ensures gas dynamic stability of operation of the compressor.

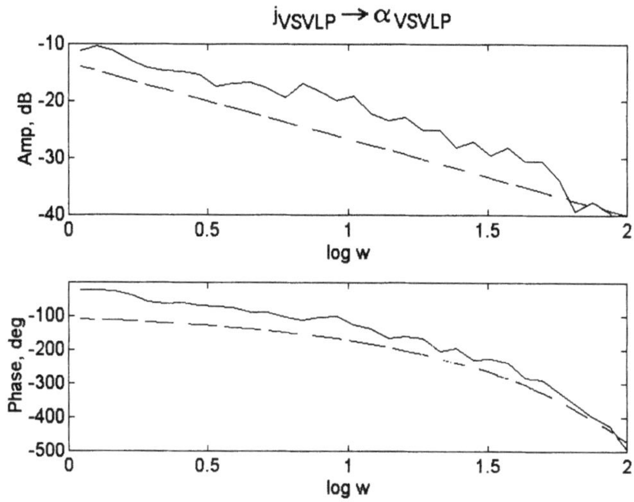

Figure 11.22. FRF of VSV LP: *a priori* model (dashed) and estimate (solid)

The VSV actuator represents a pneumatic device. The input is the time-impulse control signal j_{VSV} and the output is the angular position α_{VSV}. The linear dynamic model of the actuator is an integrator. The phase estimate in Figure 11.22 corresponds to the existing model (an integrating transfer function) with additional pure time delay, shown with a dashed line.

The FRF estimates for HP VSV substantially differ from LP estimates and correspond to the feedback model shown with a dashed line in Figure 11.23. In this case, it is not possible to identify the model from this particular data set.

Both VSV systems operate with closed-loop control. The LP and HP VSV control loops are similar in structure; see Figure 11.24. Comparison of signals in these two control loops reveals the reason for nonidentifiability of the HP VSV actuator: there is considerable noise in the measurement of the angular position of HP VSV.

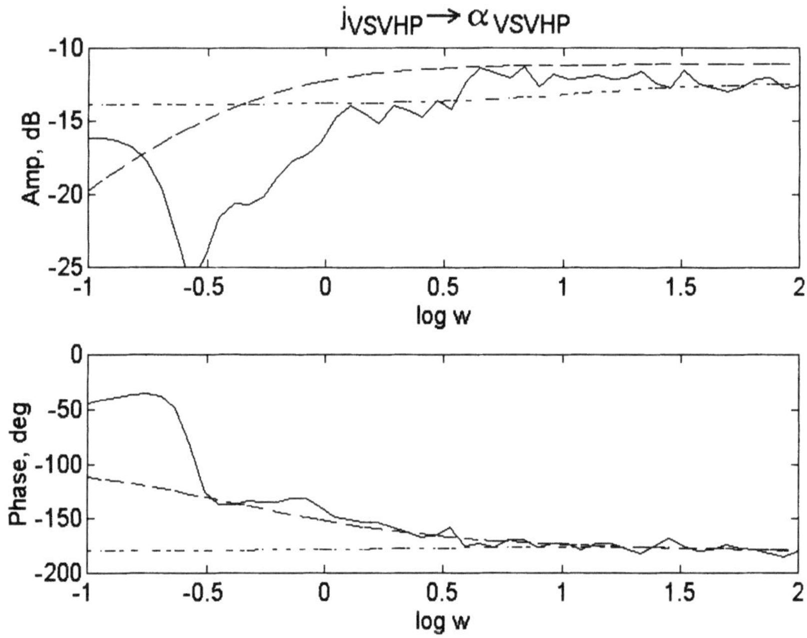

Figure 11.23. FRF of VSV HP: *a priori* plant model (dashed), feedback model (dashed) and estimate (solid)

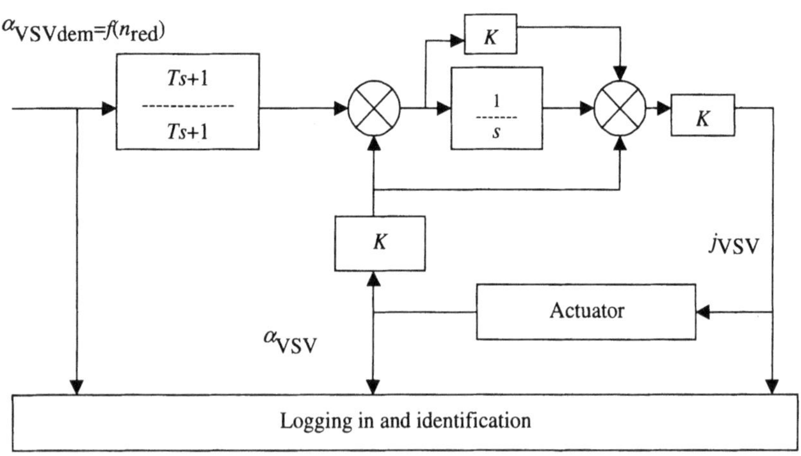

Figure 11.24. Structure of LP and HP VSV control

Demanded and measured signals of α_{VSV} are presented in Figures 11.25 and 26 and in Table 11.1. STD of $\alpha_{VSV.LP.dem}$ is much greater than STD of $\alpha_{VSV.LP.dem}$. However, STD of $\alpha_{VSV.LP}$ is less than STD of $\alpha_{VSV.HP}$. Table 11.1 also includes estimates of STD of additive output noise, calculated from spectral estimates:

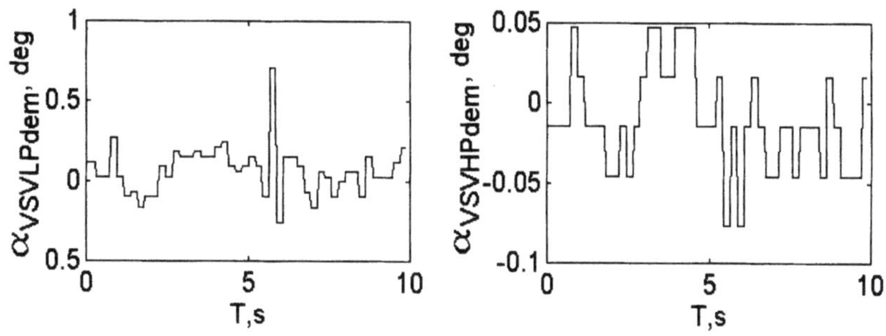

Figure 11.25. Demanded values of VSV LP and HP

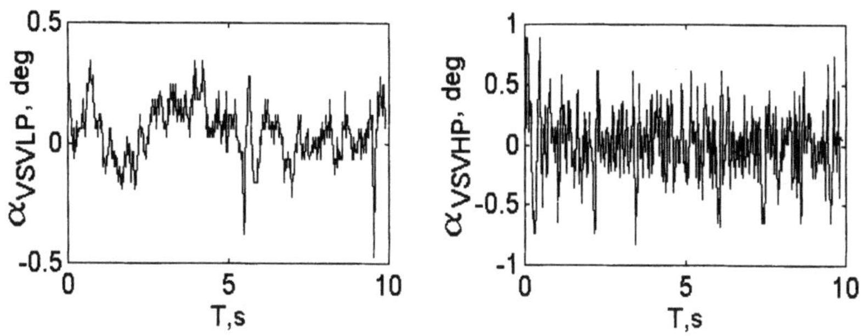

Figure 11.26. Measured values of VSV LP and HP

Table 11.1. Mean and STD of signals in LP and HP VSV control loops

Signal	Mean	STD
$\alpha_{VSV.LP.dem}$, deg	–14	0.210
$\alpha_{VSV.HP.dem}$, deg	–1.8	0.057
$\alpha_{VSV.LP}$, deg	–14	0.205
$\alpha_{VSV.HP}$, deg	–1.8	0.337
$v_{VSV.LP}$, deg	0	0.093
$v_{VSV.HP}$, deg	0	0.296

$$S_{vv}(\omega) = S_{yy}(\omega) - \frac{S_{yx}(\omega)S_{xy}(\omega)}{S_{xx}(\omega)} \tag{11.5}$$

The variance of the noise is estimated as follows:

$$D_v = \frac{1}{2\pi}\int_0^{\pi T_0} S_{vv}(\omega)d\omega \tag{11.6}$$

where T_s is the sampling interval, and S is single-sided spectral power density. STD is obtained as the root-squared variance.

Moreover, the actual measured signal $\alpha_{VSV.HP}$ is not similar to the general low-frequency shape of its demand $\alpha_{VSV.HP.dem}$, being opposite to the signals within the LP control loop; see Figures 11.25 and 11.26. Another difference is observed in the mean values; see Table 11.1. Over the range of experiments from 30% of shaft speed to take-off, the mean value of $\alpha_{VSV.LP}$ is –26 through –14 degrees, whereas $\alpha_{VSV.HP}$ is –10 through –1.8 degrees.

Analysis of the spectral densities of the signals and the coherence function presented in Figures 11.27–11.30 allows the following to be concluded: both HP and LP VSV follow their demanded values. However, HP VSV also contains powerful high-frequency noise, absent in LP VSV.

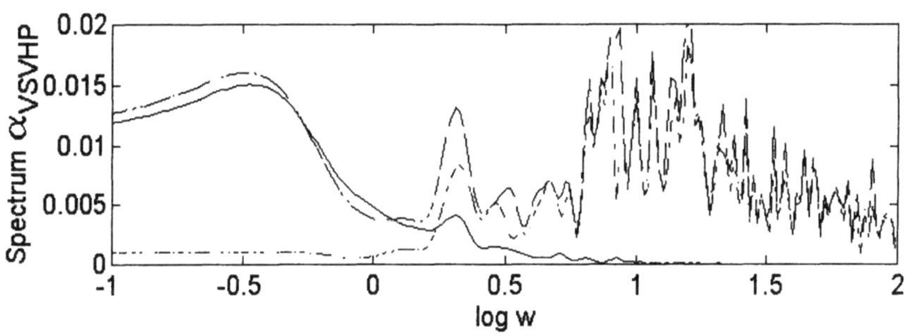

Figure 11.27. VSV HP spectra: demand (solid), measured (dashed) and noise (dash-dotted)

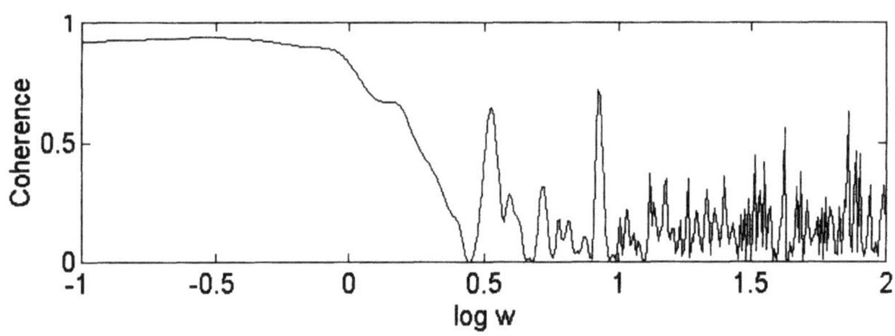

Figure 11.28. Coherence function in VSV HP identification

Taken together, all the facts discussed above in this section lead to the following explanation for the origins of the measurement noise. Measurements of the angular positions are performed using sine–cosine rotating transformers. The control computer analyses sine and cosine signals and computes the angle estimate. If the angle is close to zero, then computational errors rise because sin(0) = 0. In this situation, the computational errors create the measurement noise, when measuring near-zero angles.

In the case of LP VSV control, the random process in the demanded value of the VSV position is the most powerful disturbance in the control loop. This creates closed-loop identifiability. In the HP VSV loop, nonidentifiability stems from the powerful measurement noise. LP VSVs are more controllable in the stochastic sense, as compared with the HP VSVs. This controllability is apparently connected with closed-loop identifiability.

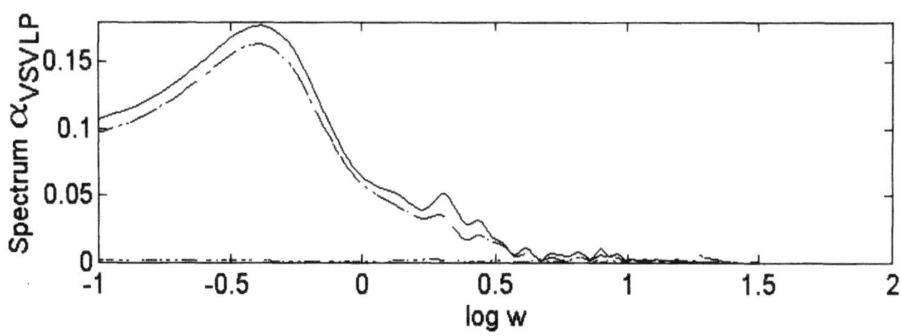

Figure 11.29. VSV LP spectra: demand (solid), measured (dashed) and noise (dash-dotted)

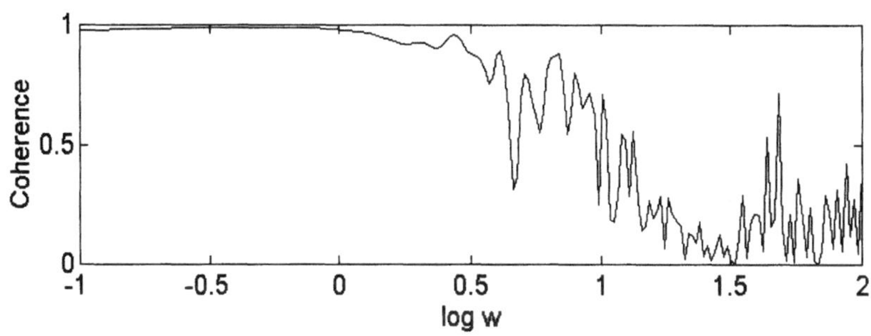

Figure 11.30. Coherence function in VSV LP identification

A possible improvement is to shift the initial angle of the transducer. Then the operating zone of the transformer will lie far from zero angles. This would result in much less measurement noise.

11.2.8 Identification of Step Drives in Prop Fan Controller

The prop fan operates under two independent control loops: forward and backward propeller control. These two control loops are coupled through the plant because the propellers are connected with the free power turbine via the differential reduction gear. Demanded values of the prop fan speeds are determined as a function of the PLA.

Each step drive operates with a local feedback; see Figure 11.31. The output of the step drive is its angular position α_{step}. The angle is proportional to the number of "steps," which are small fixed angular revolutions. The drive input is the control code U_{step} being the required number of rotation steps. The drive rotates with constant speed of steps per second to reach the demanded angular position. Within a zone of linear operation, the step drive can be considered an integrator.

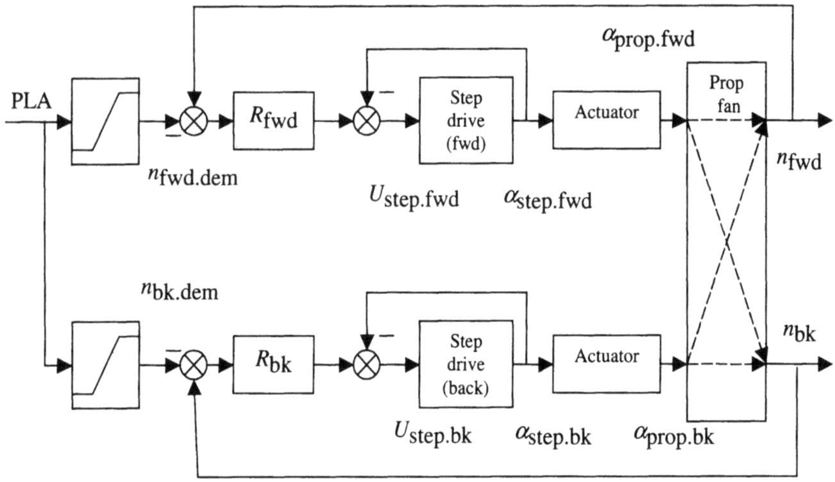

Figure 11.31. Prop fan controller

FRF estimates of the forward step drives are shown in Figure 11.34. These correspond to an integrator with time delay, similar to the existing models shown with dashed lines. The experimental model estimates agree with the expected step drive dynamics.

Successful identification of the step drives confirms identifiability of these devices. Therefore, major disturbances in these control loops are situated outside these local loops and are not applied to the measured step drive positions.

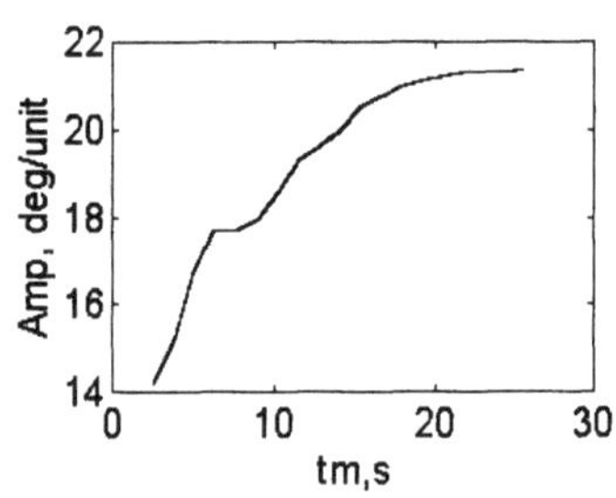

Figure11.32. Choice of resolution

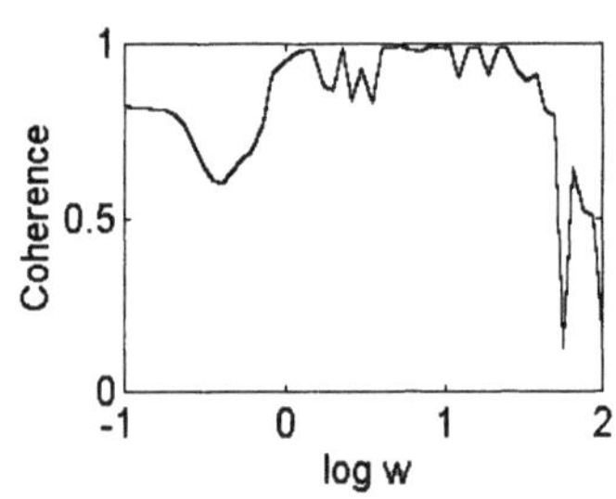

Figure 11.33. Coherence function

11.2.9 Identification of Prop Fan Dynamics

Both the forward and backward prop speed control loops are similar in structure, see Figure 11.31. The demand n_{dem} is subtracted from the measured propeller speed n: $\varepsilon = n - n_{dem}$. This differs from the conventional feedback control, where the control error is obtained via the difference $\varepsilon = n_{dem} - n$. In the prop fan controller, the sign inversion takes place within the plant: an increase in the propeller angular position results in deceleration of the propeller rotation. The transfer functions of both prop fan controllers have identical structure:

$$R(s) = K \frac{(Ts+1)(Ts+1)}{(Ts+1)(Ts+1)} \tag{11.7}$$

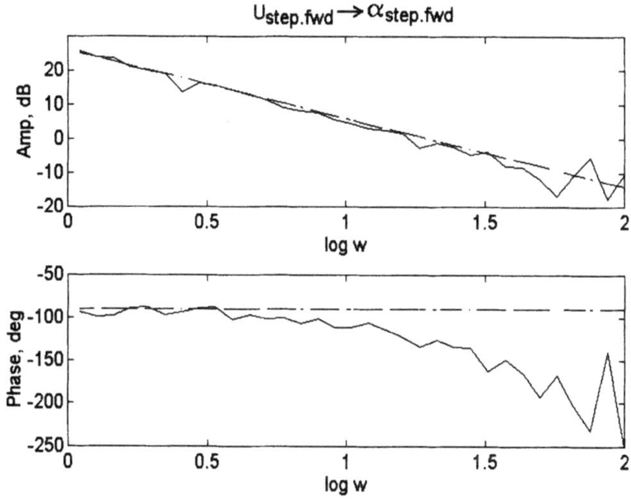

Figure 11.34. FRF of forward step drive: *a priori* model (dashed) and estimate (solid)

The angular position of the step drive α_{step} affects the propeller angle α_{prop} through the hydromechanical actuator. The actuator represents an integrator-type system described by the model:

$$R(s) = \frac{K}{s(Ts+1)} \tag{11.8}$$

The propeller angle is not measured, so the plant model being identified includes the actuator and propeller dynamics Equation (11.7) and Equation (11.8).

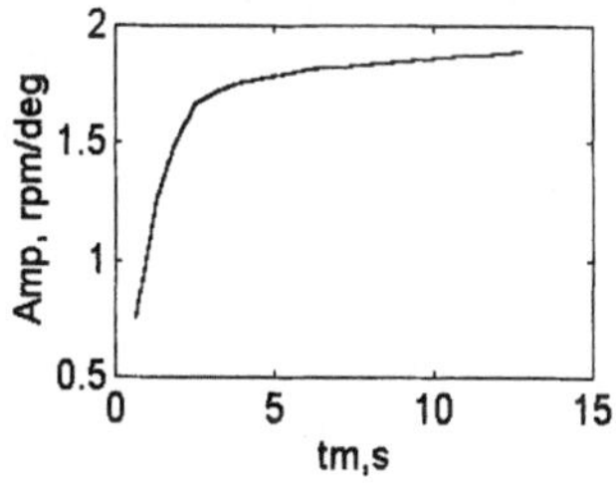

Figure11.35. Choice of resolution

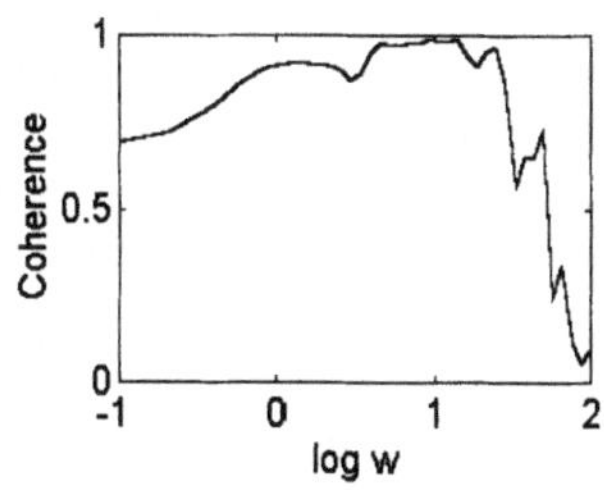

Figure 11.36. Coherence function

FRF estimates of both propellers correspond to the feedback models known *a priori*. In this case, the situation of nonidentifiability takes place; see, for example,

Figure 11.37. One of the reasons is that the demanded propeller speeds are constant at the considered operating conditions. Hence, the speed demands contain no disturbance helpful for identification. Major disturbances are most probably applied to the outputs of the identified plant, being the measured prop speeds. Here, the disturbance sources are the hydromechanical actuators.

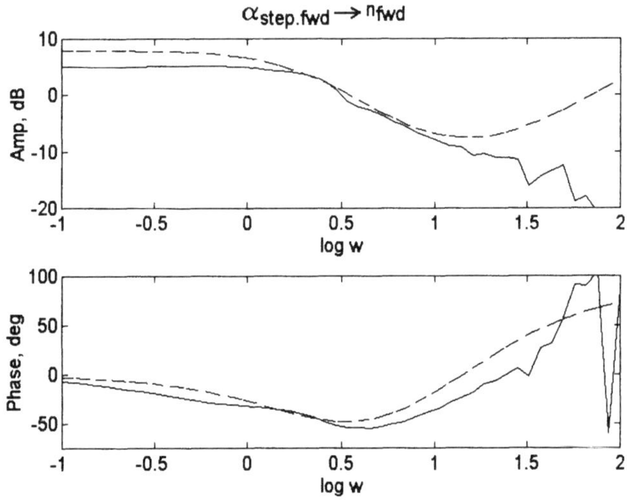

Figure 11.37. FRF of forward prop: *a priori* feedback model (dashed) and estimate (solid)

11.3 Closed-Loop Identifiability Analysis

The engine controller operates with five closed control loops. The system controls the compressor pressure ratio, the position of the variable stator vanes (VSV) in both stages of the compressor and the forward and backward propeller speed. Also, there are three local control loops, which are the fuel metering valve and step drives involved in forward and backward prop speed control. The FRF estimates correspond to the feedback models in three cases: for the HP variable stator vanes (VSV) and both the prop fans.

Analysis of Table 11.2 shows that simultaneous violation of the proposed conditions on asymmetry and excess (Chapter 7) corresponds to a closed-loop non-identifiable system. These inequalities are violated for two loops: HP VSV and the forward propeller; see grey-coloured rows. In both cases, feedback models were estimated by direct identification techniques.

The last case of identification of the "backward step drive to prop speed" model represents another combination of factors. Both asymmetry and excess inequalities are fulfilled, whereas the plant model is nonidentifiable. This determines the necessity for further research in closed-loop identifiability.

Table 11.2. Engine model identifiability in closed-loop control

Model input	Model output	$\lvert A_x \rvert > \lvert A_y \rvert$	$\lvert E_x \rvert > \lvert E_y \rvert$	Plant model estimated
Fuel flow	LP shaft speed	Yes	Yes	Yes
Fuel flow	HP shaft speed	Yes	Yes	Yes
Fuel flow	LP turbine temperature	Yes	Yes	Yes
Fuel flow	HP compressor pressure	Yes	Yes	Yes
PI control	Metering valve angle	Yes	Yes	Yes
Metering valve angle	Fuel flow	Yes	Yes	Yes
PI DC output	Metering valve angle	No	Yes	Yes
HP VSV control	HP VSV position	No	No	No
LP VSV control	LP VSV position	No	Yes	Yes
Forward control code	Forward step drive angle	No	Yes	Yes
Backward control code	Backward step angle	No	Yes	Yes
Forward step drive angle	Forward prop speed	No	No	No
Backward step drive angle	Backward prop speed	Yes	Yes	No

11.4 Concluding Remarks

The dynamic model identification of a turbo prop fan engine has demonstrated the viability of the optimal resolution technique and the identifiability verification approach proposed in previous chapters. The work demonstrated the possibility of direct passive identification in closed-loop control under normal system functioning. The identification technique can be used as the means for the engine condition monitoring.

Analysis of situations of closed-loop identifiability and nonidentifiability enabled identification of major sources of random disturbance in the system. A useful outcome from this was the introduction of appropriate random generators in a system test facility. This has provided realistic operating conditions for hardware-in-the-loop testing of control systems.

Based on analysis of random phenomena, an improvement was introduced into the controller allowing a substantial decrease in random deviation of compressor pressure at the steady-state conditions.

Chapter 12
Stochastic Gas Turbine Engine Models

12.1 Introduction

Previous chapters gave an overview of conventional methods for modelling of gas turbines for control purposes. This chapter provides basics of stochastic modelling using controllable Markov chain techniques. Accounting for stochastic properties is essential for engine modelling at system test facilities, where the real-life environment is simulated. In addition, the Markov modelling technique is a promising tool for condition monitoring and optimal control of aero engines. This chapter also introduces a novel fuzzy Markov modelling technique to further improve the modelling performance.

12.2 Disturbances Affecting Control Systems of Aero Engines

Various types of random disturbances originating from internal physical processes and the environment can influence operation of aero engine control systems. First, the build tolerances for the engine and controller elements cause differences in characteristics over the fleet. The variation can be up to 1-2 % appearing as random deviations of the engine parameters. Fuel quality has a similar effect on the engine characteristics. Second, combustion processes, temperature instability, gas turbulence in the flowing part, mechanical vibrations, and noise produced by the actuators and measurement systems affect input–output coordinates in a wide frequency range. The third group consists of environmental factors and noise in measurements of the inlet pressure and temperature, flight speed and height.

Internal disturbances in the engine are usually considered stationary. External factors are mostly nonstationary but often can be considered quasi stationary under some assumptions. Random influences are investigated in various ways: changes in the engine characteristics, identification of the engine model parameters, control quality, *etc*. A simplified classification of disturbances influencing the gas turbine engine and its control system is suggested in Table 12.1.

A study of measurement accuracy reveals that biased measurements of mean values have no effect on random components of engine parameters. Furthermore, random deviations of input–output coordinates are often measured with sufficient accuracy for the dynamic model identification of the engine and its control system.

Table 12.1. Classification of disturbing factors

Factors	Regular	Random
External	H, V, p_{in}, T_{in} Wind pulsation	p_{in}, T_{in} Wind gust
Internal	Operating point change Air bypass Wearing and deterioration Temperature nonstationarity	Gas/air flow turbulence Measurements, quantisation Spectral folding Fuel quality, actuators

12.3 Applications of Stochastic Modelling in Aero Engine Control

Modelling of random disturbances is helpful at various stages of the control system life cycle. Major application areas for stochastic modelling in aero engine control are as follows:

- signal filtering in the engine control unit;
- systems identification for control design and condition monitoring;
- optimal control of the power plant;
- stochastic simulation using computers and hardware-in-the-loop test beds.

These areas are discussed in more detail in the following sections.

12.3.1 Signal Filtering

Processing of digital signals from transducers is important in control systems design. Signal filtering can implement software or hardware tools. Currently, digital filters are implemented in software due to its flexibility. Design of digital filters consists of design specifications and creation of a transfer function fulfilling the requirements. Filtering in control systems separates the useful part of the signal from the noise of measurements and data conversion [1].

Band-pass filters are appropriate where the signal and noise spectra are not overlapped. The control-related spectrum of engine parameters occupies the frequency range of 3-15 Hz. This band is usually selected in low-pass filter design for aero engine controllers. Recent research has addressed the digital filtering problem in aero engine control using Chebyshev polynomials and Kalman filters.

The problem becomes more complicated when the signal and noise spectra are overlapped. For example, this takes place when the system controls compressor pressure ratio π_c by affecting fuel flow. Intake air pressure p_{in} strongly influences π_c as low-frequency noise because of the relation:

$$\pi_c^* = \frac{p_c^*}{p_{in}^*} \qquad (12.1)$$

Filtering of intake air pressure requires information about the spectrum of the random disturbance p_{in}. On the one hand, the intake pressure noise influences the engine control loop resulting in random deviations in fuel flow. The fluctuation amplitude can rise up to 3% of the mean fuel flow at steady-state conditions, which creates fluctuations in engine output parameters, including thrust. On the other hand, the low-frequency noise facilitates real-time identification of engine models during its normal operation. This enables in-flight engine condition monitoring to be carried out and engine performance degradation to be investigated without special experimentation at on-ground test bed facilities.

Digital filters in control systems are also responsible for monitoring of signals gathered from sensors by the control computer, which reduces the failure probability at the expense of "time redundancy." This considerably complicates the problem and requires nonlinear filter design.

The low-frequency noise p_{in} occupies the control-related frequency range, which prohibits complete compensation of the random disturbance. Hence, both the frequency range of the engine operation and spectral densities of signals must be considered during filter design.

Another important problem is testing of the developed filters. Such testing can be performed at system test facilities, which include random disturbance simulation. A less expensive means of filter testing is computer simulation, which requires models of the engine and the random disturbances affecting the engine operation.

12.3.2 Identification for Control Systems Design and Condition Monitoring

Systems identification is a well-established area of the control theory. Theoretical aspects of identification problems are widely discussed in [1-6]. General problems of systems identification in aero engine control are described in [7].

The identification problem consists of two parts: nonparametric identification (estimation of frequency response, step response, *etc.*) and parametric identification (model parameter estimation). Moreover, system identification is based on signal identification. A key problem in parametric identification is the model structure selection, which is performed via simplification and linearisation of performance-based thermodynamic models. Estimation of parametric models with known structure is less complicated; this type of identification is conducted using various optimisation techniques.

The most labour-intensive and expensive stage of control system design is experimentation with the controller on the engine test bed. The time spent on this stage can be reduced by means of more precise estimation of static and dynamic parametric models of the engine through identification. Design of control algorithms is based on engine models, which are refined using experimental data. Information about the actual engine model enables control algorithms to be improved and design specifications to be met.

Identification methods are used in engine condition monitoring and fault diagnosis. Such systems predict the engine behaviour at the expense of "algorithmic redundancy." Currently, input–output models are accurately identified only at a separate operating point and flight condition, using real-time techniques.

Experimental data for dynamic model identification of engines and hydraulic elements of its controller can be obtained from active or passive experimentation. During active experiments, various test signals are applied to the engine inputs [7]. Passive experiments are performed during normal operation of the system, logging the records of random processes.

Amplitude of test signals must exceed that of existing processes in the engine. The maximum amplitude of signals is limited in linear model identification because of the engine nonlinearity. In certain cases random processes are sufficiently powerful for identification with no probe signals. Moreover, the use of trial signals can lead to some difficulties when real-time identification is performed during the flight.

Dynamic model identification from signal perturbations around the steady state can be used for the engine condition monitoring during in-service use. Dynamic characteristics of the engine and its actuators can change because of wearing and gradual deterioration. Accounting for these changes might be helpful in optimisation of the engine operation and in predictive maintenance. Digital control systems and data logging equipment enable experimental data to be gathered in a digital form. This allows built-in monitoring to be performed using systems identification techniques.

12.3.3 Optimal Control of Power Plant

Competition in the world market for gas turbine engines drives improvements in the performance. This includes reduced fuel consumption, reduced noise level for civil engines and longer in-service time. The competitiveness of an engine can be increased through stochastic optimisation of control laws with no changes in the engine configuration. Optimum control systems with a statistical feedback can improve control accuracy and adjust control laws according to changes in the random environment.

Optimal control of a plant affected by random disturbances is still an important problem in modern control theory. Minimum variance of output coordinates of the plant is often considered to be an objective function in stochastic control systems. Such an approach to aero engine control enables fuel consumption to be reduced and the engine in-service time to be increased because of less intensive operating conditions. Analytical solutions exist if the random disturbances are precisely measured [8] or if these are stationary and the noise application points are known exactly. Note that most random factors in gas turbine engines cannot be accurately measured; the application points are known approximately. Moreover, some random disturbances are nonstationary. Stochastic control requires considerable study of random disturbances and their application points in the system. Successful handling of nonstationary random disturbances requires reliable real-time identification techniques to be developed.

12.3.4 Stochastic Simulation in Control Systems Design and Testing

Simulation test beds are widely used in design, testing, and tuning of digital control systems for aero engines. A test bed usually contains electronic simulators of sensors and actuators, the computer simulating the engine characteristics, and data logging devices; see Figure 12.1.

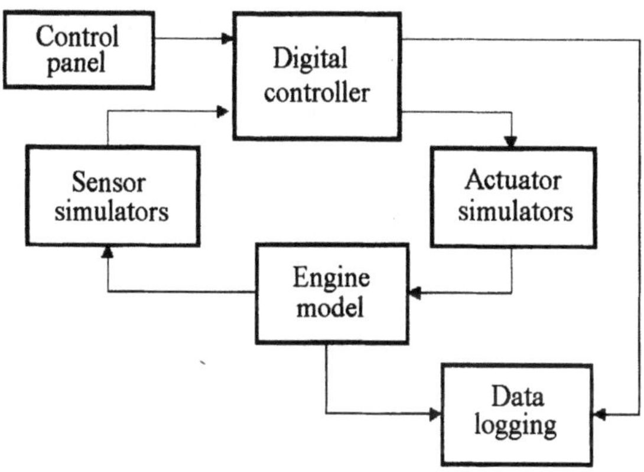

Figure 12.1. Simplified diagram of simulation test bed for digital controllers

The use of simulation test beds significantly reduces expensive testing of controllers at engine test beds and laboratory aircraft. Testing of a system at simulation test beds enables control accuracy, reliability issues, interference protection and fault tolerance of control systems to be investigated.

Deterministic engine models for simulation test beds reflect only average parameters of a general engine and do not account for individual characteristics. Sometimes, a particular digital engine control unit must be tested with an individual engine model accounting for real random disturbances. Realistic simulation includes noise of measurement and actuators, atmospheric factors, and random processes in the flowing part.

Recently, experimental data from real engines and simulation test beds from minimum to maximum thrust were studied [9]. Comparison of the data has shown that standard deviations of parameters from real engines are several times greater than those of test beds. In order to make test bed testing more realistic, random disturbances must be added.

A simple technique for stochastic simulation employs a broadband random signal generator and forming filters. Application points of the disturbances must also be determined. Experimentation with the suggested technique has proved its efficiency. Standard deviations of the simulated parameters became much closer to that of the real engine data.

Further improvements in test bed simulation will take into account individual characteristics of a particular engine. A digital controller will be tested with the model of a particular engine and tuned before installation on that engine. However, some random disturbances cannot be directly identified, thus complicating simulation with forming filters. In such a case, controllable Markov chain techniques are a promising tool for joint modelling of the deterministic and stochastic parts of the nonlinear engine model using experimental data.

12.4 Markov Modelling of Dynamic Systems

Markov modelling techniques are widely applied for exact or approximate description of various systems. Typical control applications of Markov process theory are trajectory control and reliability investigation. However, Markov modelling applications in modelling and real-time simulation of gas turbine dynamics attract insufficient attention. The essence of the Markov modelling approach is that the dynamic behaviour of a discrete linear stochastic system can be considered a Markov process [1] or Markov chain.

12.4.1 Basic Definitions

Consider four basic definitions of Markov chain theory [10, 11].

Definition 1. Let t_i be elements of the indexed set T such that $t_1 < t_2 < \cdots < t_{n-1} < t_n$. The stochastic process $\{x(t)\}$ is called a Markov process if:

$$\mathrm{Prob}\{x(t_n) \mid x(t_1), x(t_2), \cdots, x(t_{n-1})\} = \mathrm{Prob}\{x(t_n) \mid x(t_{n-1})\} \tag{12.2}$$

where $\mathrm{Prob}\{x\}$ is the probability of the event x.

Definition 2. A discrete-in-space Markov process is referred to as a Markov chain:

$$\begin{aligned}&\mathrm{Prob}\{x(t_n) = i_n \mid x(t_1) = i_1, x(t_2) = i_2, \cdots, x(t_{n-1}) = i_{n-1}\}\\ &\quad = \mathrm{Prob}\{x(t_n) = i_n \mid x(t_{n-1}) = i_{n-1}\}\end{aligned} \tag{12.3}$$

An rth-order Markov chain is characterised by the relation:

$$\begin{aligned}&\mathrm{Prob}\{x(t_n) = i_n \mid x(t_1) = i_1, x(t_2) = i_2, \cdots, x(t_{n-1}) = i_{n-1}\}\\ &\quad = \mathrm{Prob}\{x(t_n) = i_n \mid x(t_{n-r}) = i_{n-r}, \cdots, x(t_{n-1}) = i_{n-1}\}\end{aligned} \tag{12.4}$$

The current state of a simple Markov chain depends solely on its previous state. The current state of an rth-order Markov depends on r previous states only.

Definition 3. If the transition probabilities are independent of n:

$$\mathrm{Prob}\{x(t_n) = X_j \mid x(t_{n-1}) = X_i\} = P_{ij} \tag{12.5}$$

then the Markov chain is called homogeneous.

Definition 4. If there is a control parameter u discrete in space and the following relation is fulfilled:

$$\mathrm{Prob}\{x(t_n) = X_j \mid x(t_{n-1}) = X_i, u(t_{n-1}) = U_k\} = P_{ijk} \tag{12.6}$$

then this is a homogeneous first-order controlled Markov chain. Further in the text, it will be referred to as a controlled Markov chain.

12.4.2 Markov Chain Representation of Dynamic Systems

Ordinary differential and difference equations are conventional forms of dynamic systems description in control theory. Consider a process described by the stochastic differential state-space equation:

$$\frac{d\mathbf{X}}{dt} = \mathbf{AX}(t) + \mathbf{F}\xi(t) \tag{12.7}$$

where $\mathbf{X}$ is an n-dimensioned state vector, $\mathbf{A}$ and $\mathbf{F}$ are matrices of appropriate dimensions, and $\xi(t)$ is an independent random vector.

Equation (12.7) is a Markov process. This result is known as the first Doob's theorem, and its proof can be found, for example, in [11]. If $\xi(t)$ in Equation (12.7) is not white noise but a stationary Gaussian process with the correlation time much less than the system memory time, then Equation (12.7) can be approximately considered a Markov process [11].

Consider a plant described by the following differential equation:

$$\frac{dX}{dt} = AX(t) + BU(t) + F\xi(t) \tag{12.8}$$

where $X(t)$ is the state coordinate; $U(t)$ is the control coordinate; A, B, and F are constant coefficients and $\xi(t)$ is white noise.

The solution of Equation (12.8) with $X(0) = X_0$ is as follows:

$$X(t) = X_0 e^{-At} + Be^{-At} \int_{t_0}^{t} Ue^{A\tau} d\tau + Fe^{-At} \int_{t_0}^{t} \xi e^{A\tau} d\tau \tag{12.9}$$

Designate the sum of two last terms in Equation (12.9) as $C(t)$ and consider three time moments $t_3 > t_2 > t_1 > 0$. Then:

$$X(t) = e^{-A(t_3-t_2)}[X_0 e^{-At_2} + C(t_2) + C(t_3 - t_2)] = X(t_2)e^{-A(t_3-t_2)} + C(t_3 - t_2) \quad (12.10)$$

Provided $X(t_2)$ is known and the integral:

$$\int_{t_2}^{t_3} U e^{A\tau} d\tau$$

is independent of $X(t_1)$, then $X(t_3)$ is independent of $X(t_1)$. Hence, the following relation defines the state probability:

$$\text{Prob}\{X_3, t_3 \mid X_2, t_2, X_1, t_1\} = \text{Prob}\{X_3, t_3 \mid X_2, t_2\} \quad (12.11)$$

and the process Equation (12.9) is a Markov process.

Transform differential equations into difference equations and consider the following closed-loop control system:

$$\mathbf{X}(t_n) = \mathbf{AX}(t_{n-1}) + \mathbf{BU}(t_{n-1}) + \mathbf{F}\xi(t_{n-1}) \quad (12.12)$$

$$\mathrm{U}(t_n) = \mathbf{RX}(\mathrm{t}_{n-1}) \quad (12.13)$$

where $\mathbf{X}(t_n)$ is the state vector of the plant; $\mathrm{U}(t_n)$ is the control vector; $\xi(t_n)$ is the vector of random disturbances; **A**, **B**, **F** and **R** are matrices of appropriate dimension.

The system Equation (12.13) can be described as an autoregression process:

$$\mathbf{X}(t_{n-1}) = \mathbf{A}\,\mathbf{X}(t_{n-1}) + (\mathbf{B}\,\mathbf{R})\,\mathbf{X}(t_{n-2}) + \mathbf{F}\,\xi(t_{n-1}) \quad (12.14)$$

If $\xi(t_n)$ is a statistically independent signal, then Equation (12.14) can be considered a second-order Markov process. A change of variables with an increase in dimension enables the process to be reduced to a first-order complex Markov process. As a result, a dynamic system described by the difference equation Equation (12.13) with the input vector **U** and the state vector **X** is a controlled Markov process.

Level quantisation allows the Markov process Equation (12.14) to be transformed into a Markov chain. Given the random process $\xi(t)$ is stationary, then the Markov chain Equation (12.14) is homogeneous. A homogeneous Markov chain is described with the stochastic matrix of transition probabilities with the dimensions $m \times m$, where m is the number of the chain states [12]. Each element of the stochastic matrix represents the probability of the transition from the state X_i to the state X_j over the sampling period $T_s = t_n - t_{n-1}$:

$$P_{ij} = \text{Prob}\{X(t_n) = X_j \mid X(t_{n-1}) = X_i\} \qquad \sum_{i=1}^{m} P_{ij} = 1 \tag{12.15}$$

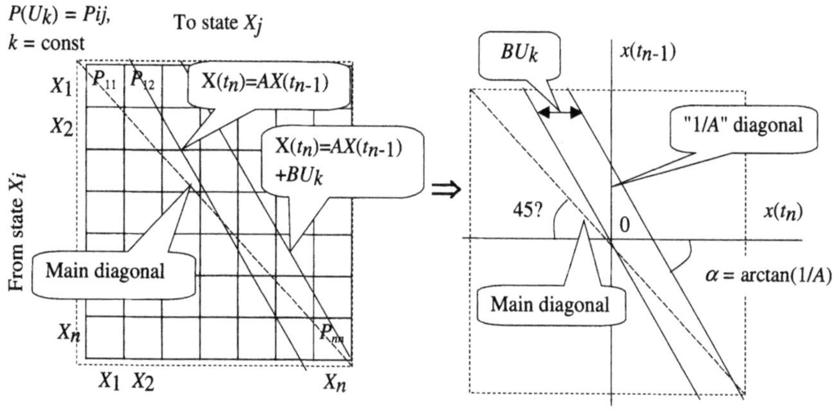

Figure 12.2. Diagram of transition probability matrix of difference equation

The stochastic matrix **P** in Equation (12.15) describes a complete set of events where the sum of probabilities is one. In SISO systems, the probability λ_i of the coordinate x being in the state x_i can be determined by the formula:

$$\lambda_i = \text{Prob}\{x \in [x_i - \Delta x/2, x_i + \Delta x/2]\} = \int_{x_i-\Delta x/2}^{x_i+\Delta x/2} P(x)dx \tag{12.16}$$

where x_i is the centre of the ith quantisation interval, and Δx is the quantisation interval.

A controlled Markov chain can be represented by a number of stochastic matrices of transition probabilities $\mathbf{P}(U_k)$,where U_k is the centre of the kth interval. In this case, each control has the corresponding transition probability matrix.

Consider a system described by a determined difference equation Equation (12.12) with $\xi(t) = 0$. Given $\Delta x \to 0$, the corresponding stochastic matrix of transition probabilities becomes a diagonal deterministic matrix, where the probabilities of state-to-state transitions are zero or one. Figure 12.2 gives a cross-sectional view of the matrix $\mathbf{P}(U)$ under fixed control $U=U_k$. The inclination angle of the unit line is defined by arctan($1/A$), where A is the autoregression coefficient; the line is shifted horizontally by BU_k.

Real physical plants are usually influenced by random disturbances from the environment and by random deviations of the system parameters. A stochastic difference equation describes a dynamic system with random transitions determined by the matrix **P** in Equation (12.15). The Markov chain of Equation (12.14) can be considered a nonparametric form for description of stochastic dynamic systems. A visual analogy for Markov modelling is a histogram, which is a nonparametric description of the probability density function of a random variable. This type of dynamic models is suitable for identification, modelling and real-time simulation of stochastic plants.

12.5 Basic Descriptive Properties

Consider basic descriptive properties derived from the representation of stochastic dynamic systems as a homogeneous Markov chain. A nonparametric Markov model can be considered partial parameterisation of experimental data.

The vector of the system state probabilities at the *i*th time moment:

$$\boldsymbol{\lambda}(t_i) = [\text{Prob}\{x(t_i) = X_1\}, \cdots, \text{Prob}\{x(t_i) = X_n\}] = [\lambda_1(t_i), \cdots, \lambda_n(t_i)] \tag{12.17}$$

is determined by the Markov formula, given the initial condition vector $\boldsymbol{\lambda}(t_0)$ and the transition probability matrix **P**:

$$\boldsymbol{\lambda}(t_i) = \boldsymbol{\lambda}(t_0) \cdot \mathbf{P}^i \tag{12.18}$$

In the case of a controlled dynamic system, the formula Equation (12.18) transforms into the following:

$$\boldsymbol{\lambda}(t_i) = \boldsymbol{\lambda}(t_{i-1})\mathbf{P}(u(t_{i-1})) = \boldsymbol{\lambda}(t_0)\prod_{n=1}^{i-1} \mathbf{P}(u(t_{n-1})) \ , \tag{12.19}$$

where $\mathbf{P}(u(t_n))$ is the transition probability matrix corresponding to the control $u(t_n)$.

12.5.1 Probability Density Function

The final stationary distribution of the state probabilities is estimated via the eigenvector of the transition probability matrix:

$$\boldsymbol{\lambda} \cdot \mathbf{P} = \boldsymbol{\lambda} \tag{12.20}$$

where $\boldsymbol{\lambda} = [\lambda_1, \lambda_2, \cdots, \lambda_n]$ is the vector of final probabilities λ_n [10]. Note that the matrix **P** in the infinite power consists of *n* vectors $\boldsymbol{\lambda}^T$ taken as columns:

$$\mathbf{P}^\infty = [\boldsymbol{\lambda}^T, \boldsymbol{\lambda}^T, \cdots, \boldsymbol{\lambda}^T] \tag{12.21}$$

In this work, a system operating at steady-state conditions is analysed, hence the state probabilities are stationary from the very first time moment of the realisation of a random process. So, the histogram estimate of the state probabilities must produce the same result as the calculation of the eigenvector $\boldsymbol{\lambda}$. This property can be used for estimating the degree of stationarity of a stochastic process.

12.5.2 Mean Value

Assume that $x(t)$ and Prob(x) are constant over the time sampling interval. Similar statements are usually assumed in zero-order interpolation of discrete systems in conventional control theory. The mean value is then expressed as follows:

$$\mu_x = E[x] = \int_{-\infty}^{+\infty} x \cdot \mathrm{Prob}(x)dx = \sum_{i=1}^{n} X_i \lambda_i = \mathbf{X}\boldsymbol{\lambda}^{\mathrm{T}} \tag{12.22}$$

where n is the number of the system states, and $\mathbf{X} = [X_1, X_2, \cdots, X_n]$ is the vector of the states, or the vector of the interval centres.

12.5.3 Correlation Function

The correlation function can be estimated from the Markov chain parameters via the following formula:

$$\begin{aligned} R_{xx}(m) &= \int_{-\infty}^{\infty}\int (x(t)-\mu_x)(x(t+m)-\mu_x)\cdot \mathrm{Prob}(x(t), x(t+m))dxdx = \\ &= \mathbf{X}\boldsymbol{\Lambda}\mathbf{P}^m\mathbf{X}^T - \left(\mathbf{X}\boldsymbol{\lambda}^T\right)^2 \end{aligned} \tag{12.23}$$

where m is the correlation time lag and $\boldsymbol{\Lambda}$ is the diagonal matrix of the final probabilities:

$$\boldsymbol{\Lambda} = \begin{pmatrix} \lambda_1 & 0 & 0 & \cdots & 0 \\ 0 & \lambda_2 & 0 & \cdots & 0 \\ \vdots & \vdots & \vdots & \ddots & \vdots \\ 0 & \cdots & 0 & 0 & \lambda_n \end{pmatrix} \tag{12.24}$$

Comparison of direct and Markov estimates of the correlation function Equation (12.23) demonstrates that dynamic properties of a first-order Markov chain are equivalent to a first-order parametric model.

12.5.4 Parameters of Autoregression Model

The autoregression coefficients A, B and F of Equation (12.14) can be obtained using the transition probability matrix $\mathbf{P}$. Multiply both parts of Equation (12.14) by $X(t)$ and find the mathematical expectation:

$$R_{xx}(1) = AR_{xx}(0) \tag{12.25}$$

where A is the autoregression coefficient.

Note that more than two points of the correlation function are available. Accounting for the whole range of the correlation lag gives $M + 1$ equations:

$$\begin{pmatrix} R_{xx}(1) \\ \vdots \\ R_{xx}(M) \end{pmatrix} = \begin{pmatrix} R_{xx}(0) \\ \vdots \\ R_{xx}(M-1) \end{pmatrix} A \tag{12.26}$$

where M is the maximum delay. Writing this in the matrix form:

$$\mathbf{F} = \mathbf{G}A \tag{12.27}$$

where $\mathbf{F} = \begin{pmatrix} R_{xx}(1) \\ \vdots \\ R_{xx}(M) \end{pmatrix}$ and $\mathbf{G} = \begin{pmatrix} R_{xx}(0) \\ \vdots \\ R_{xx}(M-1) \end{pmatrix}$

The optimum value of M must exist, ensuring the compromise between the variance and bias of the estimate A. The value of M greater than optimum does not produce further improvement of the estimate. The final expression employs the pseudo-inverse operation:

$$A = \frac{G^T F}{G^T G} \tag{12.28}$$

The use of the whole range of the lag values compensates the disadvantage of AR-estimates by the prediction error method, which is caused by greater contribution of high frequencies and insufficient contribution of a low-frequency spectrum in the statistical estimate.

12.6 Identification of Markov Chains

Application of Markov modelling allows stochastic approximation models to be created from experimental data. Systems identification using a controlled Markov chain requires the order of the process to be determined. The dimension of the transition probability matrix depends on the order of the stochastic difference equation and the width of the intervals Δx and Δu.

High-order models describe the complex plant better. However, the transition probability matrix becomes sparse due to conditional independence of subsystems, thus leading to hierarchical modelling with simple local structures. In order to optimise the model structure and parameters, nongradient genetic algorithms can be used because of the discontinuous character of the problem. An additional advantage of the multiobjective genetic algorithms (MOGA) is Pareto optimisation. The algorithms search for a set of best solutions, not a single optimal point. Finally, the best solution is chosen based on visual analysis and expert knowledge.

After determining the process order, the transition probability matrix is estimated. A first-order process gives a three-dimensional (3-D) transition probability matrix. Its elements are estimated by calculating the relative frequencies of the state-to-state transitions:

$$P_{ijk} = \frac{N_{ijk}}{\sum_{i=1}^{n} N_{ijk}} \tag{12.29}$$

where N_{ijk} is the number of the system transitions from the state X_i to the state X_j under the control U_k during a time sampling interval.

A more complex formula for estimating the transition probabilities is as follows:

$$P_{ijk} = \frac{\sum_{n} x(t_n)\varphi_j u(t_n)\psi_k x(t_{n+1})\phi_i}{\sum_{n} x(t_n)\varphi_j u(t_n)\psi_k} \tag{12.30}$$

where $x(t_n)$ and $u(t_n)$ are the current values of x and u; φ_j is the probability density function of the jth state; ϕ_i is the probability density function of the ith state; ψ_k is the probability density function for the kth control.

Simulation of the Markov chain is performed using Monte Carlo methods.

12.7 Fuzzy Markov Chains

The Markov chain described above can be considered a fuzzy system with a rectangular membership function with no overlap, as shown in Figure 12.3. The terms X_i, X_j, and U_k can be substituted with conventional linguistic variables like PB, ZO, NB, *etc*. This section focuses on general features of fuzzy Markov modelling. The results are then to be generalised for numerous fuzzification and defuzzification techniques.

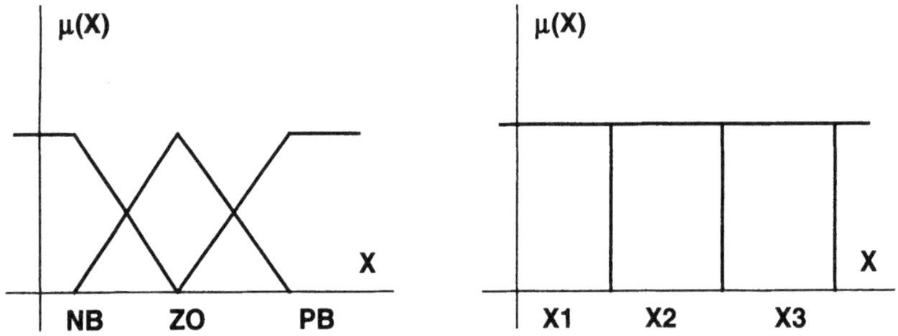

Figure 12.3. Comparison of fuzzy (left) and Markov (right) membership functions

Most fuzzy logic applications are intended for control and analysis purposes [12, 13]. Another group of applications is the system state prediction [14]. Conventional fuzzy systems cannot operate with random phenomena. On the other hand, control processes in real-life plants consist of determined and random components. Stochastic processes can be described using a Markov modelling approach [15], which provides high computational speed because it utilises only short operations of move and comparison. However, this approach allows simulation of a limited number of system states depending on state quantisation. Furthermore, the transition probability matrix must have large size to achieve high accuracy of modelling. This disadvantage can be avoided using a combination of Markov modelling with fuzzy logic.

In order to extend the application area of both techniques, a fuzzy Markov modelling approach was proposed [16]. Fuzzy systems are often referred to as "universal approximators" [17]. Therefore, fuzzy Markov systems could be used for smooth nonlinear approximation of a multidimensional probability density function. In this case, the Markov model represents a fuzzy inference system with the transition probability matrix stored within the rule base. Recently, Adaptive-Network-Based Fuzzy Inference Systems (ANFIS) were used for chaotic time series prediction [14]. Similarly, stochastic time series simulation can be carried out using fuzzy inference combined with Markov modelling.

The procedure of fuzzy Markov simulation includes four main stages: fuzzification, inference, defuzzification and randomisation; see Figure 12.4. Before performing simulation, the fuzzy system structure is chosen and the rule base is created from experimental data in the form of a multidimensional transition probability matrix **P**. Tuning of fuzzy systems often involves the use of artificial neural networks or evolutionary computation methods. Possible criteria for optimisation of a fuzzy Markov system can be the mean square errors of simulating the spectral and distribution properties compared with desired characteristics. In order to fulfil the criteria, the following degrees of freedom can be used: the order of the Markov model, the model of fuzzy inference, type, number and position of membership functions.

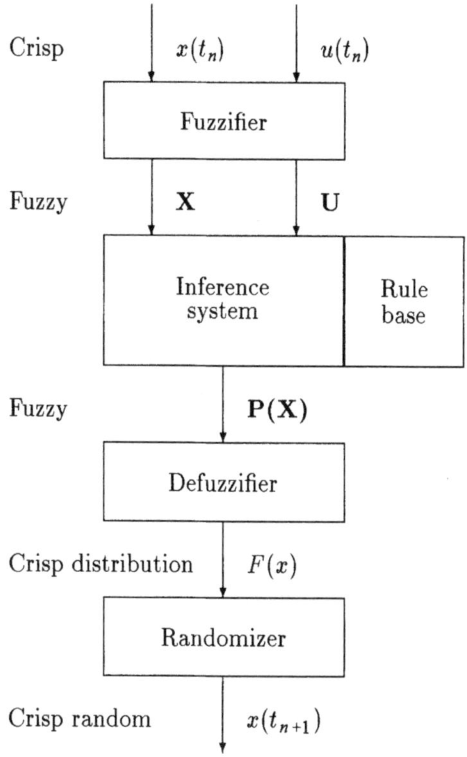

Figure 12.4. Fuzzy Markov system

A block diagram of fuzzy Markov simulation is shown in Figure 12.4. Firstly, the control input $u(t_n)$ and output $x(t_n)$ are fuzzified using the membership functions. The fuzzy values **U** and **X** with their membership degrees are used to infer the rules from the rule base. Having been extracted, the fuzzy rules are aggregated into the fuzzy probability density function **P(X)** of the output $x(t_{n+1})$.

After defuzzification, integration, and normalisation, the distribution function $F(x)$ is obtained. The normalisation places the function $F(x)$ within the interval [0, 1].

$$F(x) = \text{Prob}\{x(t+1) \leq x\} = \frac{\int_{-\infty}^{x} p(x)dx}{\int_{-\infty}^{+\infty} p(x)dx} \tag{12.31}$$

where $p(x)$ is the result of defuzzification. Finally, a random value $x(t_{n+1})$ with the desired distribution $F(x)$ is generated using a functional transform of a uniformly distributed random number y [18]:

$$x = F^{-1}(y) \tag{12.32}$$

The introduction of the modified defuzzification and randomisation stages allows fuzzy simulation of a stochastic dynamic system to be easily performed.

The rule base contains all information about the distribution and spectra in the transition probability matrix. This simple description becomes possible because the matrix simultaneously determines two types of system properties. It directly defines probabilities of transitions between states and indirectly determines final probabilities of states of the Markov chain. Conventional methods for stochastic modelling would require more complicated techniques to reflect both distribution and spectral properties.

The fuzzy Markov modelling methodology has been proposed to extend the application area of fuzzy systems. It easily describes any form of probability distribution. In addition to conventional stages of fuzzy modelling, Markov modelling also includes randomisation. This procedure transforms defuzzified "crisp" probability distribution into the random output signal. Application areas for fuzzy Markov modelling include simulation of complex stochastic systems, stationarity and stability analysis, systems identification and optimal control.

12.8 Concluding Remarks

In this chapter, the Markov modelling technique has been considered for automatic control of aero engines. The use of controlled Markov chains enables the stochastic dynamics to be incorporated into a deterministic model. Main application areas for the technique are test bed simulation, optimal control, and condition monitoring of gas turbine engines. The problem of the model structure selection can be devised using evolutionary optimisation techniques. Finally, a fuzzy Markov modelling methodology has been proposed, further improving the Markov chain techniques and extending the application area of fuzzy inference systems.

Chapter 13
Markov Modelling of Turbo Prop Fan

13.1 Introduction

In this chapter, the Markov modelling techniques developed in Chapter 12 are applied to identification and test bed simulation of a turbo prop fan engine, operating in the real-life random environment. The parameters of controlled Markov chains are identified from on-ground engine test bed data. Basic properties including standard deviations and power spectral densities are then analysed. Experiments with hardware-in-the-loop (HIL) simulation of the identified Markov models are performed.

13.2 Experimentation with Turbo Prop Fan

In this section, Markov chain techniques are applied to aero engines representing complex stochastic plants. Adequate modelling is essential for control quality assessment at the demonstration and mass production stages of the control system life cycle. The controller must effectively operate with both deterministic and stochastic dynamics of the plant. Analysis of random disturbances in Section 12.3 shows that the gas turbine engine should be considered a stochastic system.

Experimental data from a turbo prop fan engine is used in this example. The engine was tested on a ground-based test bed along with a digital controller. The prop fan engine has three shafts and is controlled by a FADEC electronic system. The experimental data was used to identify parameters of Markov models of the inlet air and engine dynamics. The obtained models were then investigated at the systems test facility.

The dynamics of the gas-producing turbine from the "minimum thrust" up to "take-off" operation points are described by the following state-space differential equations:

$$\begin{aligned} \frac{d\mathbf{X}}{dt} &= \mathbf{AX} + \mathbf{BU} + \mathbf{F}_1\xi_1 \\ \mathbf{Y} &= \mathbf{CX} + \mathbf{DU} + \mathbf{F}_2\xi_2 \end{aligned} \qquad (13.1)$$

where $\mathbf{U} = [W_f]^T$ is the control vector; $X = [n_{\mathrm{LP}}, n_{\mathrm{HP}}]^T$ is the state vector; $Y = [p_c, T_t]^T$ is the observation vector; ξ_1 and ξ_2 are the vectors of random disturbances; **A**, **B**, **C**, **D**, $\mathbf{F}_1$ and $\mathbf{F}_2$ are the matrices defined at a particular operating point.

Measured physical values of the engine coordinates are reduced to sea-level conditions where appropriate using the reduction formulae; see Table 3.1. Random processes ξ_1 and ξ_2 can be considered quasi-stationary Gaussian random processes with small correlation time of 0.025 – 0.1 s. Using level quantisation, the system is modelled with a controlled Markov chain.

The size of the transition probability matrix must provide the desired quality of the model. In this example, the quantisation level of 10% of the coordinate amplitude produces sufficient modelling accuracy because of the measurement noise.

The choice of the model order is performed using *a priori* knowledge of the engine properties. For example, the gas-producing turbine has two major accumulators of energy: two rotating turbocompressor rotors. The turbine dynamics are usually described by a second-order differential equation. It is reasonable that the Markov model should have the same structure. However, the order of the Markov model defines the size of the stochastic matrix. The choice of the Markov model order is performed in the following sequence:

- the order is chosen using *a priori* information of the engine dynamics;
- the stochastic matrix is estimated from experimental data;
- modelling accuracy is monitored via comparison of the real and simulated data;
- the model order is increased or decreased to obtain the best accuracy.

The model order reduction enables the computational load to be reduced. Results of turbo prop fan identification are given in subsequent sections. Matrix dimensions are $10 \times 10 \times 10$ for the first-order and $10 \times 10 \times 10 \times 10$ for the second-order models.

13.3 Identification of Gas Turbine Dynamics

The experiments were performed at the on-ground engine test bed under closed-loop control of compressor pressure ratio π_c at steady-state conditions. Fragments of random realisations of the signals $p_{in}, \alpha_{mv}, W_f, n_{\mathrm{HP}}, p_c, T_t$ at the operating point "85% of the maximum thrust" were chosen. The sampling frequency is 39 Hz; and the record length is 1 min, or 2340 samples of the data. First- and second-order transition probability matrices were identified from the data. Monte Carlo simulation was then conducted using the identified matrices. The realisations of the signals (in deviations from mean values) and power spectral densities of the simulated and real data are compared in Figures 13.1–13.10. Standard deviation

(STD) errors of modelling are presented in Table 13.1 in terms of absolute errors and percent of mean values.

A first-order model is sufficiently accurate for modelling of the fuel flow dynamics. In modelling of the turbine dynamics, second-order models match the real data, being in agreement with *a priori* information of the dynamic model structure.

Table 13.1. Markov modelling errors

Signal	Model order	
	First	**Second**
W_f	7.77 kg/h / 0.42%	---
n_{HP}	11.21 rpm / 0.06%	5.88 rpm / 0.03%
P_c	5.25 Pa / 0.29%	3.79 Pa / 0.21%
T_t	2.72 K / 0.35%	2.41 K / 0.31%

Example contour plots of the transition probability matrix for the fuel flow dynamics is shown in Figure 13.11. Note that maximum probability density is situated around the diagonal line representing deterministic dynamic properties of the system. Using the identified transition probability matrix, a scatter diagram "shaft speed *versus* fuel flow" is obtained via the final probabilities of the Markov chain; see Figure 13.12.

The vector of the final probabilities can be obtained using the Markov formula for homogeneous controlled Markov chains. This enables the probability density function of the engine coordinates to be determined for the time moment $t_n > t_0$ using the transition matrix and initial state. Figure 13.13 shows the 99% probability boundaries of fuel flow compared with the real data.

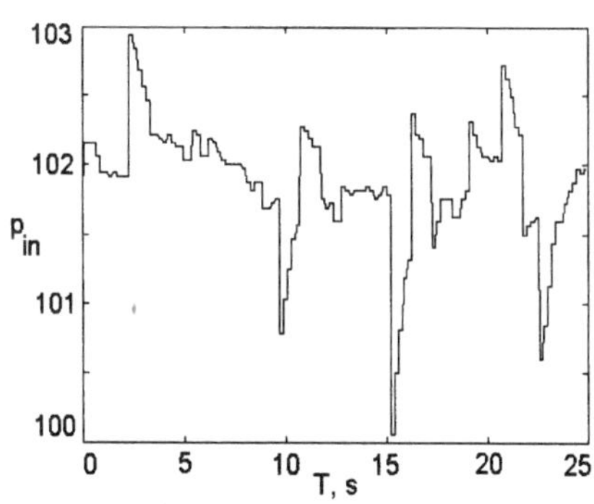

Figure 13.1. Realisation of inlet pressure p_{in}

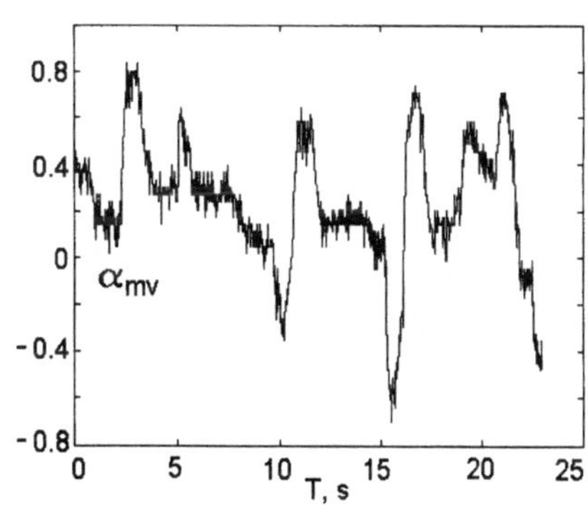

Figure 13.2. Angular position of fuel metering valve

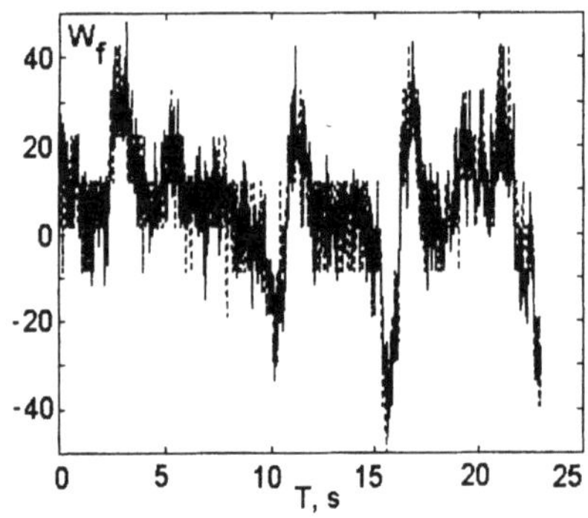

Figure 13.3. Fuel flow (solid) and first-order simulation (dotted)

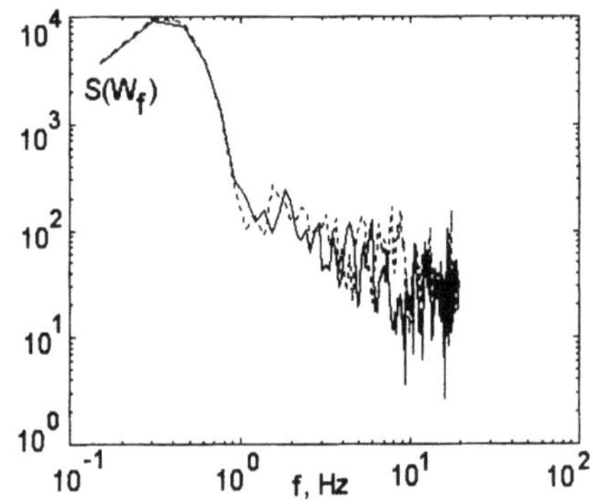

Figure 13.4. Spectral densities of fuel flow (solid) and first-order model (dotted)

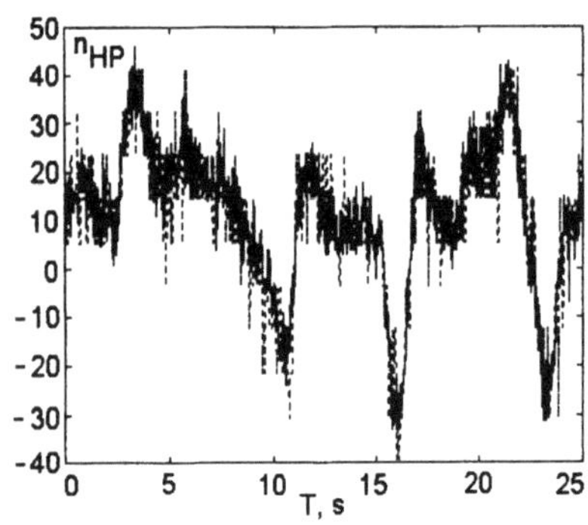

Figure 13.5. HP shaft speed (solid) and second-order simulation (dotted)

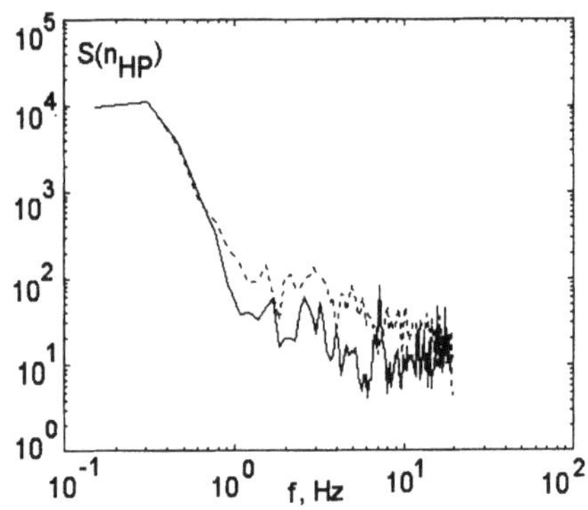

Figure 13.6. Spectral densities of HP shaft speed (solid) and second-order model (dotted)

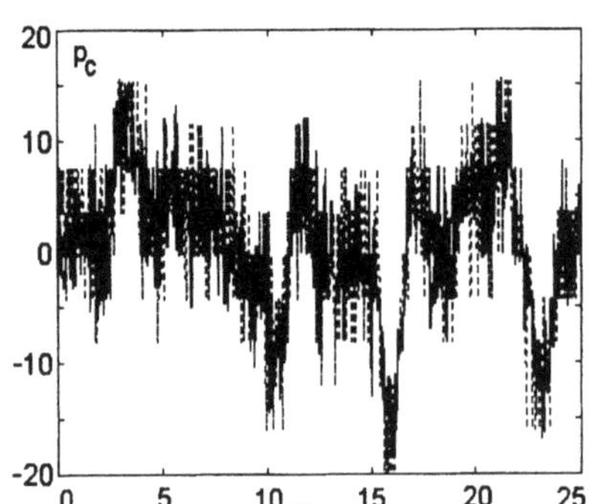

Figure 13.7. Compressor pressure (solid) and second-order simulation (dotted)

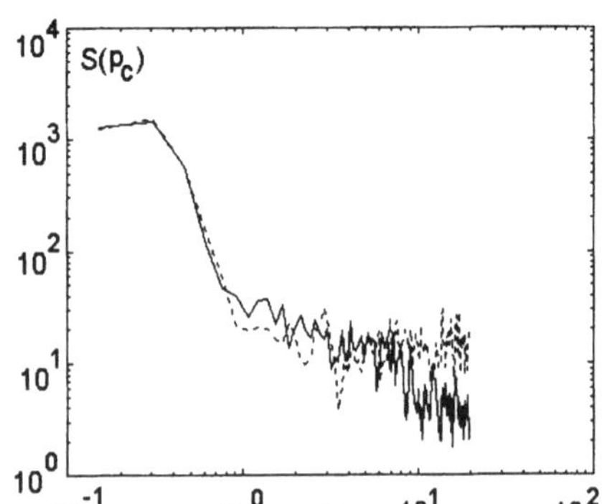

Figure 13.8. Spectral densities of compressor pressure (solid) and second order model (dotted)

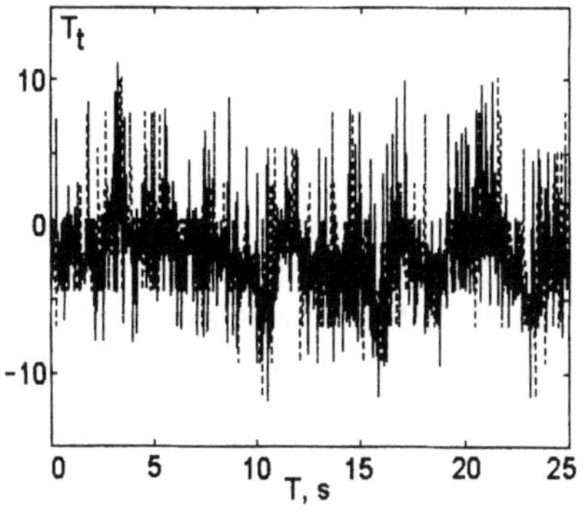

Figure 13.9. Turbine temperature (solid) and second-order simulation (dotted)

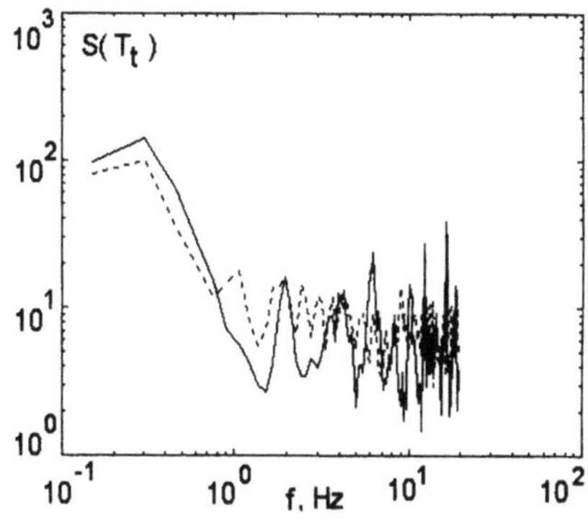

Figure 13.10. Spectral densities of turbine temperature (solid) and second-order model (dotted)

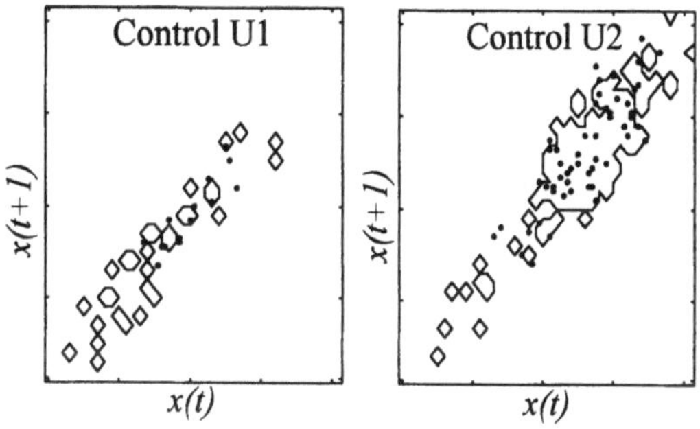

Figure 13.11. Example contour plots of transition probability matrix

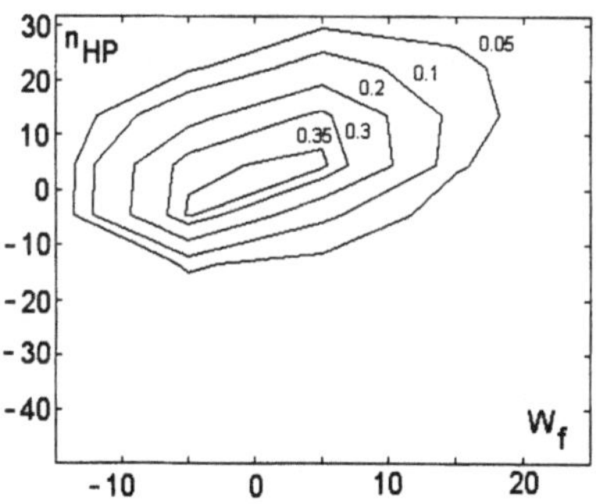

Figure 13.12. Contour scatter diagram "shaft speed *vs.* fuel flow"

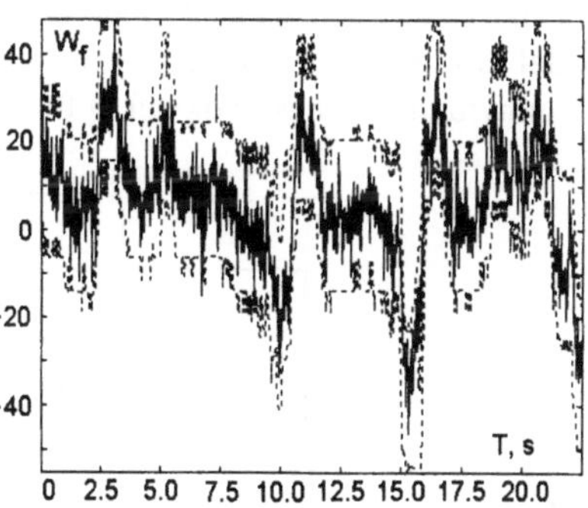

Figure 13.13. Fuel flow (solid) and 99% probability boundaries (dotted)

13.4 Simulation of Random Environment in HIL Tests of Digital Controllers

High requirements for performance of aero engines lead to increases in design quality and characteristics of control systems. In these circumstances the amount of experimentation with controllers sharply increases. Tests involving real engines are the most labour-consuming and expensive stages. The use of simulation test beds considerably reduces the number of joint tests of electronic systems with real engines at engine test beds and laboratory aircraft. In such test beds, the engine behaviour is simulated with analogue or digital electronic models. A typical test bed consists of electronic simulators of sensors and actuators of the control system, the simulating computer, and the data-logging device. The computer performs real-time simulation of the engine operation.

Most test facilities reflect the characteristics of a general engine and ideal operating conditions, neglecting the random environment and deviation of parameters of the engine and controller. However, in order to perform adequate test bed experiments, the real operating conditions of the engine and controller must be accurately simulated. This would enable the accuracy of control and reliability of the electronic unit to be investigated. In some applications, a deterministic model of the engine is added with the models of some random disturbances, which usually consist of pseudo-random generators with forming filters. Modelling of random disturbances represents a complicated problem because most of these factors are not measured directly. The problem is resolved in this chapter using stochastic models in the form of controlled Markov chains.

A general structure of a test bed with stochastic simulation is presented in Figure 13.14. The upper positions of switches 1 and 2 correspond to conventional simulation of deterministic dynamics of the plant. In practice, a particular control unit must be tested with the individual model of a concrete engine along with the model of expected random disturbances. Switching to the lower positions activates the individual stochastic model of the engine.

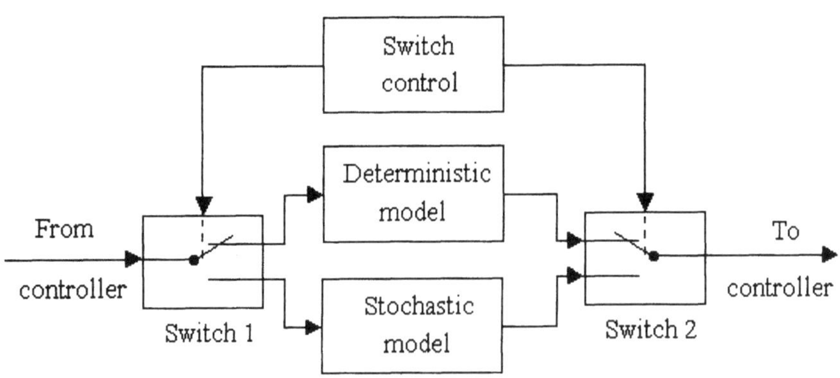

Figure 13.14. Test bed with conventional and stochastic simulation

13.4.1 Markov Simulation Technique

Simulation of the Markov models is performed using the Monte Carlo method. Consider a first-order model of the engine. Before testing of the electronic controller at the simulation test bed, the following preparations are carried out. Firstly, the vector of possible engine states $\mathbf{X} = \{X_i\}$ is chosen with M elements covering the range of the coordinate $x(t)$. Similarly, the vector of possible controls $\mathbf{U} = \{U_j\}$ is built with N elements. The transition probability matrix $\mathbf{P} = \{P_{ijk}\}$ of the dimensions $N \times M \times M$ is identified. Each element P_{ijk} is the probability of the transition from the state X_i under the control U_j to the state X_k over the sampling time. The vectors **X** and **U** are built based on *a priori* analysis of experimental data from a particular engine. Identification of the stochastic matrix **P** is performed as described in Chapter 7.

At each step of stochastic simulation, the next value of the engine state is formed using the Monte Carlo method. Generation a random value with desired distribution is performed using a uniform distribution generator and a functional transform. First, the current input $u(t)$ and output $x(t)$ are measured. Having been compared with the interval centres, the input and output are transformed to the Markov state $\{X_j, U_k\}$ via the following formulae:

$$X_j: \quad j = \arg\min |x(t) - X_j| \tag{13.2}$$

$$U_k: \quad k = \arg\min |u(t) - U_k| \tag{13.3}$$

Second, the vector **p** is extracted from the transition probability matrix **P** using the indices j and k. The elements of the vector **p** represent the corresponding state probabilities $p_i = \text{Prob}\{X_i\}$.

Finally, the obtained distribution is utilised for generating the random output x via a transformation method [1]:

$$F(x) = \int_{-\infty}^{x} p(x)dx \tag{13.4}$$

A uniform random number y is chosen between 0 and 1, as shown in Figure 13.15, and the following transform is applied:

$$x(t_{n+1}) = F^{-1}(y) \tag{13.5}$$

The use of the suggested technique for testing the engine electronic controllers at simulation test beds will provide more adequate modelling of the real operation environment.

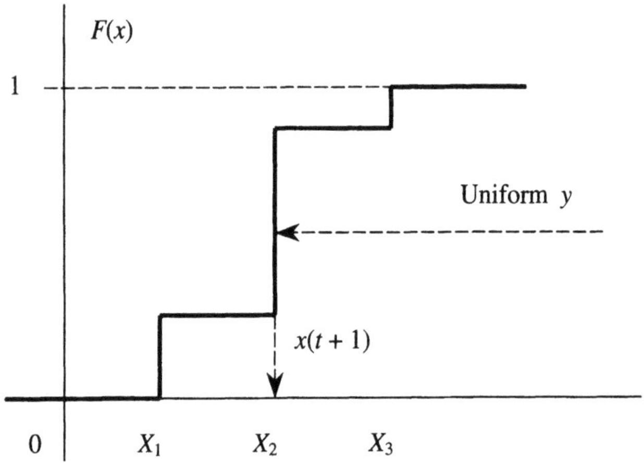

Figure 13.15. Generation of desired distribution

Computational experiments have shown that the time for calculation of the engine Markov model is comparable to that of numerical integration of differential equations. For example, Markov simulation of one minute of the engine operation in the MATLAB environment took 14 s on a Pentium PC. Runge–Kutta integration of a simplified model with differential equations describing the same engine took 11 s, without simulation of random components. This demonstrates the viability of the Markov models for applications in simulation test beds for testing electronic control units.

13.4.2 Modelling of Inlet Pressure and Temperature with Markov Chains

In this section, modelling of random components of inlet air pressure p_{in} and temperature T_{in} is considered. The usual simulation procedure [2] consists of feeding white noise to a forming filter, with parameters identified from experimental data using spectral analysis and transfer function modelling. The major disadvantage of the technique is that non-Gaussian distribution is not simulated here. The use of Markov models for stochastic simulation allows both spectral and distribution properties to be modelled as shown in Chapter 7. In modelling the processes p_{in} and T_{in} , the dimensions of the transition probability matrix **P** are 10×10.

Transition probability matrices were identified from experimental data. Figure 13.16 and 13.17 show spectral densities of the processes at the on-ground engine test bed and simulated signals. First-order Markov models proved sufficiently accurate in describing both processes.

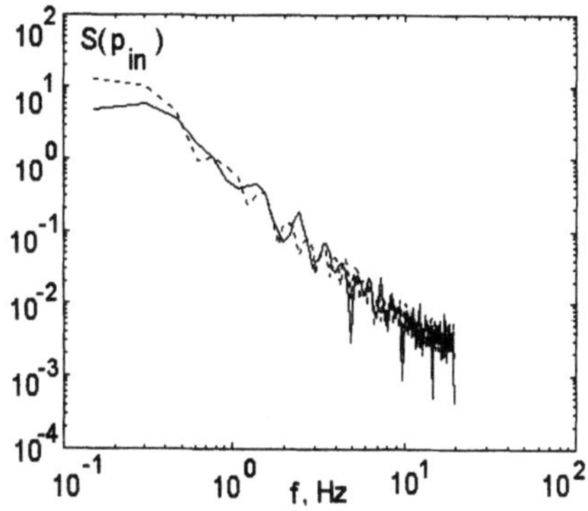

Figure 13.16. Spectral density of inlet pressure at engine test bed (solid) and Markov model (dashed)

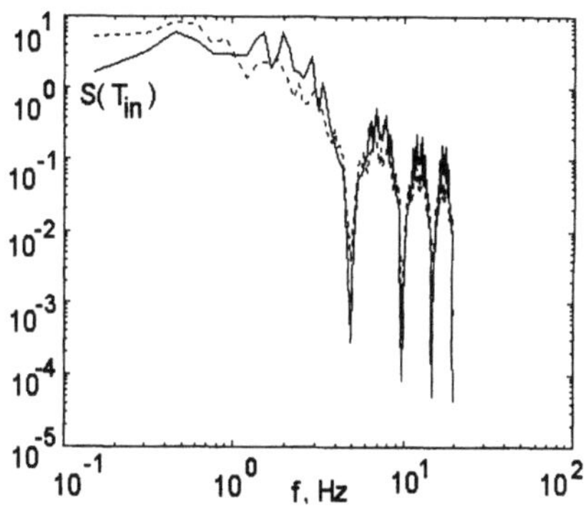

Figure 13.17. Spectral density of inlet temperature at engine test bed (solid) and Markov model (dashed)

13.4.3 Hardware-in-the-Loop Simulation Using Stochastic Models

Experimentation with Markov models of a turbo prop fan was performed at a systems test facility.

The source data are experimental records from the on-ground engine test bed at the operating point "0.85 Maximum thrust" under closed-loop control of compressor pressure ratio π_c. Markov models of the engine and hydromechanical actuators of the control system were identified. Before test bed simulation, the computational experimentation at a personal computer (PC) confirmed the accuracy of the real-time Markov models. First, experimental records of fuel flow were fed to the Markov model in open loop; closed-loop simulation was then conducted under the control laws $\pi_c = \text{const}, T_t = \text{const}$ and $n_{\text{HP}} = \text{const}$. A number of experiments with the HIL simulation test bed were then performed with closed-loop control of π_c; see Table 13.2. During the series of tests, conventional models (CM) in the form of differential equations were gradually substituted with Markov models (MM), see grey-coloured cells. Three Markov models were activated in Test D to simulate inlet pressure, engine dynamics, and fuel pump behaviour.

Power spectral densities of fuel flow and compressor pressure are shown in Figures 13.18 and 13.19. Standard deviations of signals recorded at the on-ground engine test bed and the HIL simulation test bed are presented in Table 13.3. The parameter σ_e/σ denotes the ratio of STD of real signals to that of simulated signals. Analysis of Table 13.3, Figures 13.18 and 13.19 shows that Test D is the most adequate to the real operation environment. Test D activated three Markov models of the system components: inlet pressure, gas-producing turbine dynamics and fuel pump dynamics.

Next, a series of experiments was performed with simulation of the shaft speed dynamics. Test F implemented a conventional RPLDM, whereas Test G activated a second-order Markov model "fuel flow to shaft speed." During both tests, the DECU maintained the control law $n_{HP} = \text{const}$ by affecting fuel flow. Spectral densities of the source and simulated signals are presented in Figures 13.20 and 13.21.

Table 13.2. Markov simulation experiments

Test bed	Test	Inlet pressure p_{in}	Engine: $W_f \to p_c$	Fuel pump: $\alpha_{mv} \to W_f$
HIL	A	const	CM	CM
HIL	B	MM	CM	CM
HIL	C	MM	MM	CM
HIL	D	MM	MM	MM
On-ground engine	E	Inlet air	Engine	Fuel pump

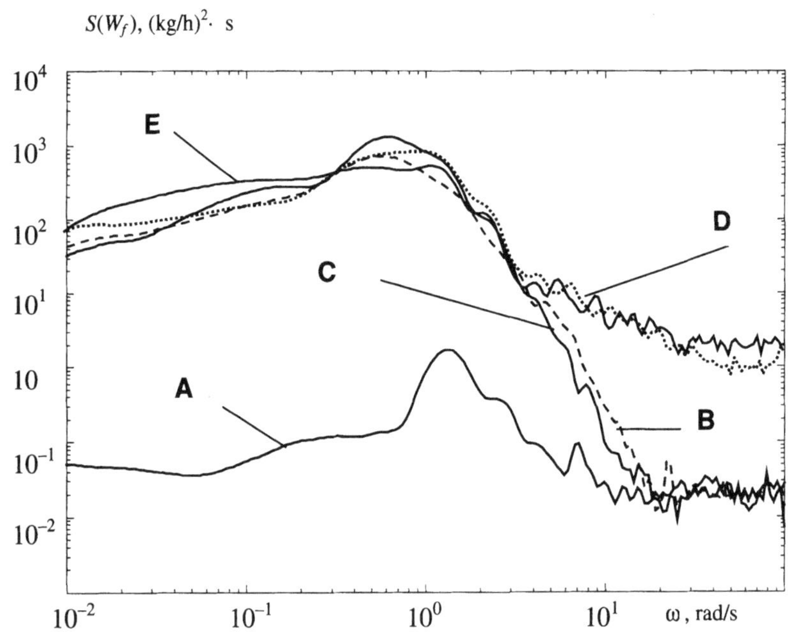

Figure 13.18. Spectral densities of fuel flow

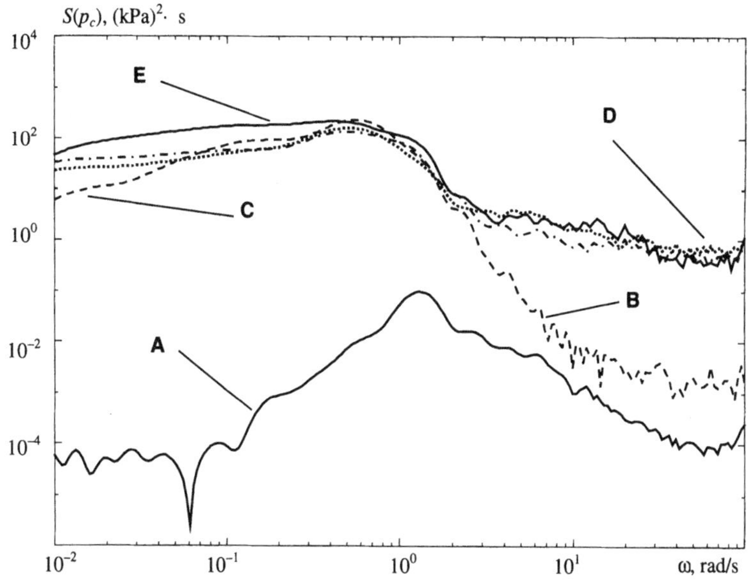

Figure 13.19. Spectral densities of compressor pressure

Table 13.3. Standard deviations at engine test bed and HIL simulation test bed

Parameter		Test				
		A	B	C	D	E
W_f	σ	0.73	14.26	17.99	20.12	15.26
	σ_e / σ	20.9	1.07	0.85	0.76	1.00
p_c	σ	0.15	5.35	6.63	7.48	6.69
	σ_e / σ	44.6	1.25	1.01	0.89	1.00
n_{HP}	σ	0.65	---	16.10	17.79	15.82
	σ_e / σ	24.33	---	0.98	0.89	1.00
n_{LP}	σ	0.49	---	16.89	18.64	16.12
	σ_e / σ	32.9	---	0.95	0.87	1.00

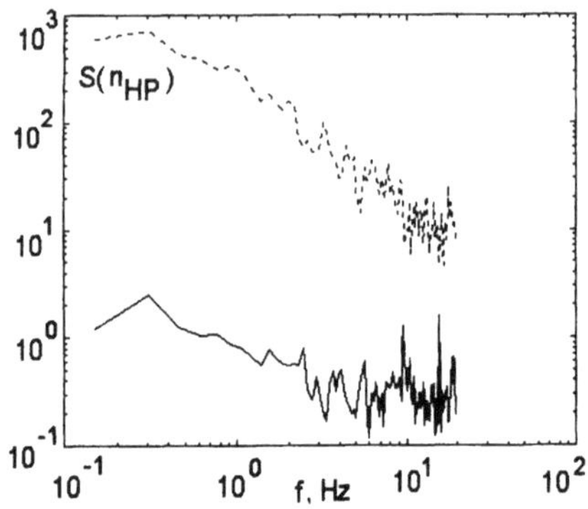

Figure 13.20. Spectral density of HP shaft speed during Test F (solid) and Test G (dotted)

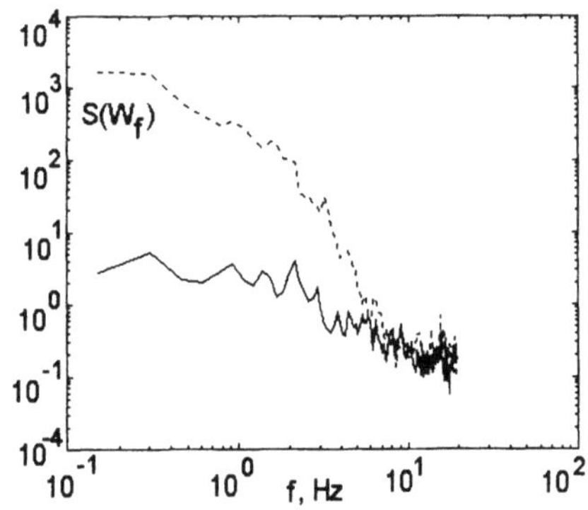

Figure 13.21. Spectral density of fuel flow during Test F (solid) and Test G (dotted)

Hardware-in-the-loop simulation of transients was also performed. A step change in the inlet pressure was fed to the test bed with the conventional and Markov models of the engine. In both cases, the controller was maintaining the law $\pi_c = \text{const}$. The diagram of the transients in p_c is presented in Figure 13.22. Note that the low-frequency component of the Markov step response is very similar to the reaction of the conventional dynamic model. In addition to modelling of the deterministic dynamics, the Markov model also simulates stochastic behaviour of the engine.

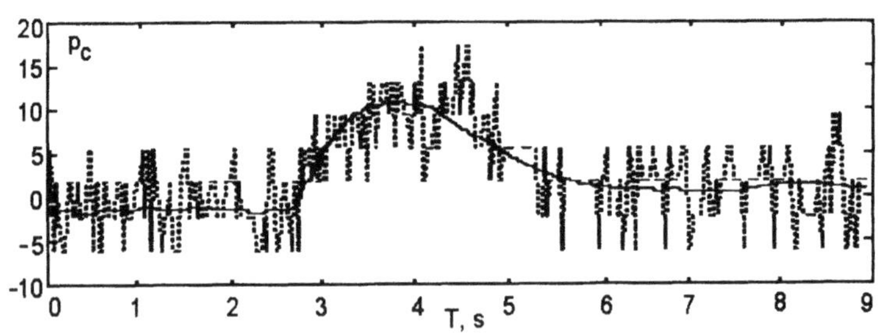

Figure 13.22. Step responses of deterministic (solid) and Markov (dotted) models of compressor pressure

13.5 Markov Modelling in Condition Monitoring of Aero Engines

The use of condition monitoring can reduce maintenance costs and improve flight safety. In condition monitoring it is necessary to consider the engine itself, and the controller that makes the power plant package. In general, direct physical and indirect model-based methods for investigating the current state of the engine components are used. These reflect the approaches of two groups of specialists: mechanical and control engineers. Direct methods include visual inspection, detecting metal particles in lubricant oil, X-ray, vibration and ultrasonic examination, and many more. Another group of methods is based on mathematical modelling of the engine operation. Some approaches cannot be so simply classified. In particular, vibration methods utilise spectral analysis for process modelling, which is an identification problem.

13.5.1 Model-Based Approach to Condition Monitoring

Mathematical modelling methods can detect abrupt and, more importantly, gradual changes in the engine performance due to wearing and deterioration of its components [3-5]. Additional maintenance following the changes detected can prevent serious flight accidents due to equipment failures. Most current engine condition monitoring systems employ static models for diagnosis. However, precise real-time condition monitoring must account for the nonlinear, stochastic nature of the engine. Markov models are particularly interesting for describing the engine dynamics as they lend themselves to simulation of the random processes in the engine. This would lead to far more accurate health monitoring [6, 7].

The problem of parameter estimation for linear stochastic systems has been considered in [8-11] utilising various assumptions. A nonparametric state-space Markov modelling approach has been suggested in [12] for stochastic systems identification. In this section, an example application of a Markov-model-based method is considered for condition monitoring of a gas turbine engine [13]. The Markov model is built using experimental data from the engine test bed. The model is then applied to a real-time simulation. A comparison of the Markov model performance with nonlinear engine models is also conducted.

13.5.2 Condition Monitoring in Dual-Lane Control

In digital controllers of aero engines, dual-lane redundancy is often introduced. These systems include two identical sets of transducers, wiring, A/D and D/A converters and control computers. This hardware redundancy improves the reliability of the information measurement channels in the case of a single fault. Both lanes simultaneously measure engine parameters during the controller operation and mutually exchange the current data. This enables condition monitoring to be conducted via the comparison of the measurements of engine parameters gathered from both lanes. One lane is controlling the engine operation, whereas another channel is waiting in "hot back-up."

Consider a dual-lane control system with two identical lanes. Experimental data is obtained from two information channels which measure the same input and output signals. This test was performed at the on-ground engine test bed. Amongst the variables measured, fuel flow W_f and HP and LP shaft speeds n_{HP} and n_{LP}

are considered. The measurement scheme is given in Figure 13.23, with no feedback shown.

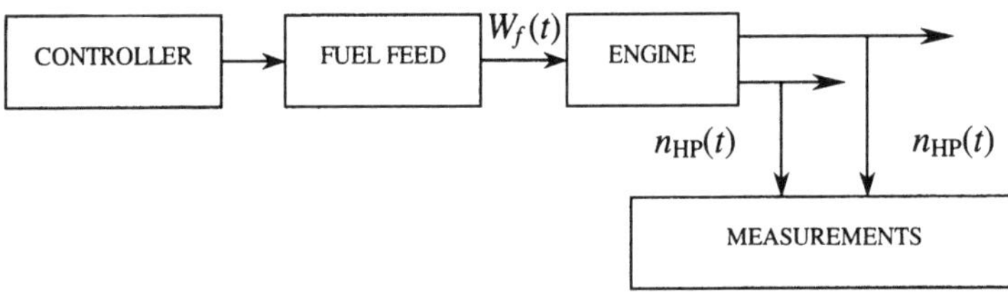

Figure 13.23. Signal measurement scheme

Since two independent transducers measure the same variable, information redundancy exists. Figure 13.24 shows HP shaft speed measurements (in deviations from mean value) from both lanes and the difference between them.

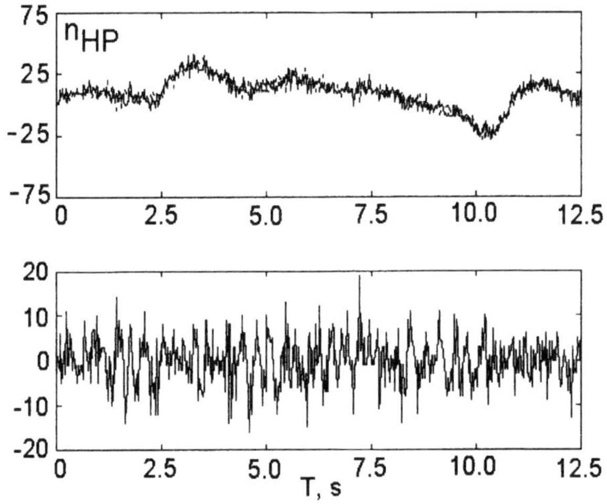

Figure 13.24. Data from measurement channels (top) and difference (bottom)

In order to detect single-sensor faults like drift and malfunction, the maximum allowed measurement difference is considered in two lanes, as demonstrated in [13]. However, this method does not provide sufficient information to detect the faulty transducer and faults in the engine itself. This problem is resolved via accurate modelling of the engine and its controller to organise the third "virtual" lane in a majority-voting scheme. Explicit modelling of the gas turbine engine is rather complex, and its real-time computations are extremely intensive. Another approach is to use systems identification methods for modelling of the normal engine behaviour, as in model-based approaches. A general schematic of model-based condition monitoring is presented in Figure 13.25.

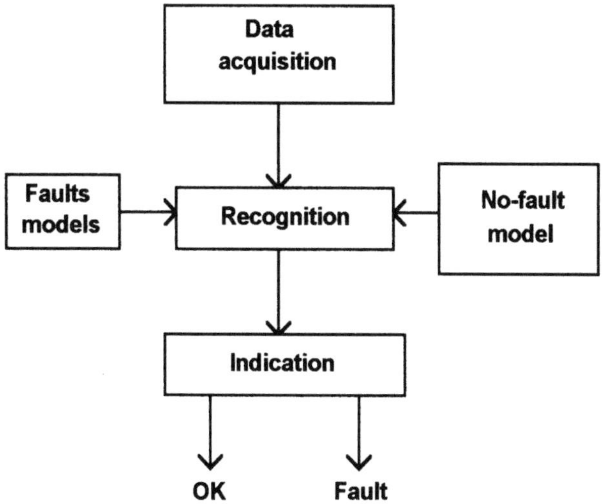

Figure 13.25. Model-based condition monitoring

A simple fault model is the signal exceeding the allowed boundaries. Figure 13.26 shows the use of on-board engine modelling in condition monitoring of a dual-lane control system and the lane-to-lane switching logic. Note that in the case of a double fault in both lanes, the engine control can be temporarily performed with the digital model.

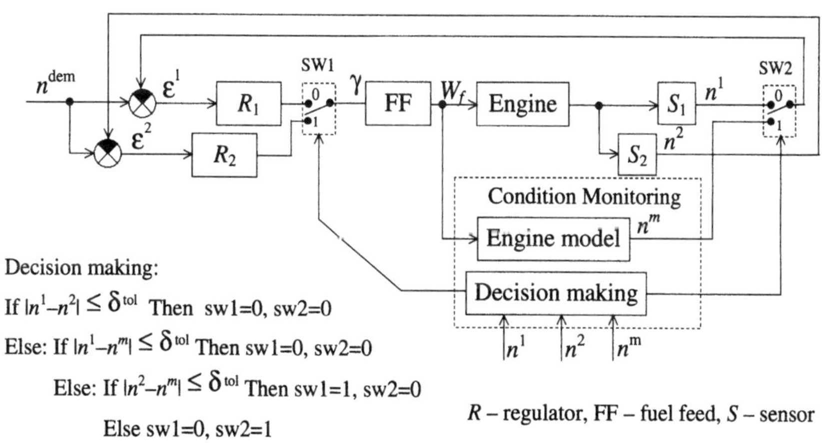

Figure 13.26. Model-based condition monitoring in dual-lane control

The total duration of fault confirmation and lane switching must not exceed the critical time limit for the system. The fault confirmation probability P_{confirm} increases with the time of the parameter being outside the allowed band. Simultaneously, the successful switching probability P_{switch} (fault accommodation) decreases; see Figure 13.27.

The engine model utilised in condition monitoring must be sufficiently accurate and operate in real time. These requirements make general engine models ineffective due to not-accounting for individual engine characteristics over the fleet. Individual modelling of flight and environmental conditions is often not feasible in real time. In order to use simplified models for condition monitoring purposes, on-line identification must be performed during the engine operation. The detailed description of Markov modelling is presented in Chapter 12. The Markov model identification represents the estimation of the transition probability matrix **P**; see Figure 13.28.

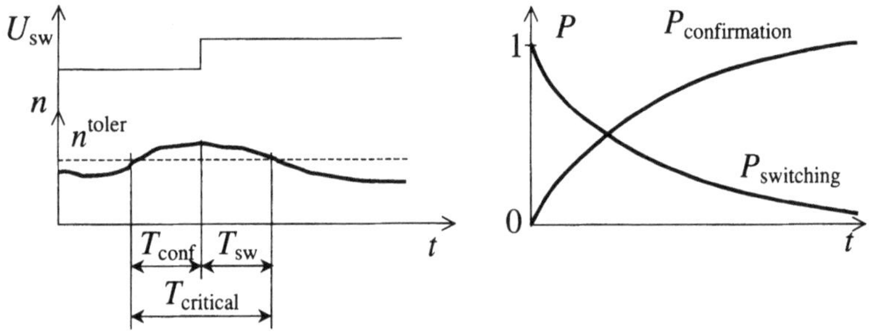

Figure 13.27. Example fault model and switching probability

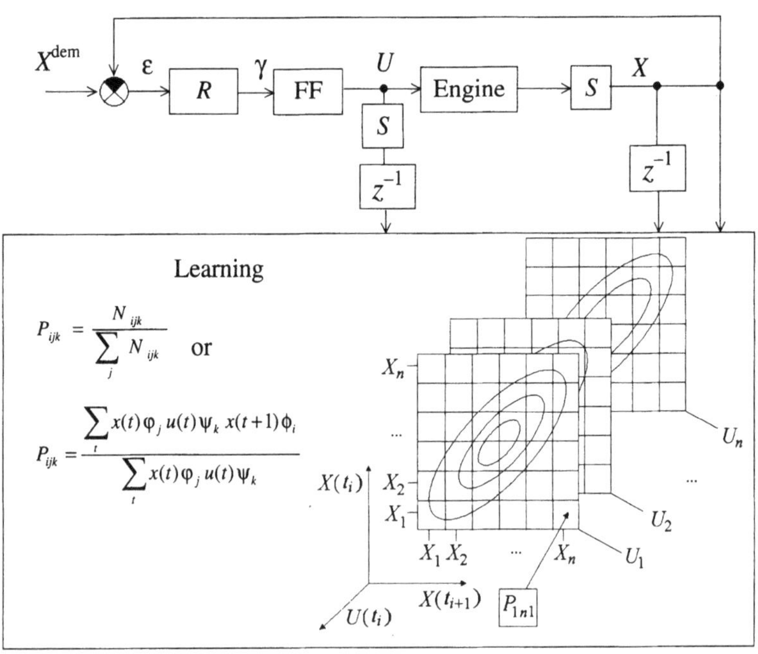

Figure 13.28. Markov model learning: regulator (R), fuel feed (FF), sensor (S) and delay operator (z^{-1})

13.5.3 Simulation Results

Markov models of both lanes of the control system of the aero engine were identified for condition monitoring purposes. The results of simulation of the fuel pump operation are shown in Figure 13.13 compared with real data. The dotted line determines 99% confidence intervals for the simulated fuel flow.

The Markov modelling technique was also applied for simulation of shaft speeds, pressures, and temperatures. The results of Markov simulation of HP shaft speed via maximal probability compared with real data and the error of simulation are shown in Figure 13.29. Maximal probability extracts the deterministic component of the engine dynamics from the Markov model.

To make a comparison, the modelling performance of radial basis function (RBF) neural networks was also investigated. The results of RBF-network simulation of HP shaft speed compared with real data and the error of simulation are shown in Figure 13.30. The mean-squared prediction error is 3.98 rpm for the Markov model approach and 4.55 rpm for RBF-network simulation.

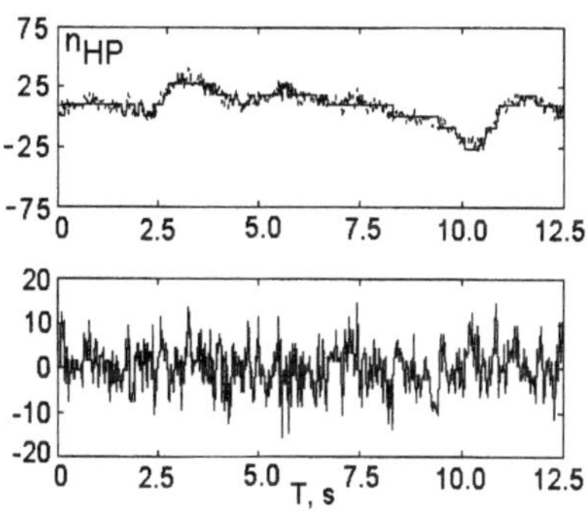

Figure 13.29. Maximum-probability Markov model simulation: real data (dotted) and simulation error (bottom)

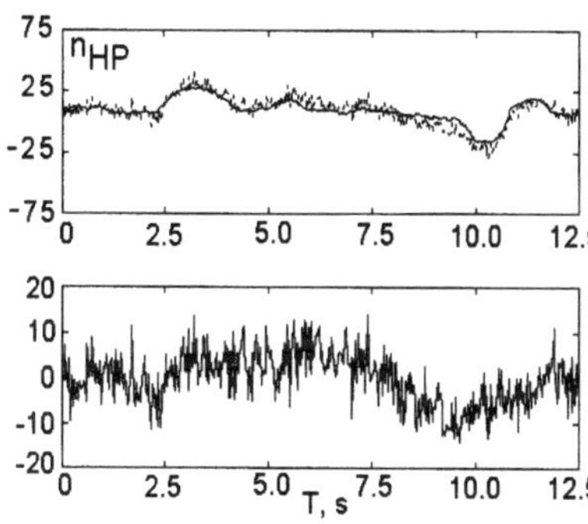

Figure 13.30. RBF-network simulation: real data (dotted) and simulation error (bottom)

An advantage of the Markov simulation technique is its high computational speed, important in real-time applications. In the practical example above, the simulation of engine operation with Markov models was about 20 times faster than that using a RBF network.

13.6 Concluding Remarks

An example of a turbo prop fan engine has been considered to demonstrate the Markov modelling technique discussed in Chapter 12. Experimental data from the real engine test bed was used to identify first-order Markov models of inlet air temperature and pressure and first- and second-order models of the engine dynamics. Simulation of the identified models at the HIL test bed facility proved the effectiveness of the Markov modelling technique. Standard deviations and power spectra of the engine parameters were much closer to the source data.

Finally, Markov modelling was considered for condition monitoring of the aero engine and its controller. The Markov model possesses the potential to be used as a third "virtual" lane for detecting faulty components in the dual-lane controller. Further refinements can improve the quality of the identified model by the parallel use of other available data and methods. The problem is then how to combine all the results in the refined model.

Chapter 14
Optimal Control of Gas Turbine Engines Using Mathematical Programming

14.1 Introduction

In previous chapters, different types of mathematical models of aero engines were described along with methods for their creation and estimation. Also, some application areas were considered, which included conventional engine control, controller testing and engine condition monitoring. In this chapter, mathematical models of aero engines are applied to design of *optimal* control. Within this approach two main issues are considered: optimal control laws based on static engine modelling and optimal control algorithms making the engine follow program trajectories with maximum accuracy. The last is based upon engine dynamic modelling. Optimal control is presented as a mathematical programming problem, which consists of an objective function, equality constraints, and inequality constraints [1]. Examples illustrate implementation of the proposed approach to solving various optimization problems in aero engine control.

14.2 Optimization of Engine Performance Through Optimal Control

To fulfill the high-performance power plant requirements for modern aircraft a greater degree of integrated control is needed [2]. This requires more complex control systems, which are multivariable (7 – 10 control loops) and multifunctional. Operation of all control loops must be coordinated across all transients and steady-state conditions.

Operational requirements become greater with demanded increases in accuracy and quality of transient performance. Effective control of power plants is also connected with integration of engine control with aircraft control. This requires new methods for analysis and design of control systems.

At present, multivariable control of a gas turbine engine (GTE) is designed using modern control theory. Note that design of linear optimal control is well investigated mostly for steady-state conditions [3-5]. Moreover, these systems cease to fulfill design specifications, when state or control variables reach their actual limits. These circumstances restrict possibilities of linear optimal control, which can be resolved using nonlinear multivariable control of GTE. The use of nonlinear control laws enables execution of any demanded trajectory of output parameters over the operational range and improvement of control quality, which can increase efficiency and service life of power plants.

Optimization of engine control for transient operation and at steady-state conditions requires solution of two major problems:

- optimal control laws (determination of the best engine state trajectory);
- optimal control algorithms (the best realisation of that trajectory).

Also, as engine characteristics differ over the fleet and also change during operation, it is necessary to solve the problem of optimization in view of their individual characteristics in real time. Such an approach can provide high operational qualities.

Therefore, the problem of design of methods for simultaneous synthesis (definition) of optimal control laws and optimal algorithms for their realisation, accounting for real GTE characteristics in real-time, is a high priority in control theory and control engineering.

For solution of problems, concerning optimal control of aircraft gas turbine engines, when a mathematical model of the plant is obviously inaccurate, it is preferable to use approximate methods. In real-time optimal control, the objective of optimization in the form of integral functionals is not feasible, because it reflects average characteristics of system quality. Instead, local quality criteria are considered for optimization, *e.g.*, specific loss functions.

The approximate character of the engine mathematical model forces the designer to pass from strict optimal analytic decisions to numeric ones. Moreover, computational speed of existing computers limits the search to optimization within a class of local-optimal control systems.

The full derivative of a Lyapunov function can be used as a local optimality criterion, which ensures asymptotic stability of the control system. Such control along with guaranteeing stability also becomes *optimal*, when additional constraints are imposed to control variables.

Multivariable local-optimal control systems of gas turbine engines possess the following advantages:

- these produce numerical solutions of the optimization problem, which extends their field of application;
- the systems asymptotically converge to strict optimal solutions, *i.e.*, they do not differ substantially from optimal systems under small perturbations;
- the synthesis procedure is based on systems analysis and enables design of multifunctional systems.

The objective of the performed work is to develop and investigate numerical algorithms of multivariable GTE control with real-time optimization of control laws and algorithms, along with accounting for individual characteristics of the engine.

14.3 Problem Formulation

The problem of multivariable GTE control can be formulated in terms of definition of local optimal control. Suppose that the controller has m independent control actions (outputs of the controller). The objective of the control system is to maintain m control laws (*e.g.*, maintenance of constant values of m engine parameters). The sampling frequency f of the digital control system is known. The sampling time period is the following: $\tau = 1/f$.

The system of differential equations describes the plant dynamics in the following form:

$$\begin{aligned} \dot{x}(t) &= A(t)\,x(t) + B(t)\,u(t) \\ y(t) &= C(t)\,x(t) + D(t)\,u(t) \end{aligned} \tag{14.1}$$

where $x(t)$ is the n-dimensioned state vector; $u(t)$ is the m-dimensioned control vector; $y(t)$ is the k-dimensioned observation (output) vector; A, B, C and D are the matrices of the dimensions $(n \times n)$, $(n \times m)$, $(k \times n)$ and $(k \times m)$ correspondingly. All elements of vectors $x(t)$ and $y(t)$ are measured by corresponding sensors. The following constraints are applied to the first derivative of the control vector $u(t)$:

$$\dot{u}_1^{\lim} \le \dot{u}(t) \le \dot{u}_2^{\lim} \tag{14.2}$$

where $\dot{u}_1^{\lim}$ and $\dot{u}_2^{\lim}$ are speed of movement limits for the actuators.

The control computer stores in its memory matrices A, B, C and D, which depend on operating and flight conditions. The computer operates with the period $\Delta t = \tau = 1/f$. At each ith cycle of its operation, current values of elements of the state vector $x(t)$ and observation vector $y(t)$ are measured, and elements of the control vector $u(t)$ are passed to actuators.

The vector $z^*(t)$ is given defining the program trajectory. The vector $z(t)$ consists of m elements of the vectors $x(t)$ and $y(t)$, where m is equal to the number of control variables. This means that the control system has to maintain m control laws simultaneously. A simple example is maintenance of constant values of m engine parameters.

Consider a vector of trajectory deviation (error) $e(t)$:

$$e(t) = z^*(t) - z(t) \tag{14.3}$$

Then the objective of control will consist of making the current vector $z(t)$ trace the demand $z^*(t)$ with maximum possible accuracy and quality:

$$\|e(t)\| = \|z^*(t) - z(t)\| \to \min \tag{14.4}$$

where $\|\cdot\|$ is the Euclidean norm of a vector. Along with delivering $\|e(t)\|\to\min$, elements of the control vector should fulfill the constraints of Equation (14.2).

The source data for a discrete model of the plant is the system of vector-matrix differential and algebraic equations represented in the normal form:

$$\begin{aligned}
&\Delta\dot{x}(t)=A(t)\,\Delta x(t)+B(t)\,\Delta u(t)\\
&\Delta y(t)=C(t)\,\Delta x(t)+D(t)\,\Delta u(t)\\
&\Delta x(t)=x(t)-x^*(t)\\
&\Delta y(t)=y(t)-y^*(t)\\
&\Delta u(t)=u(t)-u^*(t)\\
&x(t)=[x_1(t),x_2(t),\ \ldots,x_n(t)]^T\\
&y(t)=[y_1(t),y_2(t),\ \ldots,\ y_k(t)]^T\\
&u(t)=[u_1(t),u_2(t),\ \ldots,u_m(t)]^T
\end{aligned} \tag{14.5}$$

where $x(t)$, $y(t)$ and $u(t)$ are the state, observation and control vectors; and x^*, y^* and u^* are the vectors defining the demanded program trajectory of the plant. Matrices A, B, C and D have the dimensions $(n\times n)$, $(n\times m)$, $(k\times n)$ and $(k\times m)$, respectively. Their elements are functions of the engine state and atmospheric conditions. The system in Equation (14.5) is given in physical coordinates.

Digitising the equations for x and y, the system in Equation (14.5) can be written as follows:

$$\begin{aligned}
&\Delta x(i+1)=F\Delta x(i)+T\Delta u(i)\\
&\Delta y(i+1)=S\Delta x(i)+P\Delta u(i)
\end{aligned} \tag{14.6}$$

The system in Equation (14.6) consists of difference vector-matrix equations written *in deviations* from steady-state values, determining extrapolated parameters of the plant from their current values.

Consider the system in Equation (14.6) for the time moments $t = (i + 1)\tau$ and $t = i\tau$ and subtract the second system from the first one:

$$\begin{aligned}
&x(i+1)-x(i)=F(x(i)-x(i-1))+T(u(i)-u(i-1))\\
&y(i+1)-y(i)=S(x(i)-x(i-1))+P(u(i)-u(i-1))
\end{aligned}\quad\text{or} \tag{14.7}$$

$$\begin{aligned}
&x(i+1)=x(i)+F\delta x(i)+T\delta u(i)\\
&y(i+1)=y(i)+S\delta x(i)+P\delta u(i)
\end{aligned} \tag{14.8}$$

where
$$\begin{aligned}
&\delta x(i)=x(i)-x(i-1)\\
&\delta u(i)=u(i)-u(i-1)
\end{aligned} \tag{14.9}$$

Form the vector $z(i + 1)$, which describes the current state:

$$z(i+1) = \begin{bmatrix} x'(i+1) \\ y'(i+1) \end{bmatrix} = \begin{bmatrix} x(i) + H_x(F\delta x(i) + T\delta u(i)) \\ y(i) + H_y(S\delta x(i) + P\delta u(i)) \end{bmatrix} \tag{14.10}$$

where H_x and H_y are vectors of weighting coefficients, taking into account the approximate character of the mathematical model of the plant. Their values reflect the level of "uncertainty" of knowledge concerning GTE dynamics and depend on demanded engineering quality criteria.

Consider the Euclidean norm as a norm of the vector $e(t)$ in Equation (14.4):

$$J(\delta u(i)) = \sum_{j=1}^{m} \left[z_j^*(i+1) - z_j(i+1)\right]^2 \tag{14.11}$$

The expression for constraints Equation (14.2) in a discrete form is the following:

$$\dot{u}_1^{\lim} \leq \frac{u(i) - u(i-1)}{\tau} \leq \dot{u}_2^{\lim} \quad \text{or} \tag{14.12}$$

$$\begin{aligned} h_1(\delta u(i)) &= -u(i) + u(i-1) + \tau\dot{u}_1^{\lim} = -\delta u(i) + \tau\dot{u}_1^{\lim} \leq 0 \\ h_2(\delta u(i)) &= \ \ u(i) - u(i-1) - \tau\dot{u}_2^{\lim} = \ \ \delta u(i) - \tau\dot{u}_2^{\lim} \leq 0 \end{aligned} \tag{14.13}$$

At each ith step, the control vector is searched for, which provides the minimum of the objective function Equation (14.11) under the constraints Equation (14.13). In this case, optimal control represents a series of solutions of a quadratic programming problem within the control computer at discrete time instants t_i (i = 0, 1, ...) along with execution of corresponding controls $u(t_i)$. This problem can be solved using nonlinear programming methods applicable for the control computer.

Dynamic properties of the plant in a wide range of operation can be described by a set of linear models with varying coefficients. This enables engine optimal control to be built as a sequence of solutions of a quadratic programming problem in real time:

$$J = \sum_{j=1}^{m} e_j^2(\delta u(i)) \rightarrow \min \tag{14.14}$$

where the error vector e includes elements of both vectors e_x and e_y:

$$\begin{aligned} e_x &= x^* - x(i) - H_x[F\delta x(i) + T\delta u(i)] \\ e_y &= y^* - y(i) - H_y[S\delta x(i) + P\delta u(i)] \\ \delta x(i) &= x(i) - x(i-1) \\ \delta u(i) &= u(i) - u(i-1) \end{aligned} \tag{14.15}$$

with the following constraints concerning controls:

$$|\delta u(i)| \leq \tau \dot{u}^{\lim}(i) \tag{14.16}$$

14.4 Algorithm for Real-Time Resolution of a Quadratic Programming Problem

The algorithm for solution of the quadratic programming problem [6] in real-time proposed below is based upon analysis of GTE characteristics as a multivariable plant, and generalisation of requirements to digital multivariable control systems. Consider an algorithm for search for the minimum of the objective function Equation (14.14). The choice of a method for solution of the problem in Equations (14.14), (14.15) and (14.16) depends on the following factors:

- computational power necessary for solution of the problem;
- required accuracy of the solution;
- demanded accuracy of fulfilment of constraints;
- stable operation under short-time malfunctions.

Transforming Equations (14.1), (14.2) and (14.3), the problem is represented in a standard form:

$$J = Q(x) = p^T x + x^T Cx \tag{14.17}$$

$$Ax - B \leq 0 \tag{14.18}$$

where $x = \delta u(i)$ is speed of change of the control vector at the ith step of control; C is a symmetric positive defined m-dimensioned matrix; p^T and x^T are transposed column vectors.

The following solution represents the point of the minimum of the objective function Equation (14.17) if the constraints Equation (14.18) are not considered:

$$x^0 = -\tfrac{1}{2}C^{-1}p \tag{14.19}$$

If the constraints Equation (14.18) are taken into account, the problem consists of searching for the point of the allowed polyhedron Equation (14.18) lying at the level line:

$$Q(x) = p^T x + x^T Cx$$

with the minimal value of $Q(x)$:

$$\min(Q(x) \mid Ax \leq B) \tag{14.20}$$

The standard problem statement with the objective function Equation (14.14) will lead to a search for the minimum of the function at the boundary of the allowed area. This will determine a point, where reduction of control errors of some parameters is obtained only by increasing control errors in others. Therefore, in minimisation of Equation (14.17), a mechanism should be developed to exclude possible increase in absolute value of any component of the vector *e*.

A possible method for this consists of introduction into Equation (14.14) additional weighting coefficients, or "penalties," in the following form:

$$J = \sum_{j=1}^{m} \gamma_j e_j^2 \rightarrow \min \tag{14.21}$$

where γ_j are weighting coefficients calculated using special formulae.

Another approach consists of development of additional constraints, excluding increase in control errors e_j while searching for the minimum of the objective function Equation (14.1). This extends the system of limiting inequalities Equation (14.16) and makes the search procedure more complicated.

A reasonable way for solving the problem of quadratic programming is the use of a method accounting for physical sense of the objective function Equation (14.14). As it is seen from the formula Equation (14.14), the point x^* delivering the unconditional minimum of the objective function is determined by solving the system of linear equations:

$$e_j(x) = 0 \qquad j = 1, \ldots, m \tag{14.22}$$

The search for the minimum of the objective function Equation (14.14) within the allowed polyhedron Equation (14.16) starts with an acceptable point $x^0 = 0$ and then proceeds along the beam:

$$S = x^0 + \lambda x^* \tag{14.23}$$

toward the point x^* until the minimum or the boundary of the allowed area. The value of λ', where the beam crosses the boundary of the allowed area, is defined as the smallest of λ_j:

$$\lambda' = \min\{\lambda_j \mid \lambda_j > 0\} \tag{14.24}$$

where

$$\lambda_j = \frac{\delta \dot{u}_j^{\lim}}{|S_j|} \tag{14.25}$$

If the minimum of the objective function is inside of the allowed area, then obviously $\lambda = 1$. The step length, defining vector x, is selected as:

$$\lambda^0 = \min\{1, \lambda'\} \tag{14.26}$$

The advantage of the suggested method is that minimisation of a quadratic function takes only one iteration. Calculation of the function gradient and inverse matrices is not necessary for finding the unconditional minimum. The search for the minimum of a function along a straight line connecting the beginning of coordinates with the point of the unconditional minimum provides proportional reduction of control errors by all output parameters.

An easy-to-program Gaussian algorithm can be used for solving the system Equation (14.22) to define the coordinates of the point of unconditional minimum.

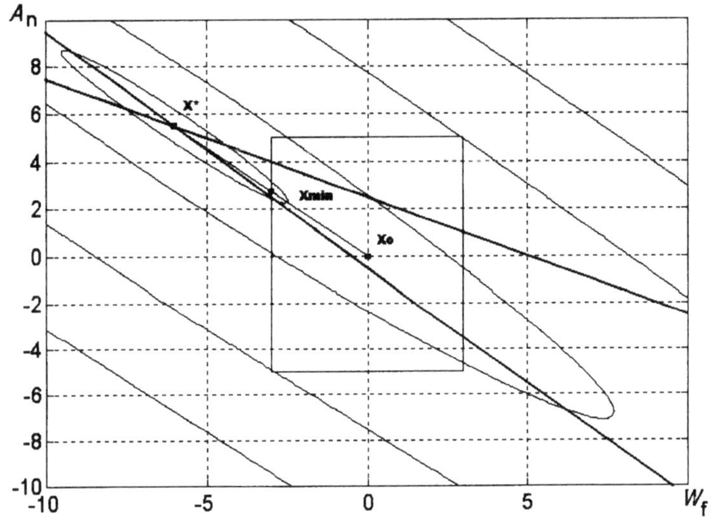

Figure 14.1. Trajectory of search for minimum of objective function

Figure 14.1 illustrates the suggested method using an example of minimisation of the following objective function:

$$J = (5 - x_W - 2x_A)^2 + (-2 - 4x_W - 4x_A)^2 \tag{14.27}$$

where $x_W = \delta u_W$ and $x_A = \delta u_A$. The constraints Equation (14.16) give the allowed area for search in the form of a rectangle with the initial point X_0 in the centre. The unconditional minimum of the objective function $J = 0$ is in the point X^* at the intersection of the straight lines $5 - X_W - X_A = 0$ and $-2 - 4X_W - X_A = 0$. These are shown by thick lines. The search for the minimum proceeds along the straight line connecting the points X_0 and X^* until the intersection with the boundary of the search area (point $X_{\min}$). The final point determines optimal control at the current step.

The sequence of calculations according to the considered method consists of the following stages:

- the point of unconditional minimum δu^* is calculated as a solution of the system:

$$e_j(\delta u(i)) = 0 \qquad j = 1,\ldots,m \tag{14.28}$$

- the parameter λ' is determined at the intersection of the beam $S = \delta u^0(i) + \lambda \delta u^*(i)$ with the boundary of the allowed area:

$$\lambda'_j = \frac{\tau \dot{u}_j^{\lim}}{|S_j|} \qquad j = 1,\ldots,m \tag{14.29}$$

- parameter λ^0 is determined according to the formula:

$$\lambda^0 = \min\{1, \lambda'_j\} \tag{14.30}$$

- the point of minimum is calculated:

$$\delta u(i) = \delta u^0(i) + \lambda^0 S \tag{14.31}$$

The developed algorithm combines high speed of calculation with low computational needs. It also provides the demanded accuracy to allow fulfilment of constraints. An additional check of whether the point found is within the allowed area Equation (14.3) introduces a degree of robustness to faults resulting from transient upsets in the computational process. This enables the algorithm to be recommended for the use in on-board digital control systems.

14.5 Example of Turbo Jet Control, Optimal by Speed

The developed technique was used for synthesis of a two-variable control system for a twin-shaft turbo jet with a variable jet nozzle. Two independent control variables (fuel flow W_f and nozzle area A_n) allow design of multivariable control.

A multivariable system was developed to control maximum values of the following parameters: low-pressure shaft speed n_{LP}, high-pressure shaft speed n_{HP} and gas temperature T_g^* behind the low-pressure turbine at maximum dry thrust operation.

The source data for control design are mathematical models of the plant and requirements on dynamic and static accuracy of control. Step responses of the output parameters should be aperiodic without overshoot. The risetime must be less than one second. Static error of shaft speed control is less than 0.3% and of temperature control is less than 5 K.

Investigation of the developed control system was performed via simulation of transients using mathematical models of the plant and controller. First, investigation was performed using linear models of the system, then with a performance-based detailed nonlinear thermodynamic model of the engine.

The program of investigation included the following points:

- response to perturbations applied to demanded values and control variables;
- influence of pure time delay;
- parametric perturbations;
- positional drift in hydraulic integrator actuators.

The mathematical model describes the plant Equation (14.5), actuators, the thermocouple and the developed controller. Actuators of W_f and A_n are modelled with integrators with maximum speed of 3000 (kg/hr)/s and 2000 cm^2/s correspondingly. The sampling frequency for transfer of control outputs to the actuators is 40 Hz. The sampling period is $\tau = 0.025$ s. Figure 14.2 shows step response of the closed-loop control system to a step of 10 rpm applied to LP shaft speed demand. The n_{LP} transient is monotonic without a peak of overshooting and lasts about 0.3 s. The T_g^* transient is flat. Reaction caused by a step change of 10^0 K in the gas temperature demand is presented in Figure 14.3. The T_g^* transitional process is stepwise function lasting around 0.125 s. The LP shaft speed has no change. Both plots show that controls are changed with maximum possible speed until gas temperature achieves its new value. This example proves efficiency of the developed method for design of control systems optimal by response speed.

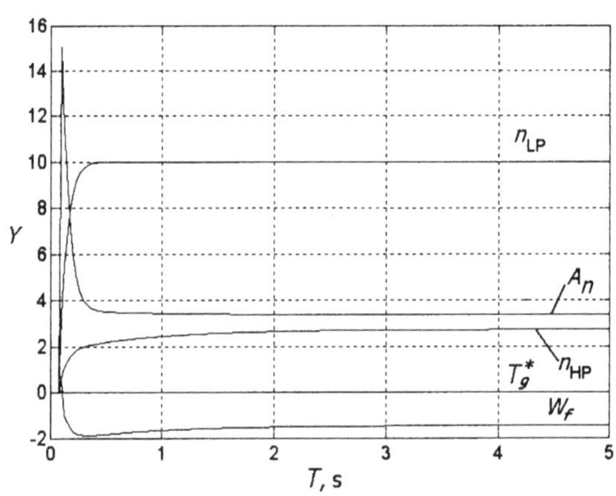

Figure 14.2. Step response of closed-loop system to $\Delta n_{LP.dem} = +10$ rpm

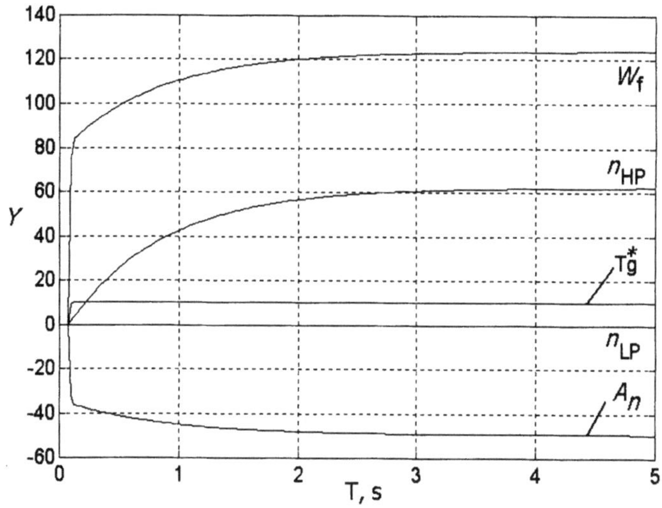

Figure 14.3. Step response of closed-loop system to $\Delta T_g{}^*{}_{dem} = +10^0$ K

14.6 Example of Turbo Jet Control, Optimal by Specific Fuel Consumption

This example illustrates the use of the proposed method for solving of a general optimization problem: how to simultaneously determine optimal control laws in real-time and their optimal realisation.

Consider a problem of optimal control of a particular power plant by a criterion of minimum specific fuel consumption at cruise flight. This problem is important because of the weight composition of modern aircraft. Fuel weight makes up to about 50-60% of take-off weight, whereas useful load is less than 5-10%.

Dry thrust of modern turbo jets at steady-state conditions is usually controlled indirectly. The controller maintains demanded values of main engine parameters, *e.g.*, shaft speed and nozzle area. An example of calculations of these parameters by the criterion of minimum specific fuel consumption (SFC) is given in Figure 14.4. Engine performance is presented as thrust F and SFC being functions of shaft speed n and jet nozzle area A.

For any particular turbo jet engine, optimal values of SFC are connected with values of shaft speed n. However, the value of shaft speed n delivering optimal SFC differs with the value of nozzle area A. Therefore, the control law should maintain minimum of SFC $= f(n)$ at each A. Optimal values of thrust and specific fuel consumption are approximated by the straight lines F_{opt} and SFC$_{opt}$ in Figure 14.4.

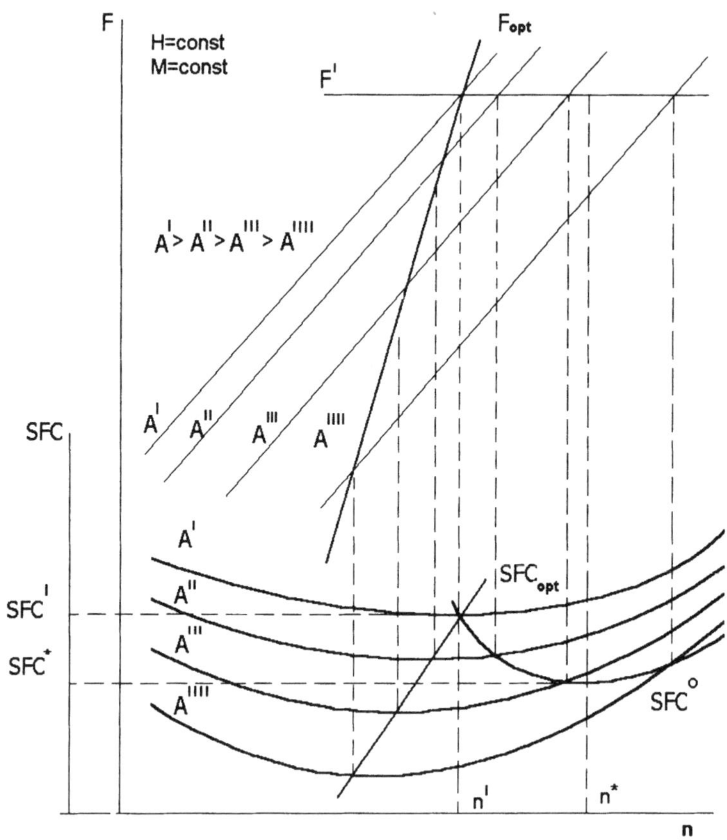

Figure 14.4. Optimal engine performance with minimum SFC

According to the control law implemented in controllers currently in service, the thrust value *F'* is achieved with shaft speed *n'* and nozzle area *A'*. In this case, specific fuel consumption SFC' is maintained. However, the same thrust *F'* can be obtained with less SFC. A number of different values of *n* and *A* are presented in Figure 14.4, which determine the same thrust *F'* (see the lines marked *A'*, *A''*, *A'''* and *A''''*).

The line SFC^0 determines corresponding values of specific fuel consumption. According to Figure 14.5, the minimum of specific fuel consumption is achieved at the point SFC*. The value SFC* is less than SFC' achieved with the general optimal program F_{opt}. Hence, the existing control law F_{opt} does not maintain the minimal possible value of SFC for demanded thrust *F*. In other words, the general law is not optimal. Note that demanded values of engine parameters are calculated from a general engine model. Also, during the engine in-service period, its characteristics are degrading because of wearing of engine elements. Therefore, fuel consumption necessary for maintaining demanded thrust differs from calculated for each engine in the fleet. Moreover, this value is different from a possible minimum.

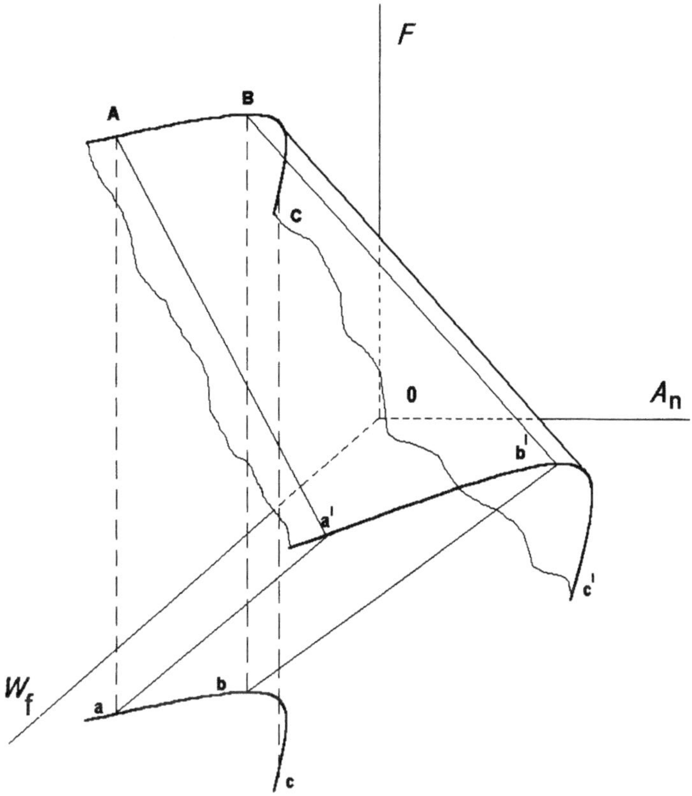

Figure 14.5. Thrust of twin-shaft bypass engine depending on fuel flow and nozzle area

This leads to solving a problem of nonsearching optimal control with the criterion of minimum SFC for each engine at current flight conditions at cruise with the required levels of stability, control quality and accuracy. A possible method for a real-time solution of the problem is described below, based on analysis of gas turbine engine characteristics.

Suppose that the control computer stores in its memory coefficients of a linear model of a turbo jet for different operating and flight conditions. At each *i*th time instant, the computer receives current values of engine state variables and output parameters and sends control signals to actuators of the system.

Thrust of a turbo jet engine is known to be characterised with the set of parameters $F = Ap_t^* / p_{\text{in}}^*$, where p_t^* is full gas pressure behind the turbine or full gas pressure in the mixing chamber, if an engine with mixture of gas flows is considered; *A* is jet nozzle area and p_{in}^* is full inlet air pressure.

A general shape of a relation between thrust, fuel consumption and nozzle area for a twin-shaft bypass engine is presented in Figure 14.7. The ABC line is the geometric place of points, where $F = 100\%$. The *a'b'c'* line is the place of points, where $F = 85\%$. The Aa' line determines the steady-state values of W_f and A given by a control law. The Bb' line is the geometric place of points, where W_f = min and F = const.

Figure 14.6 presents projections of the lines $F = 100\%$ and $F = 85\%$ at the plane $\{W_f, A_n\}$. The points **a'** and **a** show the values of W_f and A_n according to the existing control law. The points **b** and **b'** correspond to minimum fuel consumption at $F = 100\%$ and 85%. The solution of the optimal control problem can be presented as transition from the point **a'** along the line $Ap_t^*/p_{in}^* = \text{const}$ to the point **b'** and stabilisation of engine parameters there.

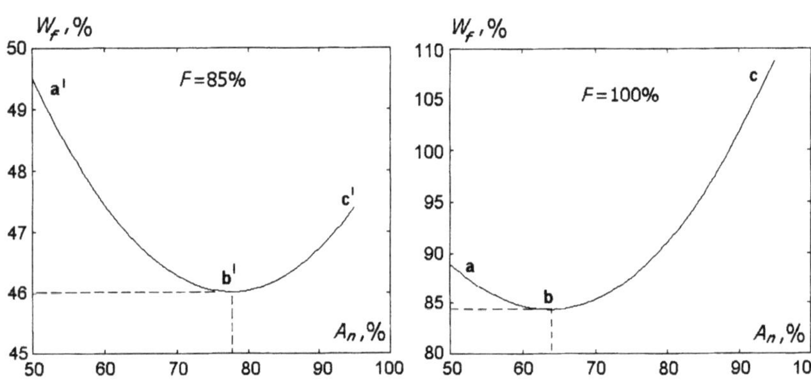

Figure 14.6. Relation between fuel consumption and nozzle area at constant thrust

Transforming the equation $Ap_t^*/p_{in}^* = \text{const}$ to a form operating with the coordinates W_f and A, using a linear discrete mathematical model of the plant:

$$\begin{aligned}
\Delta n_{\mathrm{LP},i+1} &= f_{11}\Delta n_{\mathrm{LP},i} + f_{12}\Delta n_{\mathrm{HP},i} + t_{11}\Delta W_{\mathrm{f},i} + t_{12}\Delta A_i \\
\Delta n_{\mathrm{HP},i+1} &= f_{21}\Delta n_{\mathrm{LP},i} + f_{22}\Delta n_{\mathrm{HP},i} + t_{21}\Delta W_{\mathrm{f},i} + t_{22}\Delta A_i \\
\Delta p_{\mathrm{t},i+1} &= s_{11}\Delta n_{\mathrm{LP},i} + s_{12}\Delta n_{\mathrm{HP},i} + p_{11}\Delta W_{\mathrm{f},i} + p_{12}\Delta A_i \\
\Delta n_{\mathrm{LP},i} &= n_{\mathrm{LP},i} - n_{\mathrm{LP}}^0 \\
\Delta n_{\mathrm{HP},i} &= n_{\mathrm{HP},i} - n_{\mathrm{HP}}^0 \\
\Delta P_{4,i} &= P_{4,i} - P_1^0 \\
\Delta W_{\mathrm{f},i} &= W_{\mathrm{f},i} - W_{\mathrm{f}}^0 \\
\Delta A_i &= A_i - A^0
\end{aligned} \tag{14.32}$$

The actuators are integrators and their dynamics are described by the equations:

$$\begin{aligned}
W_{\mathrm{f}} &= K_W \int U_W \, dt \\
A &= K_A \int U_A \, dt
\end{aligned} \tag{14.33}$$

where U_W and U_A are control signals passed from the control computer to corresponding actuators and K_W and K_A are integrator gains.

The control signals U_W and U_A are of impulse-width type and the sampling interval is τ. Subtract Equation (14.32) for time τ_i from Equation (14.32) for τ_{i+1}:

$$\begin{aligned}
n_{\mathrm{LP},i+1} &= n_{\mathrm{LP},i} + f_{11}\delta n_{\mathrm{LP},i} + f_{12}\delta n_{\mathrm{HP},i} + t_{11}\delta W_{f,i} + t_{12}\delta A_i \\
n_{\mathrm{HP},i+1} &= n_{\mathrm{HP},i} + f_{21}\delta n_{\mathrm{LP},i} + f_{22}\delta n_{\mathrm{HP},i} + t_{21}\delta W_{f,i} + t_{22}\delta A_i \\
p_{\mathrm{t},i+1} &= p_{\mathrm{t},i} + s_{11}\delta n_{\mathrm{LP},i} + s_{12}\delta n_{\mathrm{HP},i} + p_{11}\delta W_{f,i} + p_{12}\delta A_i \\
\delta n_{\mathrm{LP},i} &= n_{\mathrm{LP},i} - n_{\mathrm{LP},i-1} \\
\delta n_{\mathrm{HP},i} &= n_{\mathrm{HP},i} - n_{\mathrm{HP},i-1} \\
\delta W_{f,i} &= W_{f,i} - W_{f,i-1} = K_W\tau(U_{W,i} - U_{W,i-1}) \\
\delta A_i &= A_i - A_{i-1} = K_A\tau(U_{A,i} - U_{A,i-1})
\end{aligned} \tag{14.34}$$

The expression for $\delta p_{t,i+1}$ takes the form:

$$\delta p_{t,i+1} = p_{t,i+1} - p_{t,i} = p_0 + p_W\delta W_f + p_A\delta A \tag{14.35}$$

where

$$\begin{aligned}
p_0 &= s_{11}(n_{\mathrm{LP},i} - n_{\mathrm{LP},i-1}) + s_{12}(n_{\mathrm{LP},i} - n_{\mathrm{LP},i-1}) \\
p_W &= p_{11} \\
p_A &= p_{12}
\end{aligned} \tag{14.36}$$

Inlet air pressure is considered constant. Then the requirement of the operating point being at the line $F = Ap_t^*/p_{\mathrm{in}}^* = \mathrm{const}$ can be written in the form:

$$F = p_{t,i+1}A_{i+1} = (p_{t,i} + \delta p_{t,i+1})(A_i + \delta A) \tag{14.37}$$

Accounting for Equation (14.35) gives the value of thrust deviation from demand e_1. With a requirement for zero deviation:

$$e_1 = F_0 - T_W\delta W_f - T_A\delta A - P_W\delta W_f\delta A - P_A\delta A^2 = 0 \tag{14.38}$$

where:

$$\begin{aligned}
F_0 &= F - p_{t,i}A_i - p_0A_i \\
T_W &= P_W A_i \\
T_A &= p_{t,i} + P_A A_i + p_0
\end{aligned} \tag{14.39}$$

Equation (14.38) shows the relationship between deviations of fuel consumption δW_f and nozzle area δA. This information is used to maintain demanded thrust and to compensate its random deviations from demand.

Determine the conditions of the operating point where minimum fuel consumption is achieved at the required demanded thrust. This is the point **b'** in Figure 14.6, left.

The following equation determines change in pressure p_t^* at the static line with reference to the point **b'**:

$$\delta p_{t,\,i+1} = K_p^W \delta W_f + K_p^A \delta A \tag{14.40}$$

where K_p^W and K_p^A are static gains. Substituting Equation (14.40) into Equation (14.37) gives:

$$F - p_{t,i} A_i - (K_p^W A_i + K_p^A \delta A)\delta W_f - (p_{t,i} + K_p^A A_i)\delta A - K_p^A \delta A\,\delta A = 0 \ . \tag{14.41}$$

Explicit relationship between δW_f and δA is found from Equation (14.41):

$$\delta W_f = \frac{F - p_{t,i} A_i - (p_{t,i} + K_p^A A_i)\delta A - K_p^A \delta A^2}{K_p^A (A_i + \delta A)} \ . \tag{14.42}$$

Following from analysis of the graphs in Figure 14.6, minimum fuel consumption corresponds to zero of the derivative dW_f/dA_n. An expression for this derivative can be obtained by dividing both parts of Equation (14.42) by δA:

$$\frac{\delta W_f}{\delta A} = \frac{F - p_{t,i} A_i - (p_{t,i} + K_p^A A_i)\delta A - K_p^A \delta A^2}{K_p^A (A_i + \delta A)\delta A} \tag{14.43}$$

Supposing fulfillment of equality $F - p_{t,i} A_i = 0$ in Equation (14.43) gives the value of the derivative at the minimum of fuel consumption:

$$\frac{\delta W_f}{\delta A} = -\frac{p_{t,i} + K_p^A A_i + K_p^A \delta A}{K_p^A (A_i + \delta A)} \tag{14.44}$$

The condition for the derivative to be zero from Equation (14.44):

$$e_2 = p_{t,i} + K_p^A A_i + K_p^A \delta A = 0 \tag{14.45}$$

Find the limit of Equation (14.44) with $\delta A \to 0$:

$$\frac{dW_f}{dA} = -\frac{p_{t,i} + K_p^A A_i}{K_p^A A_i} \tag{14.46}$$

Hence, the condition of the derivative being zero is the following:

$$\frac{p_{t,i}}{A_i} = -K_p^A \tag{14.47}$$

Finally, the condition of minimum fuel consumption with demanded thrust is equality of turbine pressure p_t^* divided by nozzle area A to the static gain of A influence to p_t^*, taken negative.

Hence, changes of fuel consumption δW_f and nozzle area δA at each time instant must satisfy the conditions Equation (14.45) and Equation (14.38). Also, the following constraints are applied upon the values δW_f and δA:

$$\begin{aligned} &\left|\delta W_f\right| \le \dot{W}_{f.\max}\tau \\ &\left|\delta A\right| \le \dot{A}_{\max}\tau \end{aligned} \tag{14.48}$$

which are obtained from existing limitations of maximum speed of change determined by physical properties of actuators. The objective function is built by considering only linear parts of Equation (14.38) and Equation (14.45):

$$J = (F_0\gamma_1 - T_W\delta W_f - T_A\delta A)^2 + [(p_{t,i} + K_p^A A_i)\gamma_2 + K_p^A\delta A]^2 \tag{14.49}$$

The values δW_f and δA are determined as delivering minimum to Equation (14.49) and fulfilling the constraints in Equation (14.48).

The objective function J Equation (14.49) has a simple physical sense. The first term of the expression determines the difference between demanded and actual thrust. The second term reflects the "distance" from the current operating point to the point with minimum fuel consumption. The weighting coefficients γ_1 and γ_2 change the corresponding differences and so regulate dynamic properties of the system.

The proposed algorithm is able to account for individual characteristics of the engine. The engine operation with minimum SFC is determined by the coefficient of influence K_n^A. Its value is continuously obtained during the flight, accounting for individual engine characteristics.

Engine transition from demanded thrust onto more efficient operation increases efficiency of the engine as a thermal machine. This is accompanied by an increase in air pressure in flowing parts of the engine, shaft speeds and gas temperature. During transition toward minimum SFC, some critical engine parameters, *e.g.*, gas temperature, can achieve the maximum limit allowed. In this case, the control law changes and control is determined as the minimum of the following objective function:

$$J_1 = (F_0\gamma_1 - T_W \delta W_f - T_A \delta A)^2 + (T_t^{\max} - T_{t,i} - a_W \delta W_f - a_A \delta A)^2 \tag{14.50}$$

The parameter to be limited is determined by comparison of current values with their limits. The control law changes automatically using a selector. When gas temperature has reached its limit, the control law changes as follows. The values r_i are determined by the equations:

$$\begin{aligned} r_1 &= \alpha_1 \left| p_{t,i} + K_p^A A_i \right| \\ r_2 &= \alpha_2 \left| T_t^{\max} - T_{t,i} \right| \end{aligned} \tag{14.51}$$

$$r = \min(r_1, r_2) \tag{14.52}$$

If $r = r_1$, then the objective function J in Equation (14.49) is used. When $r = r_2$, the objective function J_1 in Equation (14.50) is used.

In order to make the method adaptive to variation of the engine static characteristics, recurrent identification of the parameter K_p^A is necessary.

Analysis of the engine mathematical model reveals that a 10% error in estimation of the parameter K_p^A leads to 0.82% change in fuel consumption at maximum thrust operation and 0.2% at nominal operation. This makes requirements for accurate determination of this parameter high, in particular at maximum thrust operation.

Identification of the parameter K_p^A using methods of correlation analysis is not acceptable because of high computational load and low accuracy. Previous research has shown that error of estimation of static coefficients of a gas turbine engine by methods of correlation analysis is usually greater than 12%. Passive identification of dynamic model parameters is difficult because of the feedback influence.

Methods for active identification and optimal control with test signals are not acceptable in principle. Hence, the use of other methods is necessary for taking into account individual engine characteristics. A possible solution is to monitor time drift of the extreme of the engine characteristic presented in Figure 14.6.

Using observation of the variables $W_f(A)$, an optimization algorithm is constructed, which determines a sequence $\{K_p^A, \tau\}$, $\tau = 0,1,\ldots,$ for which $\{K_{p,\tau}^A - K_{p,\text{opt}}^A\} \to \min$ if $\tau \to \infty$.

Successfully solving the problem of tracking the drift of the extreme is connected with assumptions about the drift character and properties of the function $W_f(A)$ or its gradient. The extremal relationship can often be assumed polynomial. Suppose that the following relationship fits well the extremal dependency shown in Figure 14.6:

$$W_f = a_0 + a_1 A + a_2 A^2 \tag{14.53}$$

where a_0, a_1 and a_2 are unknown coefficients to be estimated.

The coefficient K_p^A in the Equation (14.47) is determined from the ratio:

$$K_p^A = -\frac{p_{t,i}}{A_{\text{opt}}} = \frac{2a_2 p_{t,i}}{a_1} \tag{14.54}$$

Recurrent procedures for estimating the parameters a_0, a_1 and a_2 are preferable because of real-time operation. A comparative study of accuracy, convergence and computational load of recurrent algorithms for parametric identification leads to the following conclusion.

Under conditions of small observation time and highly intensive noise, all recurrent methods possess similar quality of parametric estimation, except the method of stochastic approximation. Therefore, preference is usual given to the recurrent least squares method because of its relative simplicity compared with other methods and reliable convergence of estimates.

The recurrent least squares method is applied for estimation of the parameters a_0, a_1 and a_2 in Equation (14.53). The vector of estimates $\mathbf{A} = [a_0\ a_1\ a_2]^T$ is determined

The recurrent least squares method for a twin-shaft bypass engine at maximum thrust operation gives the following initial conditions:

$$\hat{A}(i-1) = \begin{Vmatrix} 16990 \\ -11.282 \\ 0.002483 \end{Vmatrix} \qquad P(i-1) = \begin{Vmatrix} 1 & 0 & 0 \\ 0 & 1 & 0 \\ 0 & 0 & 1 \end{Vmatrix} \qquad F(i) = \begin{Vmatrix} 1 & F_i & F_i^2 \end{Vmatrix}^T \tag{14.55}$$

because the extremal characteristic $W_f = f(A)$ at maximum thrust is approximated by

$$W_f = 16990 - 11.282A + 24.83 \cdot 10^{-4} A^2 \tag{14.56}$$

Finally, the algorithm accounting for individual engine characteristics during optimal control by the criterion SFC $\rightarrow$ min is as follows:

- the engine parameters W_f, A, p_t^*, n_{LP}, n_{HP} are measured;
- the parameters a_0, a_1, a_2 are estimated;
- the value of K_p^A from Equation (14.54) is calculated.

A set of simulation experiments with the full thermodynamic model of the twin-shaft bypass turbo-jet engine and its nominal control system was done to evaluate the performance of the designed algorithm. Figure 14.7 shows transient responses of the engine fuel flow and thrust during minimisation of the specific fuel consumption of the engine at the maximum thrust operation without using the after burner.

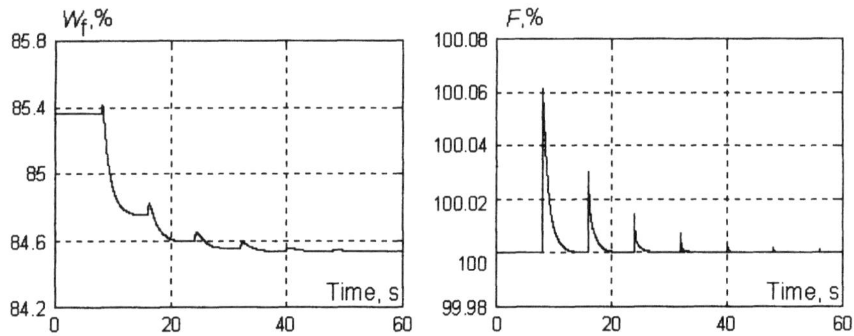

Figure 14.7. Engine fuel flow minimisation at maximum thrust operation

The nominal engine control system is in operation during the first 8 seconds. At the 8th second of simulation the designed algorithm is activated. The value of K_p^A is recalculated every 8 seconds. It can be seen from the graphs that time for transition to a new steady-state operation is approximately 30 seconds. The maximal deviation of the engine thrust from the demanded value during the transient process is 0.06%. The engine fuel flow is reduced by 0.98% of its value during the operation with the nominal control system. At cruise operation of the engine the proposed control method enabled 0.7% reduction of the specific fuel consumption of the engine.

14.7 Concluding Remarks

This chapter presented an approach to utilising mathematical models of aero engines in optimal control. Most digital on-board control systems represent mere digital realisation of analogue control laws. However, existing control computers possess enough computing power to be used in optimization of engine performance through new types of control laws and algorithms. Both types of optimal control lead to more efficient operation of the power plant via targeting to global optimization criteria and by accounting for individual engine characteristics. In the examples discussed, real-time algorithms were presented for optimization of control laws and their optimal realisation.

Chapter 15
Dynamic Model Identification of a Turbo Jet Engine

15.1 Introduction

In this chapter, the results of data analysis for Rolls-Royce SPEY engine tests are presented. This includes static and dynamic model identification for HP and LP shaft speed. Comparison with previous tests is given.

Four types of data files were processed: three-level, no test signal, multisine and closed-loop control of n_{HP} with no test signal. The types of the data are presented in Table 15.1.

Table 15.1. SPEY engine test data

Test record	Type of test	Record length, min	Sampling interval, ms
A	3-level	12	50
B	No test signal	20	40
C	Multisine	6.67	40
D	Multisine	6.67	40
E	Closed loop on n_{HP} with no test signal	20	40

15.2 Static Modelling Techniques

In this work, the gains of the dynamic and static models were compared. In fact, they represent different methods for estimating the same parameter, the dynamic gain. A static model of the engine accounts for the static nonlinear interaction between fuel flow as an input and shaft speeds as outputs. Both nonlinear static and linear dynamic models can be extracted from a thermodynamic model as they represent different aspects of the same engine behaviour [1]. The static model describes the relationship between mean values of inputs and outputs over the whole range of operating conditions from idle to maximum thrust. However, its first derivative represents a dynamic gain.

The dynamic gain can be estimated from a dynamic model through parametric fitting of FRF or using the amplitude FRF at a very low frequency. A linear dynamic model deals with the engine dynamics around a static operating point. In this case, the amplitude of signals is sufficiently low and thus makes it possible to neglect the engine nonlinearity. Hence, the static and dynamic models are closely connected to each other and to the thermodynamic model. Ideally, they should be dealt with together, as one whole model.

A static model can be fitted by a parametric function providing a smooth shape over a set of experimental mean values. In the case of a gas turbine, a logarithmic function is appropriate [2]. It provides good smooth nonlinear approximation and exploits only three parameters:

$$n_{\mathrm{HP}} = a + b \log (W_f + c) \tag{15.1}$$

Furthermore, the first derivative of Equation (15.1) incorporates only two parameters, b and c, and demonstrates good agreement with FRF estimates:

$$\frac{dn_{\mathrm{HP}}}{dW_f} = \frac{b}{W_f + c} \tag{15.2}$$

A multilevel test can provide the data describing the static model on several operating points. This was proposed for estimating the dynamic gain around one operating point (see Figure 15.1).

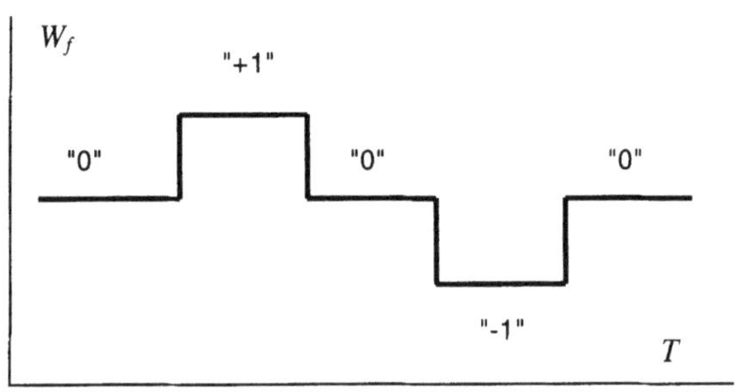

Figure 15.1. Three-level test design

A three-level test provides three static operating points. This data is used to estimate the dynamic gains by the formula:

$$K_{\mathrm{HP}}\left(\frac{W_{f1} + W_{f2}}{2}\right) = \frac{n_{\mathrm{HP1}} - n_{\mathrm{HP2}}}{W_{f1} - W_{f2}} = \frac{\Delta n_{\mathrm{HP}}}{\Delta W_f} \tag{15.3}$$

where $\{W_{f1}, n_{HP1}\}$ and $\{W_{f2}, n_{HP2}\}$ are the mean values for two static operating points; see Figure 15.2. Equation (15.3) represents a first-order linear interpolation of the static model; see Figure 15.3.

Later in this chapter, different estimates of dynamic gains will be compared. It should be noted that the estimate from the static model deals with the steady-state conditions, where all transients have finished including heat soak and other slow dynamic phenomena in the engine. This means that these estimates can slightly differ from dynamic model estimates.

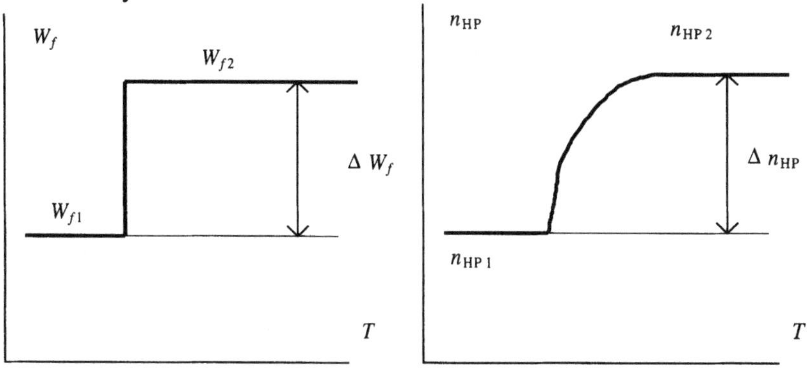

Figure 15.2. Step response in three-level test

15.3 Parametric Dynamic Modelling Techniques

The frequency response function (FRF) was estimated as described in Chapter 11. After this, a parametric model can be built in the form of a transfer function (TF):

$$H(s) = \frac{K(T_1 s + 1)e^{-\tau s}}{(T_2 s + 1)(T_3 s + 1)} \tag{15.4}$$

where K is the dynamic gain, T are time constants and τ is pure time delay.

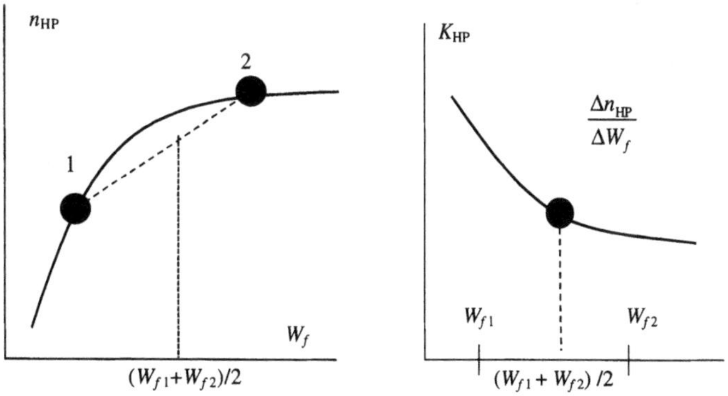

Figure 15.3. Gain estimation from static model

In order to fit the FRF by a transfer function, the least-squares method is used:

$$P = \underset{P}{\arg\min} \sum_i \{ (|H(\omega_i)| - |FRF(\omega_i)|)^2 + (\angle H(\omega_i) - \angle FRF(\omega_i))^2 \} \tag{15.5}$$

where $P = [K, T_1, T_2, T_3, \tau]$ is the vector of unknown parameters for Equation (15.4).

The SPEY engine has two turbines and compressors (LP and HP). Hence, the LP and HP shaft speed dynamics should be considered together. The interaction between two rotors is expressed in equal denominators of the TF, when using thermodynamic models or state-space methods. In the case of parametric fitting of the FRF this condition should be included in the least-square criterion Equation (15.5). Finally, the vector of the model parameters is the following:

$$P = [K_{\mathrm{HP}}, K_{\mathrm{LP}}, T_{\mathrm{HP}}, T_{\mathrm{LP}}, T_{\mathrm{den.1}}, T_{\mathrm{den.2}}, \tau_{\mathrm{HP}}, \tau_{\mathrm{LP}}] \tag{15.6}$$

Confidence intervals for the estimated coefficients are obtained using a simulation method described in [3]. It consists of the following: The variance of the FRF estimates is calculated through the coherence function for each frequency. Hence, the FRF values and their standard deviations are available for parametric modelling:

$$\begin{cases} |FRF(\omega_i)|; \sigma_{\mathrm{ampl}}(\omega_i) \\ \angle FRF(\omega_i); \sigma_{\mathrm{phase}}(\omega_i) \end{cases} \tag{15.7}$$

The probability distribution of these estimates is considered Gaussian. The next step is to perform N realisations of random numbers with the corresponding mean values and standard deviations. After simulation, the transfer function coefficients are calculated from every realisation. Finally, there are N vectors of coefficients: $\{P_1, P_2, ..., P_N\}$. Their standard deviation is computed by the formulae:

$$\sigma(P) = \sqrt{\frac{1}{N-1}\sum_{i=1}^{N}(P_i - \bar{P})^2} \qquad \bar{P} = \frac{1}{N}\sum_{i=1}^{N} P_i \tag{15.8}$$

This method provides more wide confidence intervals than the inverse Jacobian matrix obtained from gradient methods of optimization. The reason is that the simulation method accounts for a greater number of possible error sources including FRF deviation.

15.4 Static Model Identification

15.4.1 Mean Values

The mean values of variables from test records A, B, C and D are represented in Table 15.2. The data from record B has some trend, and the mean values are increasing during the test; see Figure 15.4. The mean values shown in Table 15.2 were estimated from the first 2000 points (80 s), last 2000 points (80 s), and the whole record (20 min, or 30,000 sampling points) of data B.

Table 15.2. Mean values of engine parameters

Test record	$W_{f.dem}$, cc/s	W_f, cc/s	n_{HP}, %	n_{LP}, %
A	150.14	149.85	55.111	26.362
B (first 80 s)	285.00	286.36	80.053	43.217
B (20 min)	285.00	287.17	80.199	43.317
B (last 80 s)	285.00	287.69	80.307	43.380
C	289.44	288.00	79.943	43.499
D	288.55	288.79	79.998	43.558

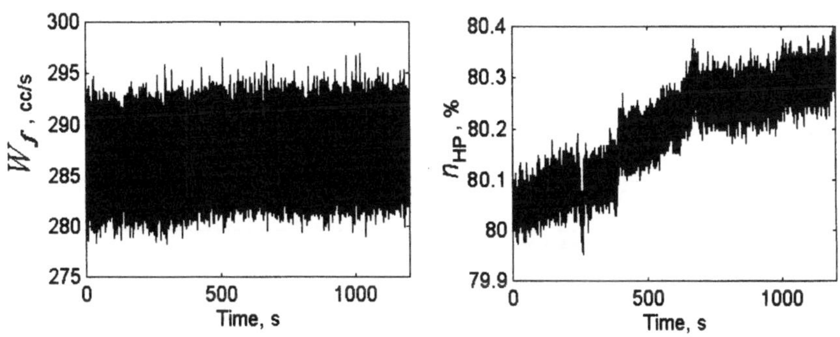

Figure 15.4. Test B: trend in fuel flow and HP shaft speed

15.4.2 "Three-Level" Test

Record A contains five steady-state sections; see Figure 15.5. The corresponding mean values are shown in Table 15.3. In fact, the data contains trends, and only small portions of data were used for estimating the static model. In this case, the trend in fuel flow can be considered the reason for the trend in shaft speeds. Moreover, the measured value of the fuel flow differs from demanded.

Table 15.3 .Test A: mean values at the steady-state conditions

Level	$W_{f.\text{dem}}$, cc/s	W_f, cc/s	n_{HP}, %	n_{LP}, %
0	260.04	251.84	74.97	40.03
+1	299.76	291.32	80.56	43.62
0	260.00	252.55	75.13	40.13
–1	225.40	218.80	69.57	36.44
0	260.00	252.09	74.98	40.06

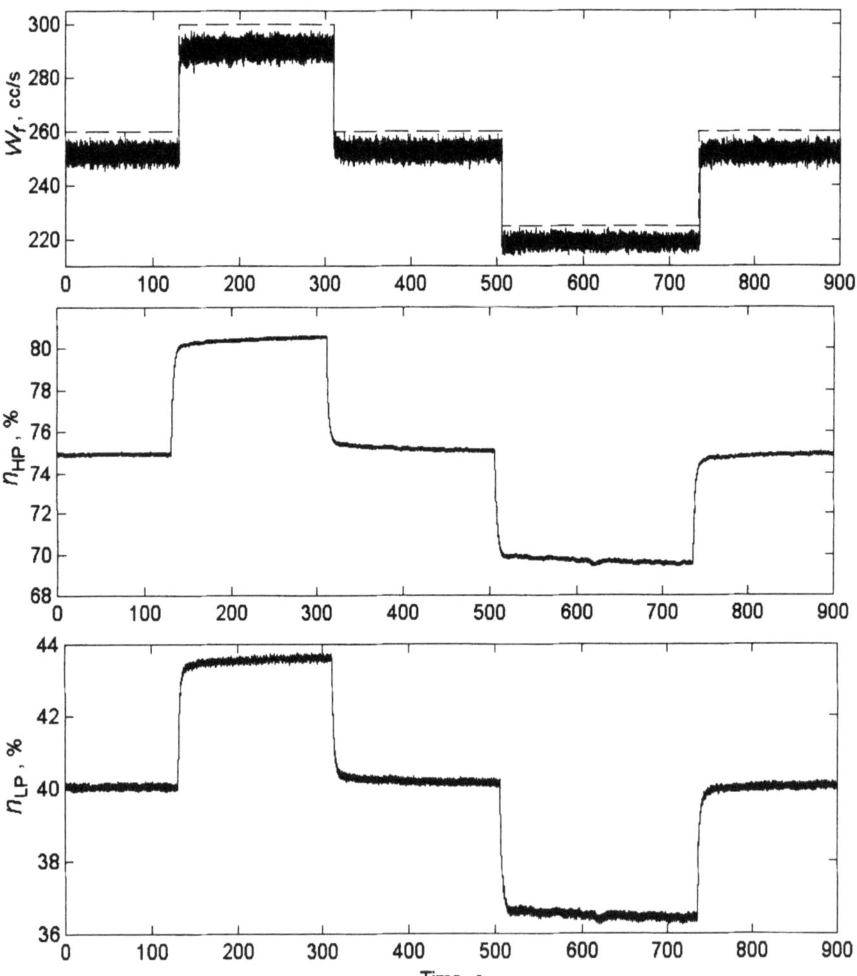

Figure 15.5. Test A: fuel flow and shaft speeds

15.4.3 Comparison with Previous Series of Tests

After the closest operating points from Sections 15.4.1 and 15.4.2 have been averaged, four operating points were obtained for estimating the static model (see Table 15.4).

The static operating points of the current series of tests represented in Table 15.4 are shown in Figure 15.6 and Figure 15.7 compared with the static model identified from previous test series. The previous series was performed at the same on-ground engine test bed facility several months before. The approximation formulae for the static model identified from the test data are as follows:

$$n_{HP} = -142.92 + 39.27 \log (W_f + 5.042)$$
$$n_{LP} = -82.58 + 22.64 \log (W_f - 26.93) \qquad (15.9)$$

Thus, the dynamic gains are described by the equations:

$$K_{HP} = \frac{39.27}{W_f + 5.042}$$

$$K_{LP} = \frac{22.64}{W_f - 26.93} \qquad (15.10)$$

Table 15.4. Mean values for static model estimation

Parameter	Point number			
	1	2	3	4
W_f, cc/s	149.85	218.81	252.16	288.22
n_{HP}, %	55.11	69.57	75.02	80.17
n_{LP}, %	26.36	36.44	40.07	43.43

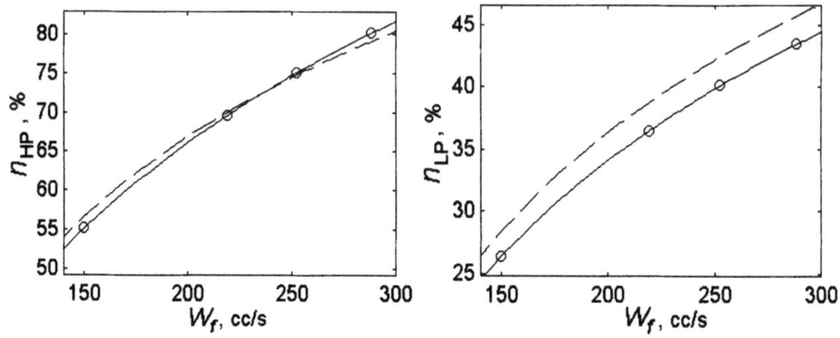

Figure 15.6. Static model: previous series (dashed) and current tests (circles and solid)

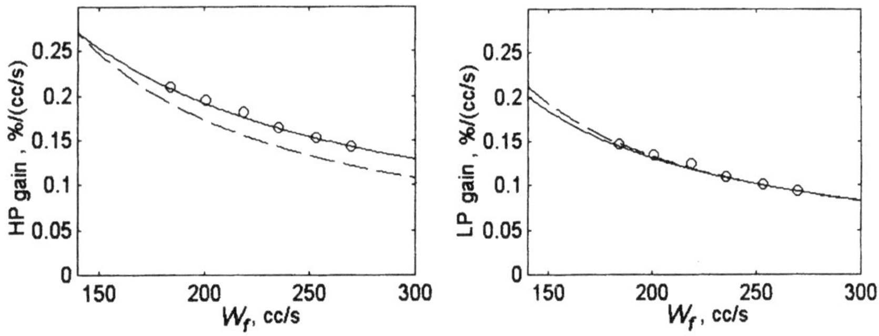

Figure 15.7. Static model gains: previous series (dashed) and current tests (circles and solid)

Static points slightly depart from the static model identified from previous tests. This might be a result of the difference in the atmospheric conditions (input temperature and pressure) or because of changes in the nozzle area or VSV and VIGV positions compared with the previous series. Also, the results might be influenced by the trend in the current data series. As a result, dynamic gains of HP have some difference from those obtained from the previous model, whereas LP gains are very close to those estimated from the previous data.

15.4.4 Notes Regarding Fuel Feed System

As shown in the previous sections, the fuel feed system in the test bed facility has some problems with providing a stable value of the fuel flow. This appears during open-loop experiments, whereas in the closed-loop test there was no trend in data. Moreover, the actual value of the fuel flow differs from demanded (see Figures 15.5 and 15.8). In future tests, this trend can be rejected by the use of closed-loop control of the fuel feed. Introducing a local feedback might improve both static and dynamic properties of the system with no changes in the existing test bed equipment. Thus, the local feedback on fuel flow can improve the results of further engine tests. A possible control structure is shown in Figure 15.9.

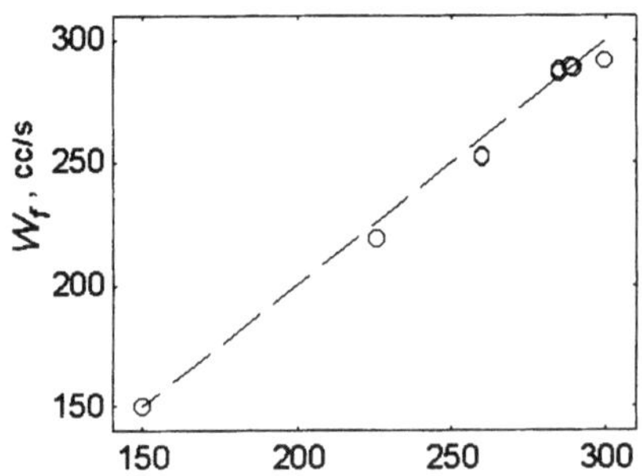

Figure 15.8. Mean values of fuel flow: measured (circles) and demanded (dashed)

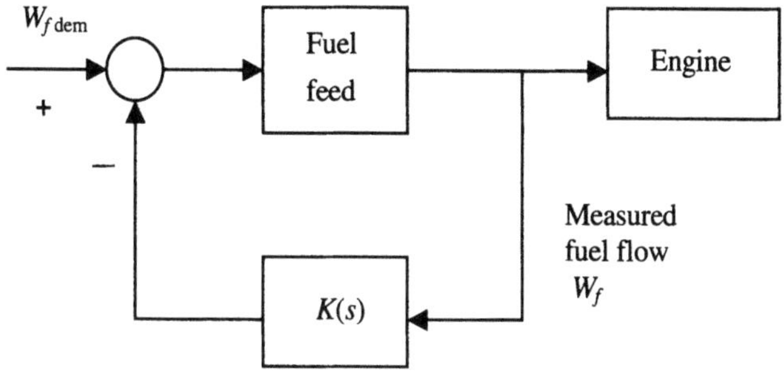

Figure 15.9. Proposed local feedback for fuel feed

15.5 Dynamic Model Identification

15.5.1 Multisine Test

The multisine test (developed by Evans *et al.* [4]) was used to estimate transfer functions (TF) of the engine and fuel feed system to perform the comparison of results obtained with different identification techniques. The engine TF estimates are shown in Table 15.5 and Figures 15.10 and 15.11. Standard deviation was estimated by the simulation method with 100 realisations. The TF estimates according to the structure Equation (15.4) are represented in the form of gains K, time constants T, and delays τ with the corresponding standard deviations (STD). Table 15.5 also contains the estimates of TF zeros Z and poles P.

Table 15.5. Engine transfer function estimates at 80% n_{HP}

Parameter	n_{HP}			n_{LP}		
	TF	STD	STD, %	TF	STD	STD, %
K	0.1174	0.0016	1.37	0.07667	0.0007	0.96
T_1	0.4617	0.0304	6.59	0.8598	0.0488	5.69
T_2	0.4524	0.0198	4.38	0.4524	0.0198	4.38
T_3	1.7996	0.0531	2.95	1.7996	0.0531	2.95
τ	0.0210	0.0023	11.1	0.0244	0.0017	7.05
Z_1	–2.1754	0.1471	6.76	–1.1668	0.0672	5.76
P_1	–2.2147	0.0993	4.48	–2.2147	0.0993	4.48
P_2	–0.5562	0.0163	2.93	–0.5562	0.0163	2.93

Figure 15.12 shows the results of dynamic model identification of the fuel feed system. A first-order TF was used to fit the experimental FRF for the use in the further analysis of the data from the closed-loop experiment:

$$H_{FF}(s) = 0.9876 / (0.1058\ s + 1) \exp(-0.0898\ s) \tag{15.11}$$

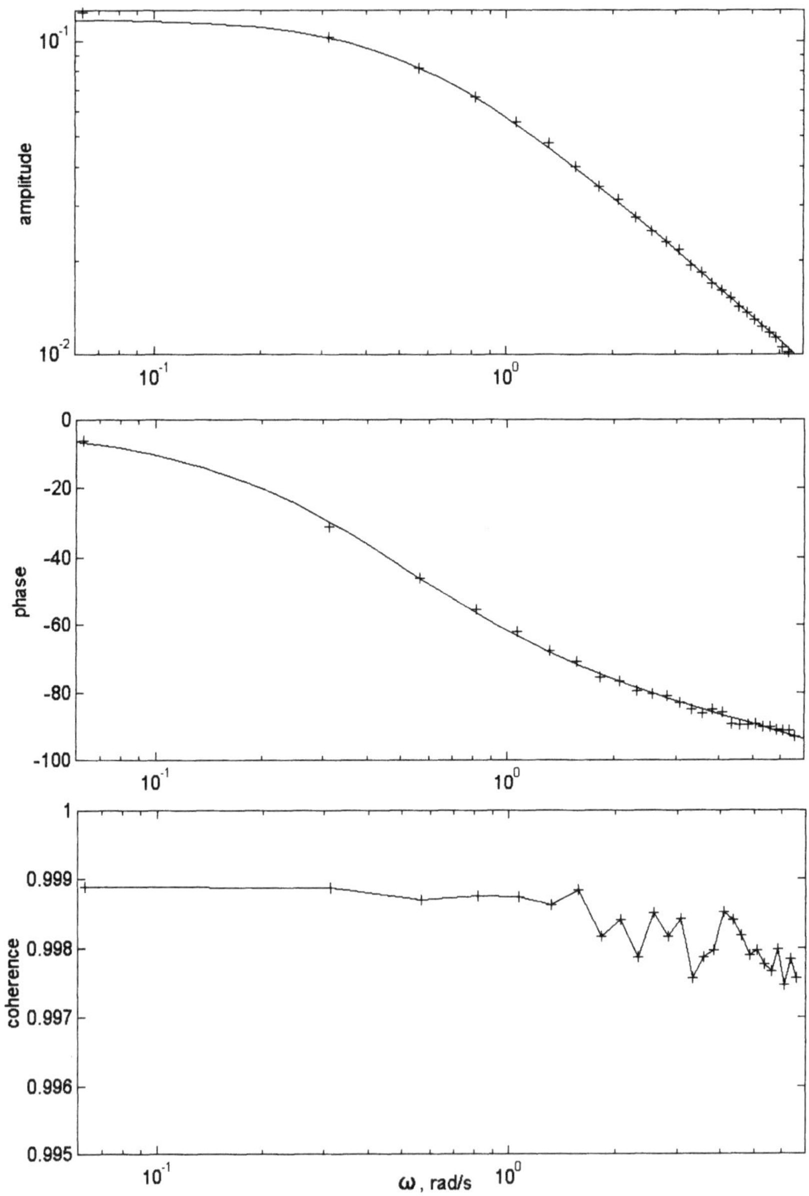

Figure 15.10. HP FRF estimates (pluses) and TF (solid)

15.5.2 Open-Loop Identification with No Test Signal

This test was carried out with no imposed probe signal. There are two sources of errors in the test data: trend and noise. The data from the experiment with no imposed test signal suffered from trend, as mentioned above in Section 15.4.1 (see Figure 15.4). However, the measurement noise is less in the current test series than in previous experimentation because an appropriate adjustment was done in the probe gearbox. This adjustment followed the analysis of shaft speed measurement errors. Basic noise properties are shown in Table 15.6. The amplitude was calculated by the formula:

$$Amp = \{\max(X) - \min(X)\} / 2 \quad (15.12)$$

The first column in Table 15.6 represent the test with no probe signal performed in 1992 with another type of n_{HP} probe. Analysis of Table 15.6 shows that the STD of the measurement noise in n_{HP} decreased after adjusting the probe. However, the noise is still greater than in 1992.

Table 15.6. Measurement noise in tests with no probe signal

	Experiment							
Parameter	**Test '92**	**Previous series**			**B (detrend)**	**E**	**A (first 2 min)**	**A (last 2 min)**
Mean W_f, cc/s	272	146	255	393	287	289	252	252
Mean n_{LP}, %	44.8	27.6	42.7	53.2	43.3	43.6	40.0	40.0
Mean n_{HP}, %	76.5	55.5	74.9	89.0	80.2	80.0	75.0	74.9
σ W_f, cc/s	1.000	1.83	2.71	6.16	2.70	2.31	2.239	2.083
σ n_{LP}, %	0.022	0.06	0.06	0.09	0.056	0.05	0.0497	0.051
σ n_{HP}, %	0.019	0.12	0.11	0.05	0.038	0.03	0.0464	0.047
Amp W_f, cc/s	3.46	4.86	6.68	13.5	9.37	7.97	7.251	6.467
Amp n_{LP}, %	0.069	0.21	0.20	0.30	0.027	0.18	0.1419	0.150
Amp n_{HP}, %	0.056	0.24	0.24	0.16	0.053	0.11	0.1124	0.126

During Test B, fuel flow increased from 286.36 to 287.69 cc/s, whereas shaft speeds increased from 43.217 to 43.380 % (LP) and from 80.053 to 80.307 % (HP), see Figure 15.4. The change of the shaft speeds divided by the change in the fuel flow gives the following gain estimates: $K_{LP} = 0.123$ and $K_{HP} = 0.151$. These figures are close to the gains estimated from the multisine test C: $K_{LP} = 0.081$ and $K_{HP} = 0.123$. The gains estimated from the static model Equation (15.2) are the following: $K_{LP} = 0.0871$ and $K_{HP} = 0.135$. Hence, the trend in the fuel flow is likely to cause the trends in shaft speeds. From the viewpoint of spectral analysis the trend should be excluded from the data. This has been done using the MATLAB function DETREND.

HP and LP FRF estimates are shown in Figures 15.13 and 14. There are two frequency ranges where HP coherence is quite high: $\omega < 0.03$ rad/s and $15 < \omega < 25$ rad/s. In these regions, amplitude and phase FRF are very close to those estimated from the multisine test C.

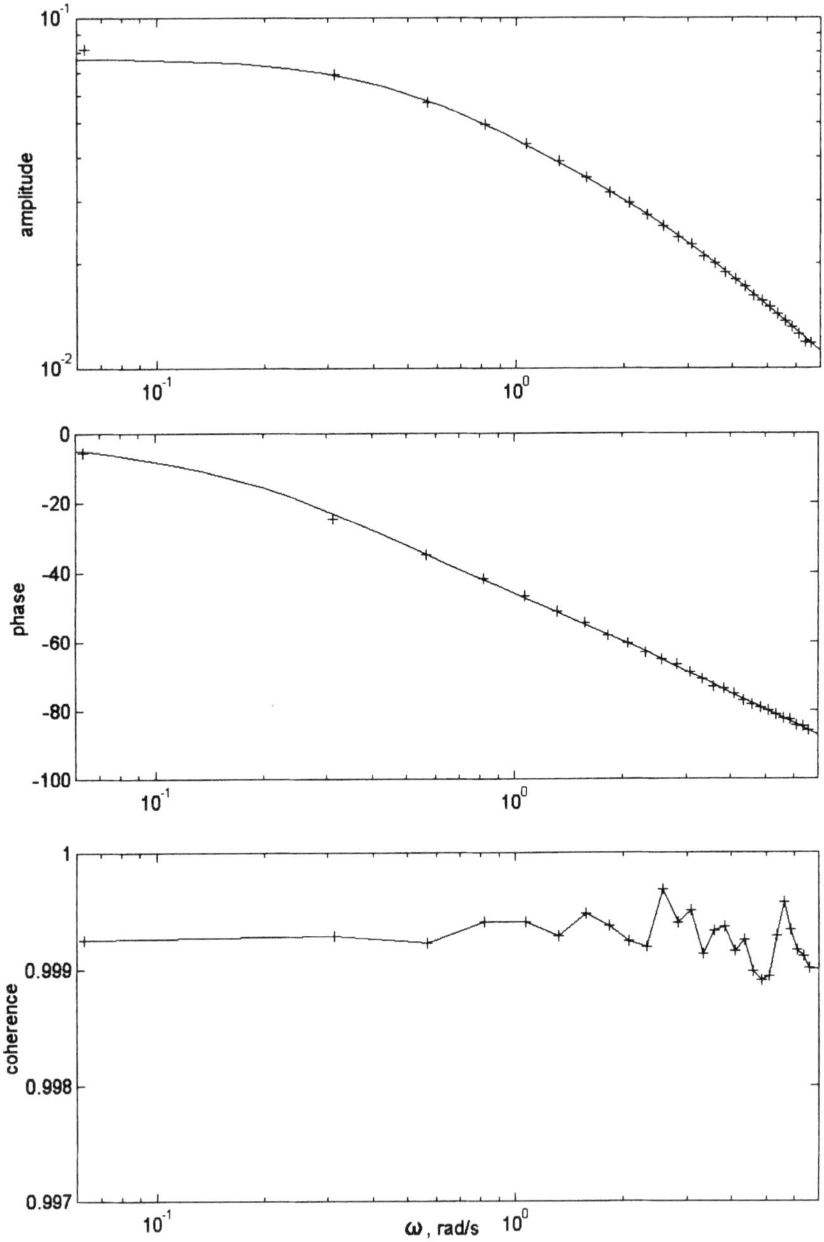

Figure 15.11. LP FRF estimates (pluses) and TF (solid)

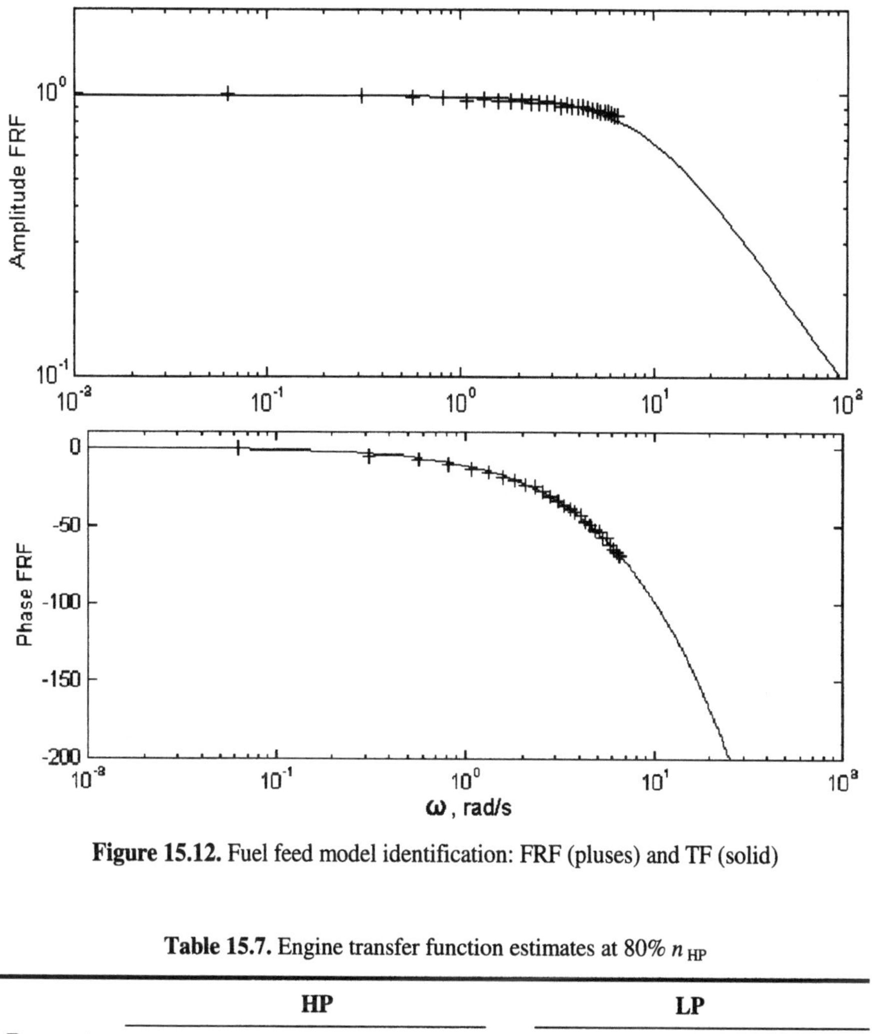

Figure 15.12. Fuel feed model identification: FRF (pluses) and TF (solid)

Table 15.7. Engine transfer function estimates at 80% n_{HP}

Parameter	HP			LP		
	TF	STD	STD, %	TF	STD	STD, %
K	0.08894	0.001425	1.60	0.06080	0.001722	2.83
T_1	0.9061	0.1188	13.1	1.2890	0.2401	18.6
T_2	1.2941	0.1046	8.08	1.2941	0.1046	8.08
T_3	1.2941	0.1046	8.08	1.2941	0.1046	8.08
τ	0.01650	0.0001965	1.19	0.01672	0.000574	3.43
Z_1	1.121	0.1404	12.5	0.7999	0.1357	16.9
P_1	0.7775	0.06059	7.79	0.7775	0.06059	7.79
P_2	0.7775	0.06059	7.79	0.7775	0.06059	7.79

The dynamic gain obtained as the amplitude FRF on a low frequency is $K_{HP} = 0.1223$, which is close to the static estimate (0.1345) and multisine estimate

(0.1174). Dynamic gains for LP are: $K_{LP} = 0.0779$ compared to the static 0.0871 and multisine 0.07667. Note that dynamic gain estimates are greater than obtained from FRF. Estimated coefficients for TF with equal denominators and their STD are shown in Table 15.7.

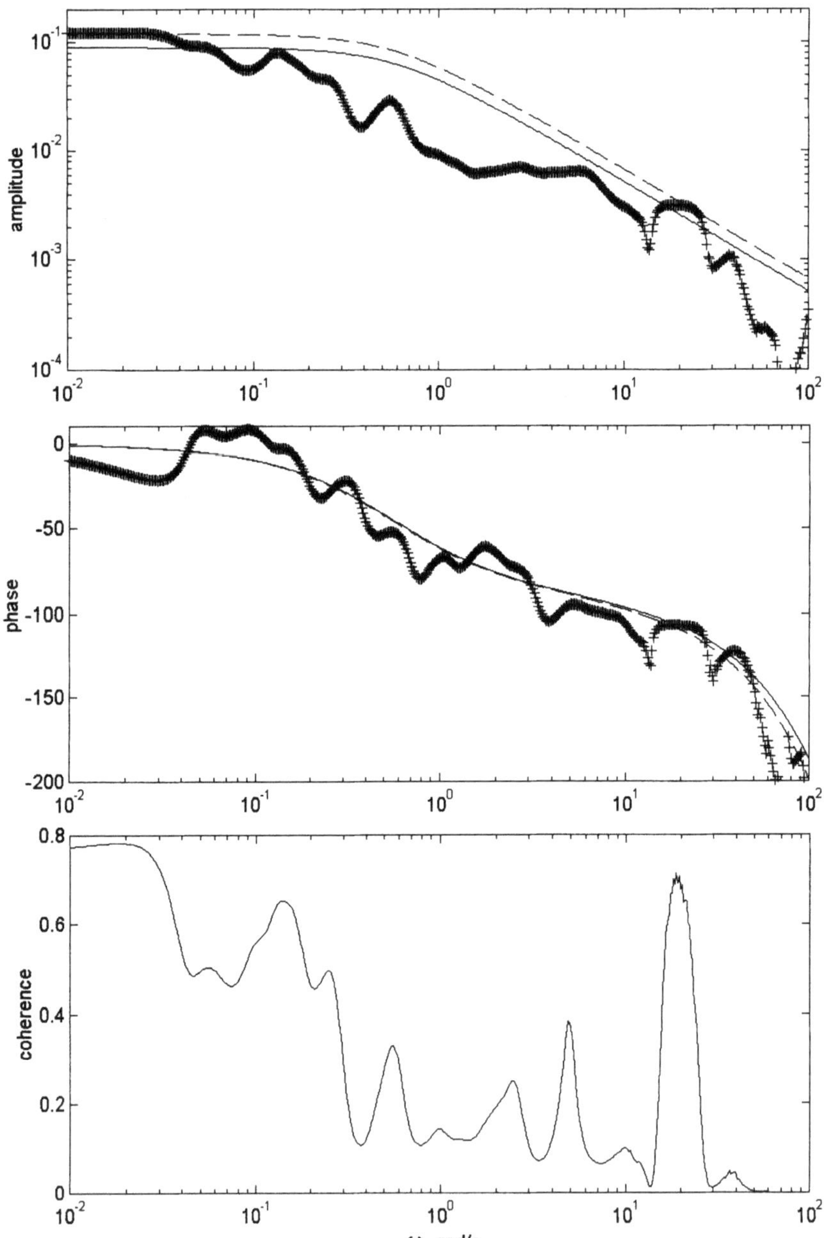

Figure 15.13. HP FRF estimates (pluses), TF (solid) and multisine TF (dashed)

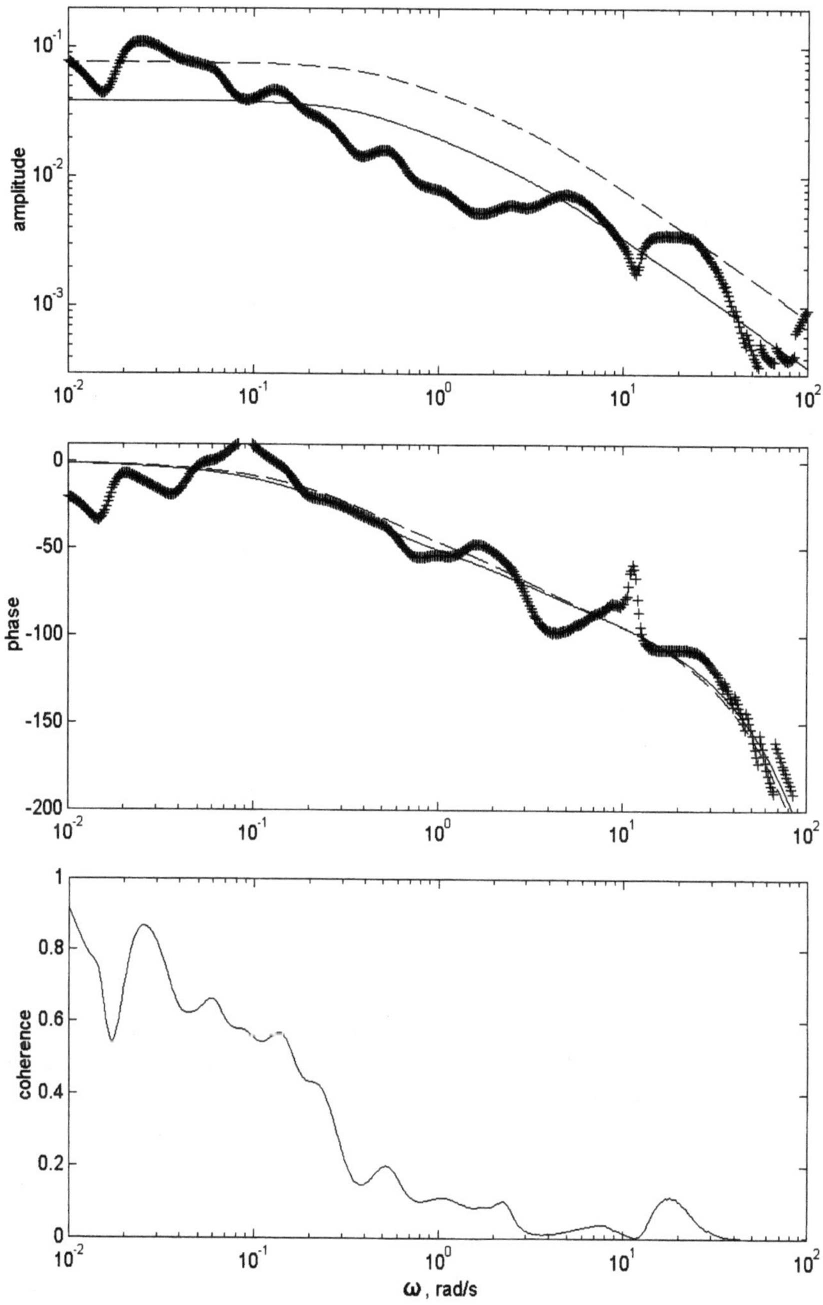

Figure 15.14. LP FRF estimates (pluses), TF (solid) and multisine TF (dashed)

15.5.3 Closed-Loop Identification with No Test Signal

Test E was performed with no imposed test signal under closed-loop control of n_{HP}. The experiment demonstrates the type of problems arising in closed-loop identification. A diagram of the closed-loop control is shown in Figure 15.15. There are three main elements in this closed loop: the controller, the fuel feed system, and the engine. This system suffers from two sources of disturbances: n_{HP} measurement (probe and A/D converter) and the fuel feed noise. In addition, some measurement noise can corrupt signals when they are going through the data acquisition system. One signal in this loop certainly has no noise: this is the fuel demand $W_{f.dem}$. It is obtained directly from the controller in the digital form and is not a result of measurement.

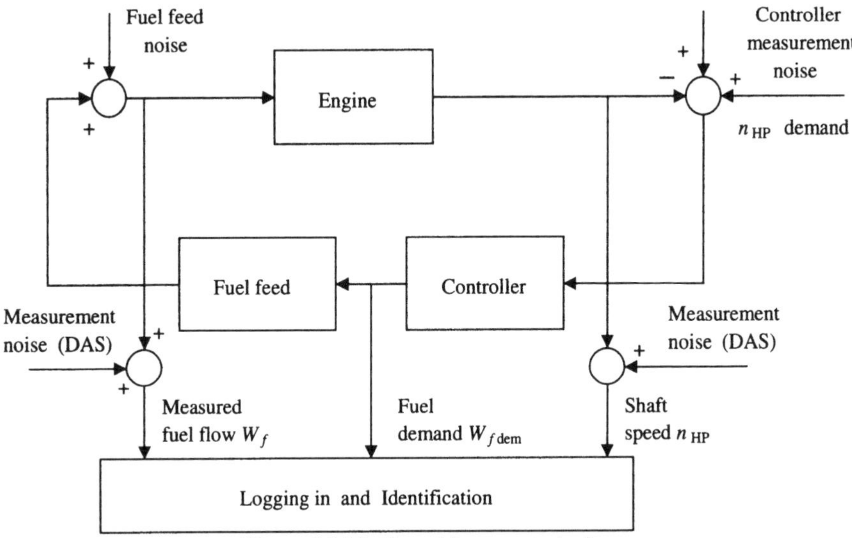

Figure 15.15. Closed-loop control of n_{HP}

A simplified diagram of the closed-loop control is shown in Figure 15.16. This includes three elements: $H_e(s)$, $H_c(s)$ and $H_f(s)$ which represent the corresponding transfer functions of the engine, controller, and fuel feed system.

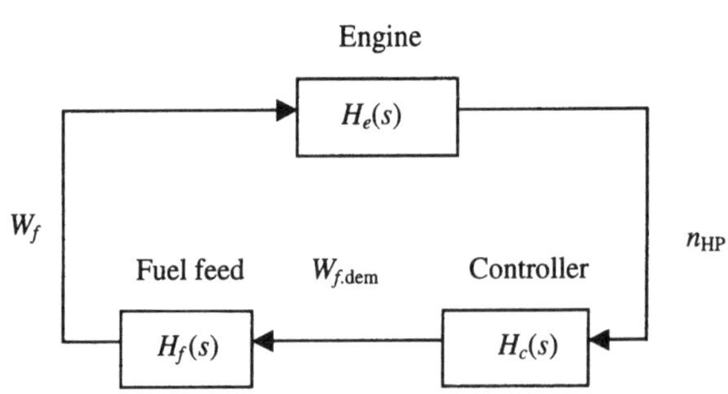

Figure 15.16. Simplified diagram of closed-loop control

The identifiability problem for closed-loop control has been investigated in [5-7]. The main approaches to the problem were discussed in Chapter 7. As regarding direct spectral methods for identification, two extreme cases exist: identifiability and nonidentifiability. The term "direct methods" designates an approach where the input and output signals are treated as if the system was in an open loop [6]. In simple words, identifiability represents such a case when the identified FRF corresponds to the plant model. A nonidentifiability case results in direct estimation of the inverse feedback model only. The type of the identifiability situation taking place during the experiment depends on which disturbance in the closed-loop is more powerful. If the measured fuel flow is more powerful than the measurement noise in n_{HP}, then the engine model is identified by direct methods.

Possible results of direct identification are considered in Table 15.8. The last two columns represent model estimates for identifiability and nonidentifiability cases.

Table 15.8. Identifiability cases

Identified plant	Input	Output	Identifiability	Nonidentifiability
Controller	n_{HP}	$W_{f.dem}$	H_c	$1/(H_f H_e)$
Fuel feed	$W_{f.dem}$	W_f	H_f	$1/(H_e H_c)$
Engine	W_f	n_{HP}	H_e	$1/(H_c H_f)$

The main problem of closed-loop identification is the lack of information about disturbances. Moreover, experiments with open-loop and closed-loop control cannot always be considered equivalent in terms of disturbances. In particular, the noise in the fuel feed and n_{HP} measurement channels may not be considered as purely additive and estimated precisely before closed-loop experimentation.

The results of direct identification from the test data are shown in Figures 15.17-15.20. These graphs demonstrate the expected identifiability and nonidentifiability models from Table 15.8. The transfer functions chosen for the comparison are the following:

- the controller model (obtained from the test bed program listing):

$$H_c(s) = \frac{-42.673(0.82018s+1)}{s} \tag{15.13}$$

- the engine model (identified from the multisine test):

$$H_e(s) = \frac{0.119\exp(-0.019s)}{1.79s+1} \tag{15.14}$$

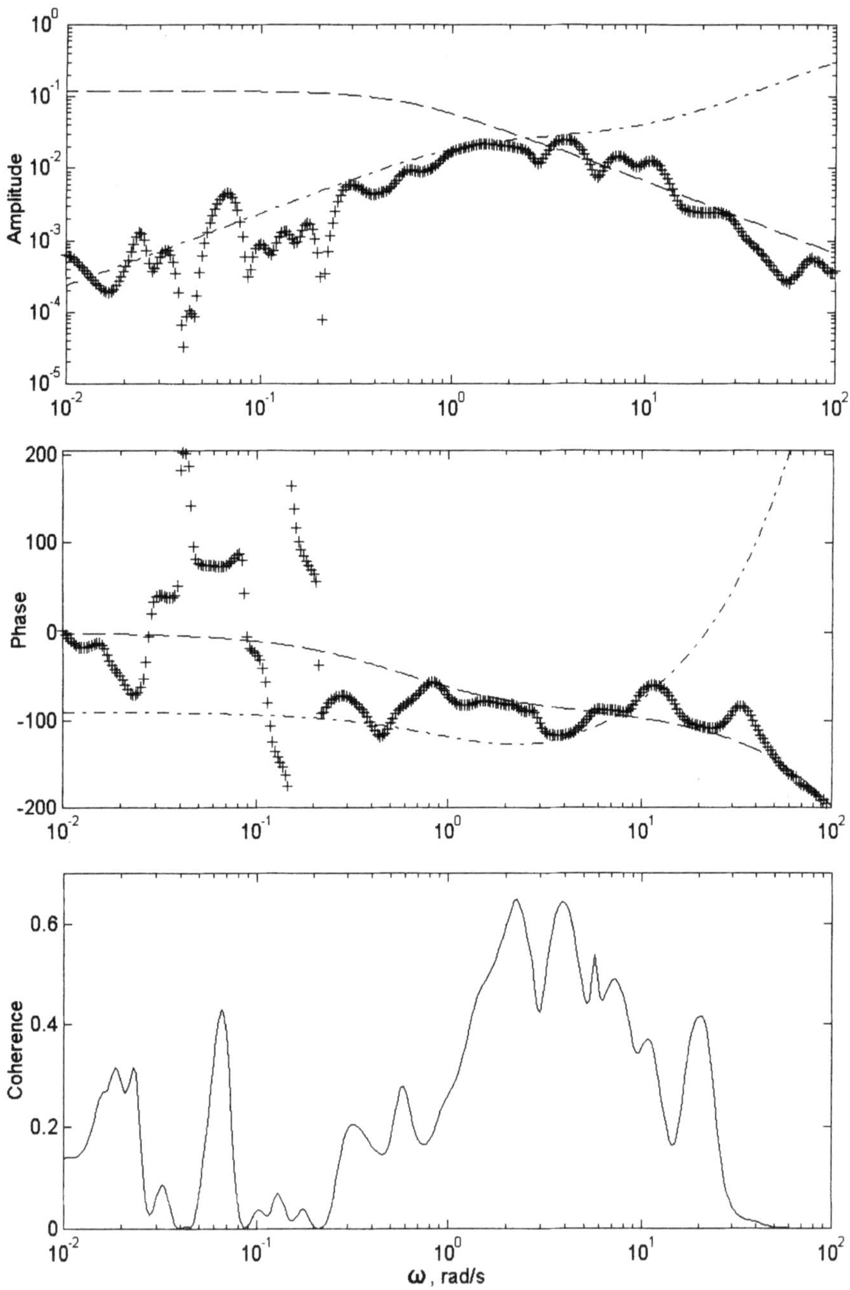

Figure 15.17. n_{HP} FRF and coherence estimates (pluses, solid), *a priori* models for identifiability (dashed) and nonidentifiability (dash-dotted)

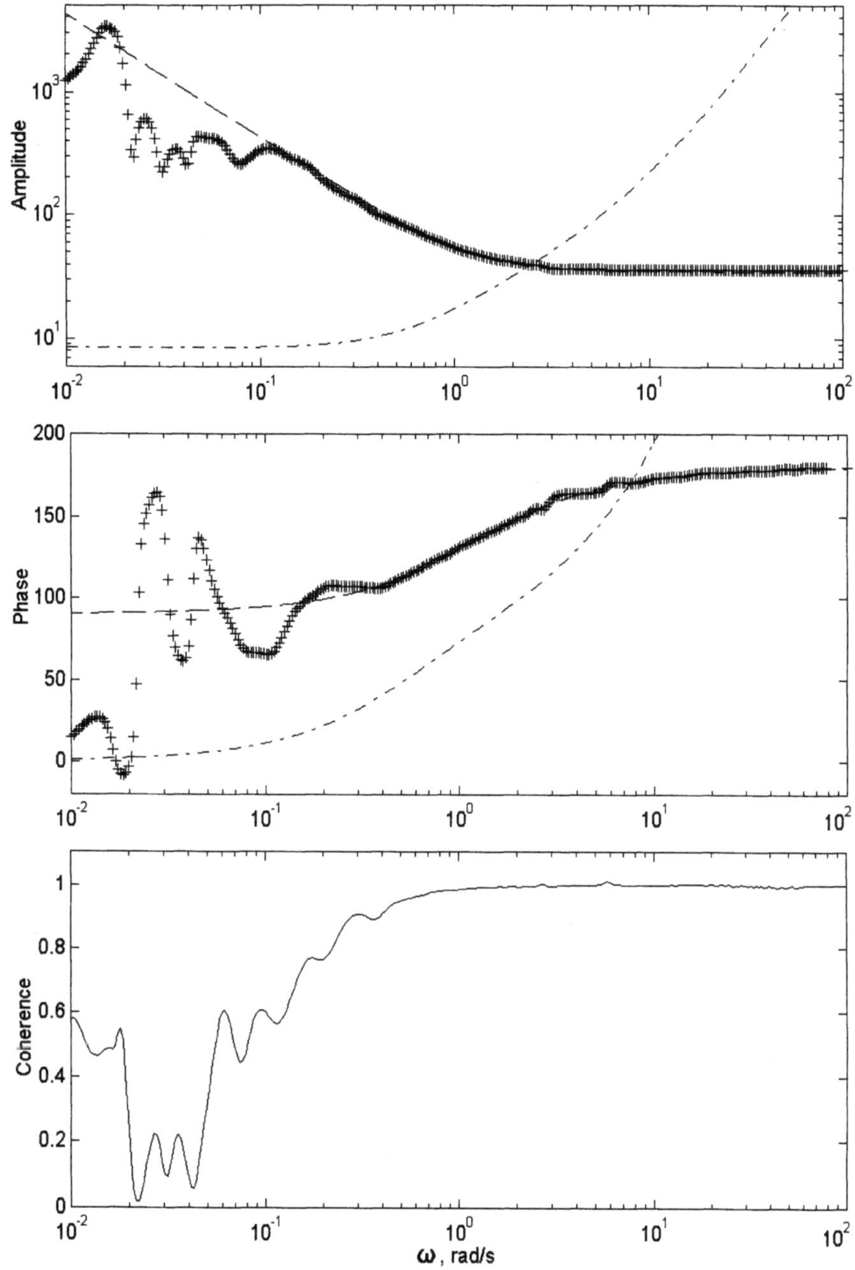

Figure 15.18. Controller FRF and coherence estimates (pluses, solid), *a priori* models for identifiability (dashed) and nonidentifiability (dash-dotted)

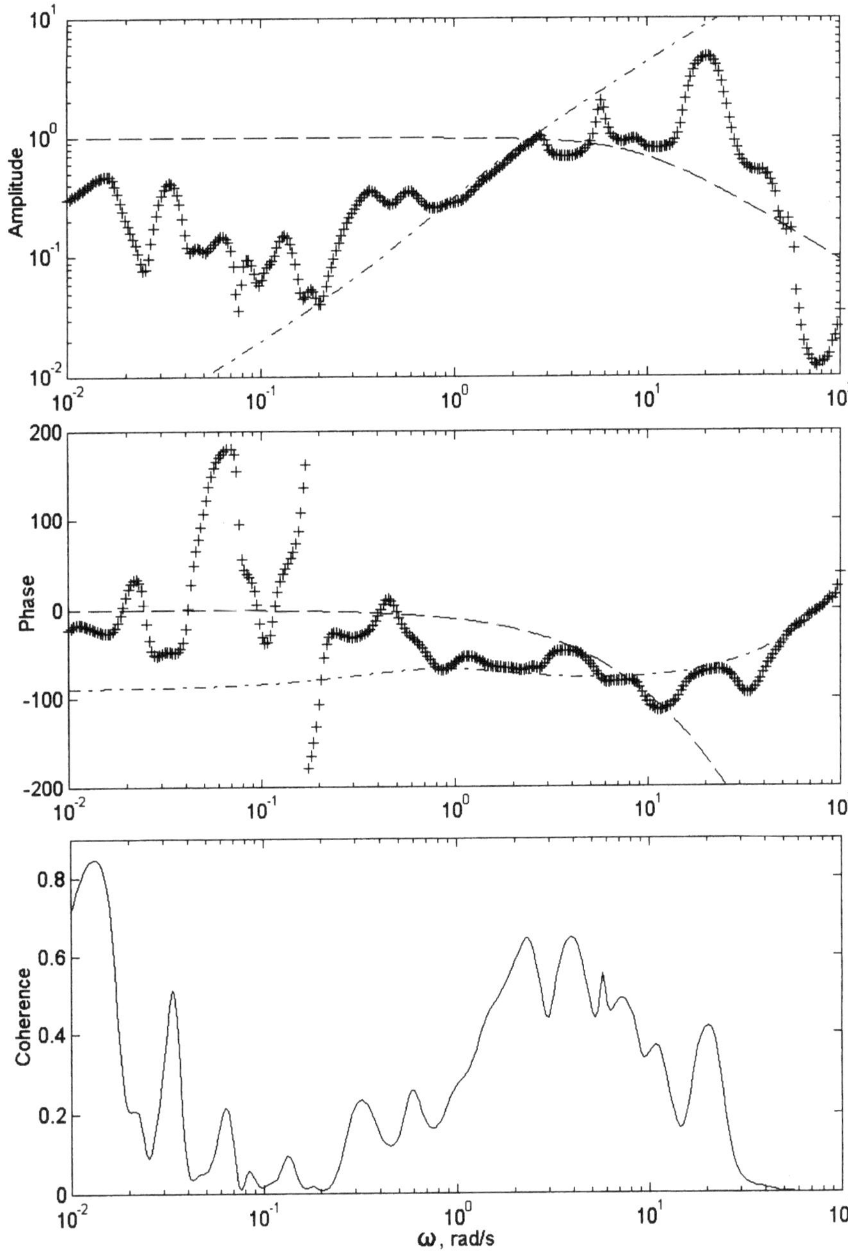

Figure 15.19. Fuel feed FRF and coherence estimates (pluses, solid), *a priori* models for identifiability (dashed) and nonidentifiability (dash-dotted)

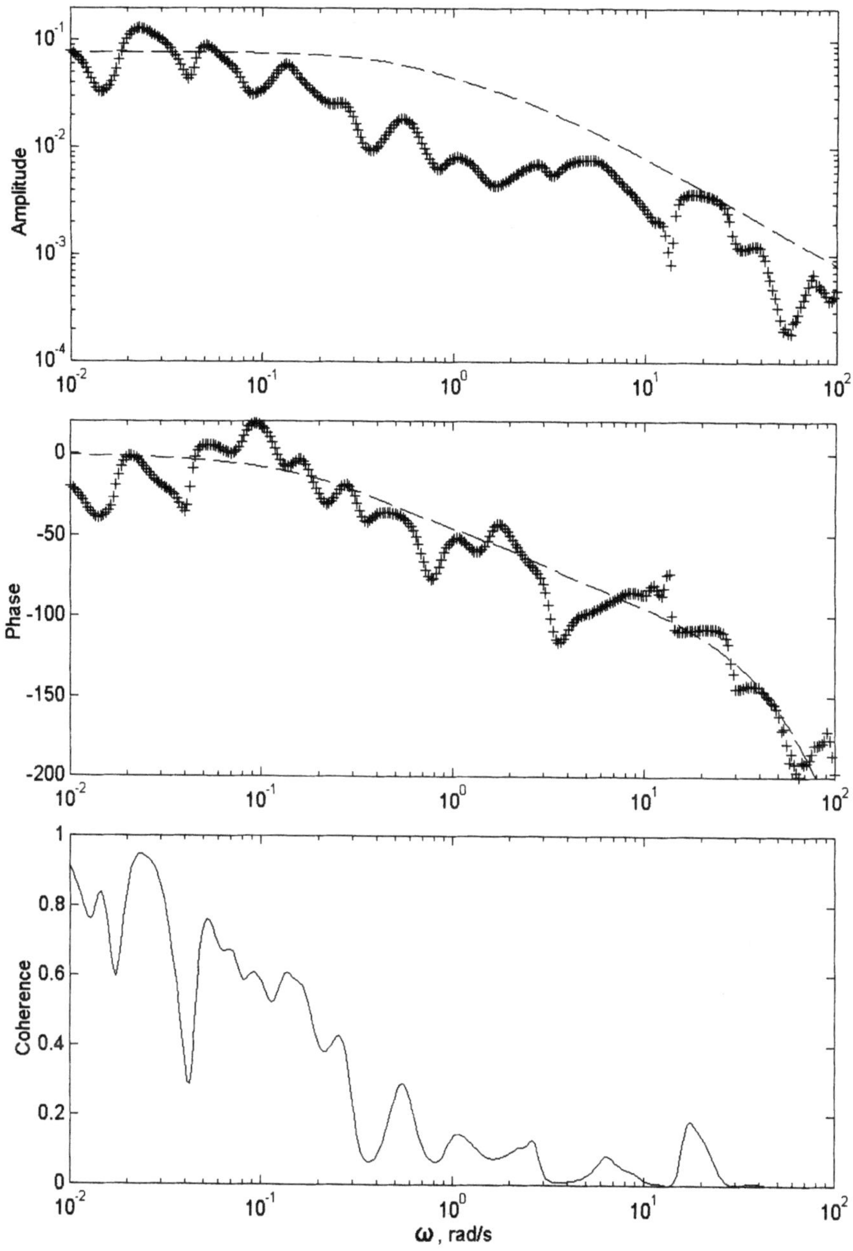

Figure 15.20. Engine n_{LP} FRF and coherence estimates (pluses, solid) and model identified from multisine test C (dashed)

- the model of the fuel feed system (identified from the multisine test):

$$H_f(s) = \frac{0.9876 \cdot \exp(-0.089s)}{0.1058s + 1}. \qquad (15.15)$$

Analysis of the FRF estimates from the closed-loop experiment shows that there are frequency ranges of identifiability for models of n_{HP} ($\omega > 2$ rad/s) and controller ($\omega > 0.1$ rad/s). The fuel feed system represents a nonidentifiability case. This means that the noise in the measured fuel flow is more powerful than the random process in the fuel demand. LP shaft dynamics can be treated as an open-loop data; the FRF estimate is in good agreement with the multisine model estimate.

15.6 Concluding Remarks

Methods for static and dynamic model identification have been demonstrated in this chapter using experimental data from Rolls-Royce SPEY engine tests. Dynamic models were identified in the form of frequency response and transfer functions using the optimal resolution technique and static modelling was proposed for cross-validation. The estimates with no test signal are in good agreement with multisine test results and the static model.

The real-life example of a turbo jet engine modelling has shown practical problems arising in systems identification. As opposite to "pure" simulation data, some additional factors intrude which are usually not expected. This results in a series of identification tests with step-by-step improvement of the experiment design. In the fuel feed system, a difference between the demanded and measured fuel flow was detected. A local feedback was then suggested to improve static and dynamic properties of the fuel feed system, with no changes in the existing equipment.

References

Chapter 1

1. Cohen H, Rogers GFC, Saravanamuttoo HIH. *Gas Turbine Theory*. Harlow: Addison-Wesley-Longman, 1996.
2. Cherkhasov BA. *Automatica and Control of Jet Engines.* (In Russian). Moscow: Mashinostroenie, 1988.
3. Dutton K, Thompson S, Barraclough B. *The Art of Control Engineering*, New York: Addison-Wesley, 1997.
4. McGlone ME. Transition of a technology base for advanced aircraft gas turbine control systems. *ASME Trans. Journal of Engineering for Gas Turbine and Power*, 1998;12;3: 437–441.
5. Kulikov GG. General schematic of gas turbine engines and their control programs. In: Shevyakov AA, Martyanova TS, editors. *Optimisation of Multi-variable Control Systems of Aero Gas Turbine Engines.* (In Russian). Moscow: Mashinostroenie, 1989; 16–23.
6. Shevyakov AA, editor. *Integrated Automatic Control of Aero Power Plants.* (In Russian). Moscow: Mashinostroenie, 1983.
7. Thompson HA. *Parallel Processing for Jet Engine Control.* Advances in industrial control Series. London: Springer-Verlag, 1992.
8. Kulikov GG, Arkov YuG, Arkov VYu. Method for control of radial clearance in turbine of turbo prop engine and device for its implementation. Patent No.2052644 (RF), 1996.
9. Kikstra JF, Roffel B, Schoen P. Model predictive control of a combined heat and power plant using local linear models. *ASME Trans. J. engineering for Gas Turbine and Power*, 1998;12;4:689–693.

Chapter 2

1. Kulikov GG. Mathematical modelling of gas turbine engine as a multi-variable plant. In: Shevyakov AA, Martyanova TS, editors. *Optimisation of Multi-variable Control Systems of Aero Gas Turbine Engines.* (In Russian). Moscow: Mashinostroenie, 1989; 30–88.
2. Cohen H, Rogers GFC, Saravanamuttoo HIH. *Gas Turbine Theory*. Harlow: Addison-Wesley-Longman, 1996.
3. Shevyakov AA. *Automatica of Aero Power Plants.* (in Russian). Moscow: Mashinostroyeniye, 1965.
4. Kulikov GG. Principles of design of digital control systems for aero engines (Chapter 7). Mathematical models used in CAD systems for engine and its control design (Section 8.4). In: Cherkasov BA, editor. *Control and*

Automatics of Jet Engines. (in Russian). Moscow: Mashinostroyeniye, 1988; 253–274, 288–308. (Japanese edition is also available).

Chapter 3

1. Dorf RC, Bishop RH. *Modern Control Systems*. New York: Addison-Wesley, 1995.
2. Kuo BC. *Automatic Control Systems*. Englewood Cliffs, NJ: Prentice-Hall, 1995.
3. Kulikov GG. Nonlinear performance-based static and dynamic models of gas turbine engines. In: Shevyakov AA, Martyanova TS, editors. *Optimisation of Multivariable Control Systems of Aero Gas Turbine Engines*. (In Russian). Moscow: Mashinostroenie, 1989; 30–35.
4. Kulikov GG. Dynamic characteristic of gas turbine engine. In: Shevyakov AA, Martyanova TS, editors. *Optimisation of Multivariable Control Systems of Aero Gas Turbine Engines*. (In Russian). Moscow: Mashinostroenie, 1989; 35–41.
5. Shevyakov AA. *Automatica of Aero Power Plants*. (in Russian). Moscow: Mashinostroyeniye, 1965.
5. Friedland B. *Advanced Control Systems Design*. Englewood Cliffs, NJ: Prentice-Hall, 1996.
6. Graupe D. *Identification of Systems*. New York: Robert Krieger Publishing, 1976.
7. Kulikov GG. Methods for creation of linear mathematical models of gas turbine engines, In: Shevyakov AA, Martyanova TS, editors. *Optimisation of Multivariable Control Systems of Aero Gas Turbine Engines*. (In Russian). Moscow: Mashinostroenie, 1989; 41–81.

Chapter 4

1. Kulikov GG Real-time dynamic model of gas turbine engine with variable coefficients, In: Shevyakov AA, Martyanova TS, editors. *Optimisation of Multivariable Control Systems of Aero Gas Turbine Engines*. (In Russian). Moscow: Mashinostroenie, 1989; pp. 81–88.
2. Nise NS. *Control Systems Engineering*. New York: Benjamin & Cummings, 1995.
3. Dutton K, Thompson S, Barraclough B. *The Art of Control Engineering*. New York: Addison-Wesley, 1997.
4. Kerr LJ, Nemec TS, Gallops GW. Real-time estimation of gas turbine engine damage using a control-based Kalman filter algorithm. *ASME Trans. J. of Engineering for Gas Turbines and Power*, 1992;114:187–195.
5. Halasz N, Davidson P, Abu-Hakima S, Phan S. JETA: a knowledge-based approach to aircraft of gas turbine engine maintenance. *J. of Applied Intelligence*, 1992;2:25–45.
6. Milne R, Trave-Massuyes L, Quevedo J. TIGER: numeric and qualitative model-based diagnosis. In: *Proc. 13th IFAC Triennal World Congress*, 1996; 127–132.

7. Isermann R, Balle P. Trends in the application of model-based fault detection and diagnosis of technical processes. In: *Proc. 13th IFAC Triennal World Congress*, 1996; 1–12.
8. Merrington GL. Fault diagnosis of gas turbine engines from transient data. *ASME Trans. J of Engineering for Gas Turbines and Power*, 1989; 111:237–243.
9. Milne R, Trave-Massuyes L. Real-time model-based diagnosis of gas turbines. In: *Proc. IMeC Workshop on Quantitative and Qualitative Methods for Fault Diagnosis in Process Control*, 1994; 28-41.
10. Torella G, Lombardo G. Neural Networks for Diagnostics and Trouble-Shooting of Aero Engines. In: *Proc. 5th AIDAA/AAAF/DGLR/RAeS European Propulsion Forum*, EPF-95, Paper 95-03, 1995.

Chapter 5

1. Ljung L. *System Identification: Theory for the User*. Englewood Cliffs, NJ: Prentice Hall; 1987.
2. Godfrey KR. Correlation methods. *Automatica*, 1980;16:527–534.
3. Ljung L. *System Identification Toolbox for Use with Matlab*. Natick, MA: Mathworks, Inc., 1995.
4. Norton JP. *An Introduction to Identification*. London: Academic Press, 1986.
5. Brillinger D. *Time Series: Data Analysis and Theory*. San Francisco: Holden-Day, 1981.
6. Schoukens J, Renneboog J. Modeling the noise influence on the Fourier coefficients after a discrete Fourier transform. *IEEE Trans. Instrumentation and Measurement*, 1986;35:278–286.
7. Gade S, Herlufsen H. Use of weighting functions in DFT/FFT analysis. Parts I and II. *Brüel and Kjaer Technical Review 3 and 4*, 1987.
8. Bendat JS, Piersol AG. *Engineering Applications of Correlation and Spectral Analysis*. New York: Wiley-Interscience, 1980.
9. Guillaume P. Identification of multiinput multioutput systems using frequency-domain methods. Ph.D. dissertation. Vrije Universiteit Brussel, Department ELEC, Belgium, 1992.
10. Schoukens J, Guillaume P, Pintelon R. Design of broadband excitation signals. (Chapter 3). In: Godfrey K, editor. *Perturbation Signals for System Identification*. Englewood Cliffs, NJ: Prentice-Hall, 1993.
11. Pintelon R, Guillaume P, Rolain Y, Verbeyst F. Identification of linear systems captured in a feedback loop. IEEE Trans. *Instrumentation and Measurement*, 1992;41:747–754.
12. Schoukens J, Pintelon R, Renneboog J. A maximum likelihood estimator for linear and nonlinear systems — a practical application of estimation techniques in measurement problems. *IEEE Trans. Instrumentation and Measurement*, 1988;37:10–17.
13. Kollár I. *Frequency-Domain System Identification Toolbox for Use with Matlab*. Natick, MA: Mathworks, Inc.,1994.
14. Kollár I. On frequency-domain identification of linear systems. *IEEE Trans. Instrumentation and Measurement*, 1993; 42:2–6.

15. Pintelon R, Schoukens J. *System Identification: A Frequency-Domain Approach*. IEEE Press, 2001.
16. Evans C. Identification of linear and nonlinear systems using multisine signals, with a gas turbine application. Ph.D. dissertation. University of Glamorgan, School of Electronics, UK, 1998.
17. Schoukens J, Dobrowiecki T, Pintelon R. Parametric and nonparametric identification of nonlinear systems in the presence of nonlinear distortions – A frequency-domain approach. *IEEE Instrumentation and Measurement Technology Conference, IMTC/98*, St. Paul, USA, 1998;43:176–190.
18. Schoukens J, Pintelon R, Van Hamme H. Identification of linear dynamic systems using piecewise constant excitations: Use, misuse and alternatives. *Automatica*, 1994; 30:1153–1169.
19. Hill DC. Identification of gas turbine dynamics: time-domain estimation problems. ASME Gas Turbine Conference, *97-GT-31*, 1997:1–7.
20. Godfrey KR. *Perturbation Signals for System Identification*. Englewood Cliffs, NJ: Prentice-Hall, 1993.
21. Van den Bos A. Estimation of parameters of linear system using periodic test signals. Ph.D. dissertation. Technische Hogeschool Delft, Netherlands, 1970.
22. Schroeder MR. Synthesis of low peak-factor signals and binary sequences of low auto-correlation. *IEEE Trans. Information Theory*, 1970; 16:85–89.
23. Guillaume P, Schoukens J, Pintelon R, Kollár I. Crest factor minimisation using nonlinear Chebyshev approximation methods. *IEEE Trans. Instrumentation and Measurement*, 1991;40:982–989.
24. Kollár I, Pintelon R, Schoukens J. Frequency-domain system identification toolbox for Matlab: A complex application example. *Prepr. 10th IFAC Symp. on System Identification*, Denmark, 1994;4:23–28.
25. Schoukens J, Pintelon R, Vandersteen G, Guillaume P. Frequency-domain system identification using nonparametric noise models estimated from a small number of data sets. *Automatica*, 1997;33:1073–1086.
26. Evans C, Rees D, Jones L. Identifying linear models of systems suffering nonlinear distortions, with a gas turbine application. *IEE Proc. Control Theory and Applications*, 1995;142:229–240.
27. McCormack AS, Godfrey KR, Flower JO. The detection of and compensation for nonlinear effects using periodic input signals. *IEE International Conference "Control 94,"* University of Warwick, 1994:297–302.

Chapter 6

1. Hill DC. System identification of gas turbine engines. Ph.D. dissertation. University of Birmingham, School of Electronic and Electrical Engineering, UK, 1994.
2. Evans C. identification of linear and nonlinear systems using multisine signals, with a gas turbine application. Ph.D. dissertation, University of Glamorgan, School of Electronics, UK, 1998.
3. Evans C, Borrell A, Rees D. Testing and modelling gas turbines using multisine signals and frequency-domain techniques. *ASME J. of Engineering for Gas Turbine and Power*, 1999;121:451–457.

4. Haber R. Nonlinearity tests for dynamic processes. In: *Proc. 7th IFAC Symp. Identification and System Parameter Estimation*, York, 1985;409–414.
5. Hill DC. Identification of gas turbine dynamics: time-domain estimation problems. *ASME Gas Turbine Conf.*, paper 97-GT-31,1997:1–7.

Chapter 7

1. Kuo BC. *Digital Control Systems*. London: Saunders College Publishing, 1992.
2. Åström KJ, Wittenmark B. *Computer-Controlled Systems: Theory and Design*. London: Prentice-Hall, 1997.
3. Rohrs CE, Melsa JL, Schultz DG. *Linear Control Systems*. New York: McGraw-Hill, 1993.
4. Ljung L. *System Identification: Theory for the User*. Englewood Cliffs, NJ London: Prentice-Hall, 1987.
5. Eykhoff P. *System Identification: Parameter and State Estimation*. London: Wiley-Interscience, 1974.
6. Ljung L, Glad T. *Modeling of Dynamic Systems*. Englewood Cliffs, NJ: Prentice-Hall, 1994.
7. Söderström T, Stoica P. *System Identification*. Englewood Cliffs, NJ: Prentice-Hall, 1989.
8. Isermann R. *Digital Control Systems*. New York: Springer-Verlag, 1981.
9. Otnes RK, Enochson L. *Applied Time Series Analysis: Basic Techniques*. New York: Wiley-Interscience, 1978.
10. Marple L. *Digital Spectral Analysis with Applications*. Englewood Cliffs, NJ: Prentice-Hall, 1987.
11. Eisen M. *Introduction to Mathematical Probability Theory*. Englewood Cliffs, NJ: Prentice-Hall, 1969.
12. Rozanov IA. *Introductory Probability Theory*. Englewood Cliffs, NJ: Prentice-Hall, 1969.
13. Bendat J, Piersol A. *Random Data: Analysis and Measurement Procedures*. New York: Wiley, 1986.
14. Forssell U, Ljung L. Closed-loop identification revisited. *Automatica*, 1999; 35;7:1215–1241.

Chapter 8

1. Schetzen M. *The Volterra and Wiener Theories of Nonlinear Systems*. New York: Wiley-Interscience, 1980.
2. Leontaritis IJ, Billings SA. Experimental design and identifiability for nonlinear systems. *Int J. Systems Science*, 1987;18:189–202.
3. Narendra KS, Gallman PG. An iterative method for the identification of nonlinear systems using a Hammerstein model. *IEEE Trans. Automatic Control*, 1966;11:546–550.
4. Billings SA, Tsang KM. Spectral analysis of block-structured nonlinear systems. *J. Mechanical Systems and Signal Processing*, 1990;4:117–130.

5. Weiss M, Evans C, Rees D. Identification of nonlinear cascade systems using paired multisine signals. *IEEE Trans. on Instrumentation and Measurement*, 1998;47:332–336.
6. Vandersteen G, Schoukens J. Measurement and identification of nonlinear systems consisting of linear dynamic blocks and one static nonlinearity. *IEEE Trans. Automatic Control*, 1999;44:1266–1271.
7. Leontaritis IJ, Billings SA Input-output parametric models for nonlinear systems. Part I: deterministic nonlinear systems. *Int. J. Control*, 1985;41:303–328.
8. Billings SA, Voon WSF Piecewise identification of nonlinear systems. *Int. J. of Control*, 1987;46:215–235.
9. Johansen TA, Foss BA Constructing NARMAX models using ARMAX models. *Int. J. Control*, 1993;58:1125–1153.
10. Sjöberg J Zhang Q, Ljung L, Benveniste A, Delyon B, Glorennec P, Hjalmarsson H, Juditsky A. Nonlinear black-box modeling in system identification: A unified overview. *Automatica*, 1995;31:1691–1724.
11. Ljung L. Black-box models from input–output measurements. In: *Proc. IEEE Instrumentation and Measurement Technology Conf., Budapest*, 2001;138–146.
12. Ljung L. *System Identification: Theory for the User*. Englewood Cliffs, NJ: Prentice-Hall, 1987.
13. Chen S, Billings SA, Luo W. Orthogonal least squares methods and their application to nonlinear system identification. *Int. J. of Control*, 1989; 50:1873–1896.
14. Korenberg MJ, Billings SA, Liu YP, Mollroy PJ. Orthogonal parameter estimation algorithm for nonlinear stochastic systems. *Int. J. of Control*, 1988; 48:193–210.
15. Aguirre LA. Some remarks on structure selection for nonlinear models. *Int. J. Bifurcation and Chaos*, 1994;4:1707–1714.
16. Liu YP. Identification of nonlinear systems: The NARMAX polynomial model approach. Ph.D. dissertation. University of Sheffield, Department of Automatic Control and Systems Engineering, UK, 1988.
17. Billings SA, Voon WSF. Structure detection and model validity tests in the identification of nonlinear systems. *IEE Proceedings, Pt. D.* 1983;130:193–199.
18. Aguirre LA, Donoso-Garcia PF, Santos-Filho R. Use of a priori knowledge in the identification of global nonlinear models: A case study using a buck converter. *IEEE Trans. Circuits and Systems*, 2000;47:1081–1085.
19. Nøgaard M, Ravn O, Poulsen KN, Hansen LK. *Neural Networks for Modelling and Control*. London: Springer-Verlag, 2001.

Chapter 9

1. Bäck T. *Evolutionary Algorithms in Theory and Practice: Evolution Strategies, Evolutionary Programming, Genetic Algorithms*. Oxford: Oxford University Press, 1996.

2. Rechenberg I. *Evolutionsstrategie: Optimierung Technischer Systeme nach Prinzipien der Biologischen Evolution*. Stuttgard: Frommann-Holzboog, 1973.
3. Fogel LJ, Owens AJ and Walsh MJ. *Artificial Intelligence Through Simulated Evolution*. New York: Wiley, 1966.
4. Holland JH. *Adaptation in Natural and Artificial Systems*. Ann Arbor: The University of Michigan Press, 1975.
5. Koza JR. *Genetic Programming: On the Programming of Computers by Means of Natural Selection*. Cambridge, MA: MIT Press, 1992.
6. Baker JE. Reducing bias and inefficiency in the selection algorithm. In: *Proc. 2nd Int. Conf. on Genetic Algorithms, Cambridge, MA, USA*, 1987; 14–21.
7. Goldberg DE. *Genetic Algorithms in Search, Optimisation and Machine Learning*. Reading, MA: Addison-Wesley, 1989.
8. Hwang CL, Masud ASM. *Multiple Objective Decision Making, Methods and Applications*. New York: Springer-Verlag, 1979.
9. Fonseca C, Fleming PJ. Genetic algorithms for multiobjective optimization: Formulation, discussion and generalisation. In: *Proc 5th Int. Conf. on Genetic Algorithms*, 1993; 416–423.
10. Evans DC, Rees D, Jones DL. Design of tests signals for identification of linear systems with nonlinear distortions. *IEEE Trans. on Instrumentation and Measurement*, 1992;41;6:768–774.
11. Billings SA, Voon WSF. Correlation based model validity tests for nonlinear models. *Int. J. of Control*, 1986;44;1: 235–244.

Chapter 10

1. Kulikov GG. Mathematical modelling of a gas turbine engine as a multivariable plant. In: Shevyakov AA, Martyanova TS, editors. *Optimisation of Multivariable Control Systems of Aero Gas Turbine Engines*. (In Russian). Moscow: Mashinostroenie, 1989; 30–88.
2. Eykhoff P. *System Identification: Parameter and State Estimation*. London: Wiley-Interscience, 1974.
3. Söderström T, Stoica P. *System Identification*. Englewood Cliffs, NJ: Prentice-Hall, 1989.
4. Ljung L. *System Identification: Theory for the User*. Englewood Cliffs, NJ: Prentice-Hall, 1987.
5. Otnes RK, Enochson L. *Applied Time Series Analysis: Basic Techniques*. New York: Wiley-Interscience, 1978.
6. Marple L. *Digital Spectral Analysis with Applications*. Englewood Cliffs, NJ: Prentice-Hall, 1987.
7. Bendat J, Piersol A. Random Data: Analysis and Measurement Procedures. New York: Wiley, 1986.
8. Ljung L, Glad T. *Modeling of Dynamic Systems*. Englewood Cliffs, NJ: Prentice-Hall, 1994.
9. Ljung L. *System Identification Toolbox for Use with MATLAB: User's Manual*. Natick, MA: MathWorks, Inc., 1991.

10. Kollár I. *Frequency-Domain System Identification Toolbox for Use with MATLAB: User's Manual.* Natick, MA: MathWorks, Inc., 1994.
11. Patel VC, Kadirkamanathan V, Kulikov GG, Arkov VYu and Breikin TV. Gas turbine engine condition monitoring using statistical and neural network methods. In: *Proc. IEE Colloquium on Modelling and Signal Processing for Fault Diagnosis.* Paper 1996/260, 1996; 1–6.
12. Kulikov GG, Arkov VY, Breikin TV. On condition monitoring of FADEC information channels. *Russian Aeronautics*, 1995;4:81–85.
13. Ivashenko NN. *Automatic Control: Theory and Elements of Systems.* (in Russian). Moscow: Mashinostroenie, 1978.
14. *The Encyclopedia of Cybernetics.* (in Russian). USE, Kiev, 1975; v.1.
15. Lotosh MM, Shuster AL. *Basics of Automatic Control Theory: Mathematical Methods.* (in Russian). Moscow: Nauka, 1992.
16. Livshits NA, Pugatchev VN. *Probabilistic Analysis of Automatic Control Systems.* (in Russian). Moscow: Sovetskoe radio, 1963.
17. Gubarev VV. *Probabilistic Models: Reference Book.* (in Russian). Novosibirsk: NETU, 1992.
18. Peterson EL. *Statistical Analysis and Optimisation of Systems.* New York: John Wiley & Sons, 1961.
19. Cooper G, McGillem C. *Probabilistic Methods of Signal and System Analysis.* New York: Holt, Rinehart & Winston, 1986.
20. Isermann R. *Digital Control Systems.* New York: Springer-Verlag, 1981.
21. Arkov V, Kulikov G. Gas turbine engine frequency response function identification using spectral methods: Optimisation approach, In: Rahnejat H, Whalley R, editors. *Multibody Dynamics: Monitoring and Simulation Techniques.* London: Mechanical Engineering Publications, 1997; 123–131.
22. Arkov VY, Kulikov GG. Dynamic model identification using spectral analysis: optimisation approach. In: Proc. *IEEE Singapore Int. Symp. on Control Theory and Applications*, 1997; 415–418
23. Arkov V, Evans DC, Fleming PJ, Hill DC, Norton JP, Pratt I, Rees D, Rodriguez-Vazquez K. System identification strategies applied to aircraft gas turbine engines. In: *Proc. 14th Triennal IFAC World Congress*, 1999; 145–152.
24. Graps AL. An introduction to wavelets. *IEEE Trans on Computational Sciences and Engineering*, 1995;2;2:50–61.
25. Jiang Q. Orthogonal multiwavelets with optimum time-frequency resolution. *IEEE Trans. on Signal Processing*, 1998;46;4:830–844.
26. Warden AT, Percival DB, McCoy EJ. Spectrum estimation by wavelet thresholding of multitaper estimators. *IEEE Trans. on Signal Processing*, 1998;46;12:3153–3165.

Chapter 11

1. Kulikov GG, Arkov VY, Breikin TV. On condition monitoring of FADEC information channels. *Russian Aeronautics*, 1995;4:81–85.
2. Ljung L. *System Identification Toolbox for Use with MATLAB: User's Manual.* Natick, MA: Mathworks, Inc., 1991.

3. Arkov V, Kulikov G. Gas turbine engine frequency response function identification using spectral methods: Optimisation approach. In: Rahnejat H, Whalley R, editors. *Multibody Dynamics: Monitoring and Simulation Techniques*. London: Mechanical Engineering Publications, 1997; 123–131.

Chapter 12

1. Isermann R. *Digital Control Systems*. New York: Springer-Verlag, 1981.
2. Bendat JS, Piersol AJ. *Random Data: Analysis and Measurement Procedures*. New York: Wiley, 1986.
3. Jenkins GM, Watts DG. *Spectral Analysis and Its Applications*. London: Holden-Day, 1968.
4. Ljung L. *System Identification: Theory for the User*. Englewood Cliffs, NJ: Prentice-Hall, 1987.
5. Marple SL. *Digital Spectral Analysis with Applications*. Englewood Cliffs, NJ: Prentice-Hall, 1987.
6. Eykhoff P, editor. *Trends and Progress in System Identification*. Oxford: Pergamon Press, 1981.
7. Dedesh VT, editor. *Identification of Control Systems of Aero Gas Turbine Engines*. (in Russian). Moscow: Mashinostroenie, 1984.
8. Åström KJ. *Introduction to Stochastic Control Theory*. New York: Academic Press, 1970.
9. Kulikov GG, Fleming PJ, Breikin TV, Arkov VY. Markov *Modelling of Complex Dynamic Systems: Identification, Simulation and Condition Monitoring with Example of Digital Automatic Control System of Gas Turbine Engine*. (in Russian). Ufa, Russia: UGATU, 1998.
10. Searle SR, Hausman WH. *Matrix Algebra for Business and Economics*. New York: Wiley-Interscience, 1970.
11. Tikhonov VI, Mironov MA. *Markov Processes*. (in Russian). Moscow: Sovetskoe Radio, 1977.
12. Zadeh LA. Outline of a new approach to the analysis of complex systems and decision processes. *IEEE Trans. on Systems, Man and Cybernetics*. 1973;3; 1:28–44.
13. Mamdani EH. Advance in linguistic synthesis with a fuzzy logic controller. *Int. J. Man-Machine Studies*. 1976;8;669–678.
14. Jang JSR, Sun CT. Predicting chaotic time series with fuzzy if-then rules. In: *Proc. IEEE Int. Conf. on Fuzzy Systems*, 1993.
15. Breikin TV, Arkov VY, Kulikov GG. On stochastic system identification: Markov models approach. In: *Proc. 2nd Asian Control Conf. ASCC'97*, 1997; 775–778.
16. Arkov VY, Breikin TV, Kulikov GG. Fuzzy Markov simulation technique for product testing equipment. In: *Prepr. 4th IFAC Workshop on Intelligent Manufacturing Systems, IMS'97*, 1997; 415–419.
17. Castro JL. Fuzzy logic controllers are universal approximators. *IEEE Trans. on Systems, Man and Cybernetics*, 1995;25;4:629–635.

18. Press WH, Flannery BP, Teukolsky SA, Vetterling WT. *Numerical Recipes in C: the Art of Scientific Computing*. New York: Cambridge University Press, 1990.

Chapter 13

1. Press WH, Flannery BP, Teukolsky SA, Vetterling WT. *Numerical Recipes in C: the Art of Scientific Computing*. New York: Cambridge University Press, 1990.
2. Dedesh VT, editor. *Identification of Control Systems of Aero Gas Turbine Engines*. (in Russian). Moscow: Mashinostroenie, 1984.
3. Koza J. Genetic Programming: *On the Programming of Computers by Means of Natural Selection*. Cambridge, MA: MIT Press, 1992.
4. Chipperfield A, Fleming P, Pohlheim H, Fonseca C. *Genetic Algorithm Toolbox for use With MATLAB: User's Guide*. Sheffield, UK: University of Sheffield, 1994.
5. Isermann R. Process fault detection based on modelling and estimation methods: A survey. *Automatica*, 1984;20;4:387–404.
6. Basseville M. Detecting changes in signals and systems: A survey. *Automatica*, 1988;24;3:309–326.
7. Isermann R. Fault diagnosis of machines via parameter estimation and knowledge processing: Tutorial paper. *Automatica*, 1993;29;4:815–835.
8. Breikin TV, Arkov VY, Kulikov GG, Kadirkamanathan V, Patel VC. On gas turbine engine and control system condition monitoring. In: *Prepr. IFAC Symp. on Fault Detection, Supervision and Safety for Technical Processes, SAFEPROCESS'97*, 1997; 66–70.
9. Green MD, Duyar A, Litt JS. Model-based fault diagnosis for turboshaft engine. In: *Prepr. IFAC Symp. on Fault Detection, Supervision and Safety for Technical Processes, SAFEPROCESS'97*, 1997; 71–76.
10. Goodwin GC, Sin KS. *Adaptive Filtering, Prediction and Control*. Englewood Cliffs, NJ: Prentice-Hall, 1984.
11. Ljung L. *System Identification: Theory for User*. Englewood Cliffs, NJ: Prentice-Hall, 1987.
12. Soderstrom T, Stoica P. *System Identification*. London: Prentice-Hall, 1989.
13. Tugnait JK. Techniques for stochastic system identification with noisy input and output system measurements. *Control and Dynamic Systems*, 1995;73:41–88.
14. Breikin TV, Arkov VY, Kulikov GG. On stochastic system identification: Markov models approach. In: *Proc. 2nd Asian Control Conf., ASCC'97*, 1997; 775–778.
15. Kulikov GG, Arkov VY and Breikin TV. On condition monitoring of FADEC information channels. *Russian Aeronautics*, 1995;4:81–85.

Chapter 14

1. Greensite AL. *Control Theory: Elements of Modern Control theory*. New York: Spartan, 1970.

2. Shevyakov AA, editor. *Integrated Control of Aero Power Plants.* (In Russian). Moscow: Mashinostroenie, 1983.
3. Singh MG, Titli A, editors. *Systems: Decomposition, Optimisation and Control.* New York: Pergamon, 1987.
4. Dutton K, Thompson S, Barraclough B. *The Art of Control Engineering.* New York: Addison-Wesley, 1997.
5. Kuo BC. *Automatic Control Systems.* Englewood Cliffs, NJ: Prentice-Hall, 1995.
6. Canon MD, Cullum CD, Polak E. *Theory of Optimal Control and Mathematical Programming.* New York: McGraw-Hill, 1970.

Chapter 15

1. Kulikov GG. Mathematical modelling of a gas turbine engine as a multivariable plant. In: Shevyakov AA, Martyanova TS, editors. *Optimisation of Multivariable Control Systems of Aero Gas Turbine Engines.* (In Russian). Moscow: Mashinostroenie, 1989; 30–88.
2. Breikin TV, Arkov VY, Kulikov GG, Kadirkamanathan V, Patel VC. On gas turbine engine and control system condition monitoring. In: *Prepr. IFAC Symposium on Fault Detection, Supervision and Safety for Technical Processes, SAFEPROCESS'97*, 1997; 66–70.
3. Press WH, Flannery BP, Teukolsky SA, Vetterling WT. *Numerical Recipes in C: The Art of Scientific Computing.* New York: Cambridge University Press, 1990.
4. Arkov V,Evans DC, Fleming PJ, Hill DC, Norton JP, Pratt I, Rees D, Rodriguez-Vasquez K. System identification strategies applied to aircraft gas turbine engines. *Annual Reviews in Control*, 2000;24: 67–81.
5. Eykhoff P. *System Identification: Parameter and State Estimation.* London: Wiley-Interscience, 1974.
6. Ljung L. *System Identification: Theory for the User.* Englewood Cliffs, NJ: Prentice-Hall, 1987.
7. Söderström T, Stoika P. *System Identification.* London: Prentice-Hall, 1989.

Index

P

Q

R

S

T

MIX
Papier aus verantwortungsvollen Quellen
Paper from responsible sources
FSC® C105338

If you have any concerns about our products,
you can contact us on
ProductSafety@springernature.com

In case Publisher is established outside the EU,
the EU authorized representative is:
Springer Nature Customer Service Center GmbH
Europaplatz 3, 69115 Heidelberg, Germany

Printed by Libri Plureos GmbH
in Hamburg, Germany